Advances in Rock Mechanics

Advances in Rock Mechanics

Chief Editor

Yunmei Lin

Northeastern University

World Scientific

Singapore • New Jersey • London • Hong Kong

Published by

World Scientific Publishing Co. Pte. Ltd.

P O Box 128, Farrer Road, Singapore 912805

USA office: Suite 1B, 1060 Main Street, River Edge, NJ 07661

UK office: 57 Shelton Street, Covent Garden, London WC2H 9HE

British Library Cataloguing-in-Publication Data
A catalogue record for this book is available from the British Library.

ADVANCES IN ROCK MECHANICS

ISBN 981-02-3621-2

Printed in Singapore by Eurasia Press Pte Ltd

Chief Editor: Yunmei Lin

International Editorial Board

Acknowledgement

The support of Prof. Xuehan Yang (The Chinese Geological University) in the process of publication is gratefully appreciated.

Preface

This is a publication presenting the state of the art in the field of rock mechanics. It includes 29 contributions which present ongoing or recently completed research in various aspects of rock mechanics, as well as examples of current practice with advanced technologies or methods.

The main topics of these contributions are: mechanical properties of rock; ultrasonic and micro-seismic measurements; intelligent rock mechanics; rock stress; numerical analysis; theoretical analysis; stability of rock engineering; rock reinforcement and rock classification, etc.

A lot of new developments have been presented in various papers, such as the developed computer program for numerical analyses in 3D-layered strata (*Russia*); the theoretical advances in anisotropic fractured porous media (*USA*); the investigation on basic fracture modes of granite by using specific testing devices (*France*); the new approach to reconstruction of large-scale *in situ* stresses (*Australia*); the manifold method; the new method for stability analysis of rock slope (*China*); the theoretical-experimental designing method (TEM) and its program "TUNNEL" (*Russia and USA*); predicting rock fracturing behaviors by using neural model NN (11, 17, 2) (*China and Japan*), etc.

An introduction of the first national enforced standard of rock classification in the world is presented at the end of the book.

On the whole, this book offers an interesting and comprehensive understanding of the new advances in rock mechanics worldwide in recent years.

Contents

Theoretical Analysis

Stability of Rock Engineering

Rock Reinforcement

Drilling Model

Rock Classification

Mechanical Properties of Rock and its Analysis

BASIC FRACTURE MODES AND BREAKAGE RESISTANCE OF GRANITE

M. P. LUONG
CNRS-LMS, Ecole Polytechnique
91128 Palaiseau Cedex, France
E-mail: luong@athena.polytechnique.fr

The paper presents an investigation on fundamental fracture modes of granite, using specific testing devices in order to highlight the salient features of fracture processes in granite such as path dependence (mode I tension fracture, mode II shear fracture and mode III torsion fracture), and shear transfers due to aggregate interlock (crack dilatancy), cross effect (frictional contact slip). Experimental results provided useful mechanical characteristics such as cracking strength, stiffness, toughness index and fracture energy according to the three basic fracture modes, and particularly the Hillerborg characteristic length of granite in tension readily derived from direct tension testing. In addition, computerized X-ray tomography analysis evidences localization of cracking in specimens subject to different fracture modes, readily detected by a non-linear analyser based on Volterra series expansion.

1 Introduction

The literature on rock excavation technology often includes such terms as grinding, wearing, cutting, breaking, shearing, scraping, melting, chipping, scabbing, slabbing, or spalling, but these terms are seldom defined. A realistic mechanical understanding of fundamental rock fracture modes is thus necessary in the search for greater efficiency and effectiveness of excavation techniques, particularly relevant in cases of rock fragmentation by blasting (Fourney [6] 1983). The majority of rock drilling and boring techniques employ principles of mechanical attack by means of which stresses are induced in the rock by impact, abrasion, wedge penetration or cutting, erosion, and sometimes by a combination of these methods.

Interest in the strength and deformation characteristics of rock materials has increased in these last decades because they are directly applicable to the design of large underground structures, and to studies of rock mining, tunnelling, drilling, cutting, crushing and blasting, and sometimes indirectly applicable to consideration of the behavior of large jointed rock masses. This is also due to the need for an understanding of rock response as a structural material to various types of loading, and because non-linear fracture mechanics has entered the field as an aid to understanding rock damage and failure. In addition, techniques for numerical analysis, such as finite element methods have been developed which require comprehensive material models.

Failure, in the form of cracks, involves not only mode I in pure tension, but also mode II and mode III shear fracture. Experimental observations show that the shear displacement or shear slip is affected not only by the shear stress, but also, by the normal displacement owing to rough asperities of aggregate or surface roughness and variable granite strength. The normal stresses across the interface in mode II shear tests and mode III torsion tests are measured and applied in different ways, which result in the different numbers of Belleville springs, ensuring different shear stiffness. These factors must be taken into account when measuring toughness indices of the brittle disordered materials, and particularly hard rocks such as granite. Many types of tests have been devised, using equipment and techniques that range from the crude and empirical, with results that are almost impossible to interpret analytically, to the theoretically elegant, which are almost impossible to execute practically (Lama and Vutukuri [15] 1978, Roberts [27] 1977, Hawkes and Mellor [7] 1970).

2 Direct tension strength

Tensile strength of quasi brittle materials is one of their most fundamental and important properties, particularly because these materials are very much weaker in tension than in compression. Fracture processes in blasting mainly involve tensile strength. Most rock materials are more or less brittle. When unconfined, test samples cannot yield plastically to relieve the stress concentrations that are produced at the localised points around the specimens, where these are gripped to be pulled apart by the testing machine. Consequently premature failure is generated from these points. Difficulties in ensuring truly axial loading also exist, so that the specimen is liable to be twisted or bent when gripped and pulled from either end. Different fracture energy values were obtained if different specimen geometries were used (Jenq and Shah [10] 1989). Does this difference result from different testing set-ups involved: different degrees of eccentricity, alignment and support conditions, etc.?

Direct tension applied to a specimen is theoretically the simplest method of determining the tensile strength. It is both practical and theoretically meaningful. The direct test is of more fundamental value because the stress field of an isotropic specimen is determined directly by the applied loading and the boundary conditions, irrespective of the material properties. indirect tests have the inherent disadvantage that a stress strain relationship must be assumed in order to obtain usable results; the usual assumption of linear elasticity, with equal moduli in compression and tension, is invalid for many rocks. However the greatest difficulty is the gripping the specimens and applying a load parallel to the axis of the specimen.

The proposed test specimen (Luong [18] 1989) is a cylindrical tube (Figure 1) easily prepared, with two parallel flat ends and two inverted tubular coaxial borings. The external surface requires no particular preparation. This configuration converts the applied load on the specimen into a tension $\sigma_I = F / \pi(r_e^2 - r_i^2)$ applied on the intermediate tube, defined by internal radius r_i and external radius r_e. Thus the usual compressive test machine can be used. The cylindrical symmetry permits self alignment of loads parallel to the axis of the specimen. The test requires no special device for specimen gripping, and compressive loads for example can be applied without any precaution. There will be no tendency to cause bending so that abnormal stress concentrations are avoided. The main advantage of this direct tensile test is evidenced by the uniformity of uniaxial tension stress throughout the test volume. Thanks to its cylindrical symmetry, there is no bending or torsion stresses, no stress concentrations arising from geometrical irregularities and no end restraint effects perturbing the stress field: most of observed failures occur in the central part of the specimen. The complete stress-deformation properties of granite (Figure 2) can be described by means of two relations, one stress-strain relation (σ - ε), including unloading branches, which is valid for all the material, and one stress-displacement relation (σ - w), also including unloading branches, which gives the additional deformation due to fracture zone (Hillerborg [8] 1989).

3 Direct shear strength

Shear strength is often used to cover several different concepts (Everling [5] 1964) such as (i) strength against pure shear, (ii) shear stress required for failure without normal stress, (iii) shear diagram on solid interface depending upon normal stress and (iv) Mohr's stress envelope. Several methods have been devised for measuring the shear strength of rock: torsion test, single shear test, double shear test, punch test, shear loaded beam with starter notches according to mode II of fracture mechanics. The main requirements of a mode II test specimen must be: simple compact geometry, ease of preparation, simple loading system and stress conditions little affected by extremely small geometric alterations. Attempts to realize mode II crack propagation on a laboratory scale usually fail because mode I growths become predominant. On the other hand, traces of mode II growth are often detected at or

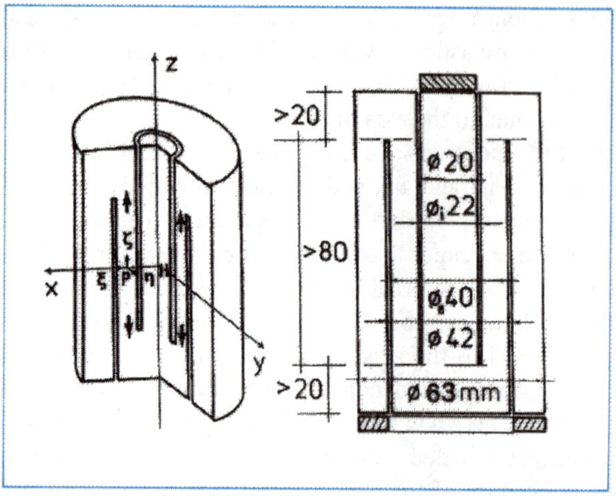

Figure 1: Direct tension (mode I) test specimen
(Young's modulus E_T = 80 GPa, Poisson's ratio ν = 0.27 and Tensile strength σ_I = 6.6 MPa).

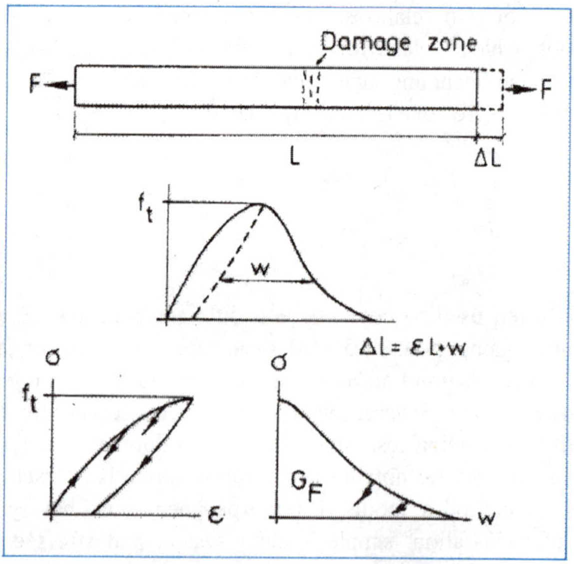

Figure 2: The basis of the fictitious crack model
(Fracture energy G_F = 10 J.m^{-2}, Characteristic length ℓ_c = 20 mm).

after earthquake slipping. The present paper uses a direct shear test (Luong [19] 1990a) that maximizes mode II conditions of crack propagation.

The axial load F is applied on the central cylinder of the specimen, which is retained by the four concentric tubular parts (Figure 3). The notches define the sheared zone S = 4s. The specimen geometry has been optimized by a numerical analysis of stress intensity factors in mixed modes problems by path independent integrals J_I and J_{II} (Bui [3] 1982). A criterion for deciding whether the crack will grow in mode I by formation and extension of a kink or in mode II in the original crack direction, was established similarly as done by Melin [26] (1986), by comparing the stress intensity factors K_{Imax} for a small kink with K_{IImax} for the main crack. If K_{Ic} and K_{IIc} denote the critical stress intensity factors for mode I and II respectively, mode II growth will be preferred to mode I if

$$\chi = K_{IImax} / K_{Imax} > K_{IIc} / K_{Ic} = \chi_c \qquad (1)$$

It is commonly considered that the strength of rock will become constant if sufficient relative movement is developed (Barton [2] 1976). Some authors have suggested that the displacement necessary to develop the residual state increases with rock strength (Krsmanovic et al.[14] 1966). Various rocks behave differently with regard to relations of the peak and the residual strengths.

The main parameter is the stiffness of the normal loading, able to partially prevent the aggregate interlock breakdown, which softens the mechanical response of the materials. When the tangential displacement or shear slip δ_t occurs along the crack interface, the shear stress τ working parallel to the crack is induced and is accompanied by the compressive stress σ and the normal displacement δ_n or crack width normal to the crack plane. These four parameters define the deformation characteristics of the cracked interface (Yoshikawa et al.[32] 1989).

4 Toughness index

A simple method of evaluating the post-cracking performance of rock materials is based on a toughness index that ideally should be a material characteristic, independent of test specimen geometry, dimensions and loading conditions. For ease of comparison and quality control, it should be readily obtained from a load-deflection curve of simple tests. Some toughness indices reported in literature (Barr

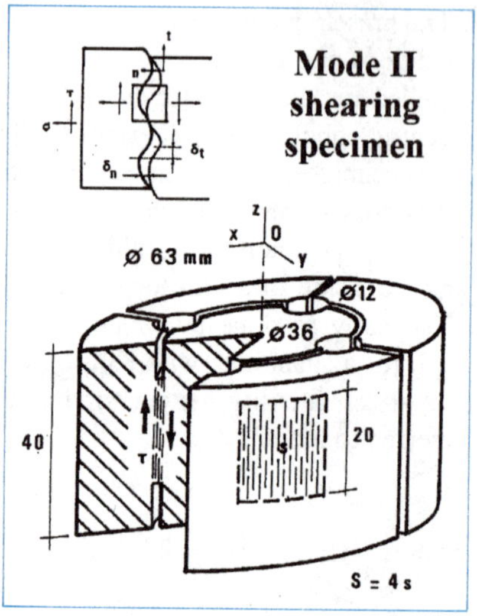

Figure 3: Direct shear (mode II) test specimen.

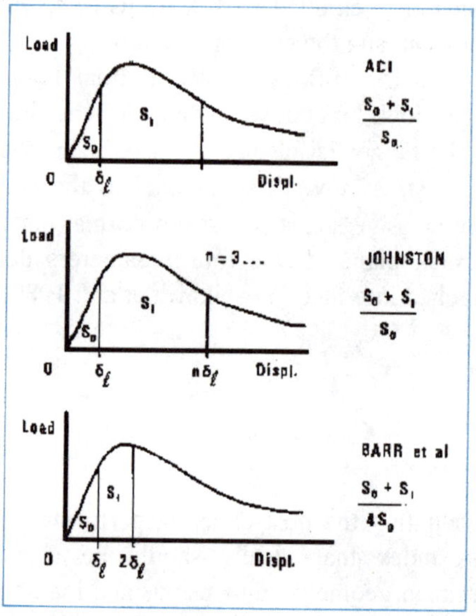

Figure 4: Toughness indices reported in literature.

et al.[1] 1982, Wang and Backer [30] 1989) are defined in Figure 4. The toughness index is a measure of the energy absorbed during fracture in the non-linear portion of the load-deflection curve. It gives the numerical value to the toughness and is independent of the units used. The index is particularly useful in analyzing the rheological behavior of rock materials. Using the so-defined toughness indices based on the load-deflection curves, dilatancy effects can be evaluated for mode II and mode III conditions as a function of the normal stiffness (Luong [22] 1991b).

Several series of tests under direct shear loading were performed on specimens of quasi-brittle materials subject to diverse values of normal stress and different types of loading on the sheared interface, in order to analyze the dilatancy effects. The nominal mode II shear strength is given by $R_{II} = F / 4s = \tau$ in relation with the normal stress $\sigma = F_n / s$ where F_n denotes the lateral force directly applied on the failure surface using a variable number of coned discs or Belleville springs for the control of normal loading stiffness.

Table 1: Toughness indices of granite subjected to direct shear loading.

Normal load	2.5 MPa		5.0 MPa	
Normal stiffness	1.1 kN.mm^{-1}	2.2 kN.mm^{-1}	1.1 kN.mm^{-1}	2.2 kN.mm^{-1}
ACI	4.00		3.90	4.00
Barr et al.	0.52	0.50	0.50	0.46
Johnston	2.50	2.40	2.30	2.30

These experimental values show that toughness indices of granite, subjected to direct shear loading, decrease slowly when normal load and normal stiffness increase, demonstrating that dilatancy effects decrease with the occurrence of plasticity, generated by increased mean stresses.

5 Direct torsion strength

When studying the torsion responses of members, it is convenient to identify two different modes of torsion resistance: (i) circulatory torsion and (ii) warping torsion (Krpan and Collins [13] 1981). In the former case, the torsion shear stress remains constant in magnitude from point to point around the cross section and there is no change in magnitude of the longitudinal stresses. In the latter case, torsion shear stresses are not uniform and account must be taken of the variable effects of dilatancy in regard to the difference of normal stiffness.

Conventional techniques of determining shear strength of rock materials used indifferently tests that promote predominantly mode II or mode III of fracture mechanics (Lama and Vutukuri [15] 1978). In this study, a tubular cylindrical specimen for direct anti-plane shear or torsion testing (Figure 5) is proposed for the determination of mode III fracture strength of quasi-brittle materials (Luong [21] 1991a). It is a square cross section prism with a thin cut turned in its central part. It is assumed that, in spite of the low tensile strength of granite, fracture will occur from shearing stresses in mode III, since the screw-shaped surface of tensile fracture would be much larger than the thin round neck under torsion (Luong [23] 1992).

$$R_{III} = T / \pi r^3 (1 - m^4) \qquad (2)$$

The fracture surfaces appear complex, consisting of inclined facets, known as factory roof fracture surfaces by metallurgical workers, formed by the intersection of a large number of independently initiated mode I cracks. Thanks to the thin round neck, mode III or torsion fracture has been preferred as shown by the fracture surfaces, mode I fracture being controlled by the stiffness of normal stress loading or dilatancy effect. The normal stress $\sigma = F_n / \pi r^2 (1 - m^2)$ is applied on the specimen by means of a long screw. The stiffness of normal stress loading on the failure surface is controlled by one or several Belleville springs. The dilatancy effect is evidenced by an increase of nominal torsion strength with the normal loading stiffness.

6 Fracture energy

The majority of rock drilling and boring techniques employ principles of mechanical attack by means of which stresses are induced in the rock by impact, abrasion, wedge penetration or cutting, erosion, and sometimes by a combination of these methods. Thus there is a need to evaluate energy spent for fragmentation of rock (Luong and Hoteit [25] 1997). Under torsion loading, Figure 6 presents the relationship between fracture energy and strength. The performed tests in this work readily provide energy characteristics to bring the rock materials to fracture in different modes (Figure 7).

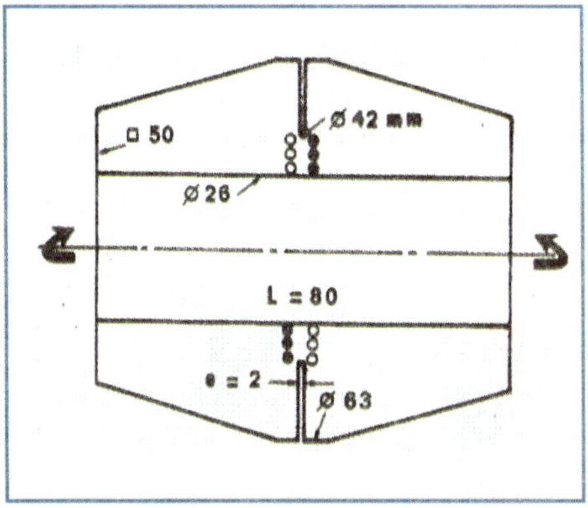

Figure 5: Direct torsion (mode III) test specimen.

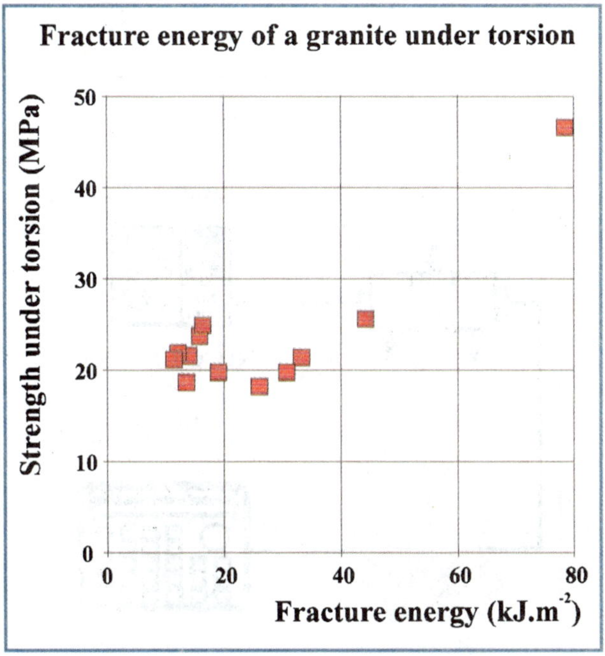

Figure 6: Fracture energy of granite
subject to torsion loading.

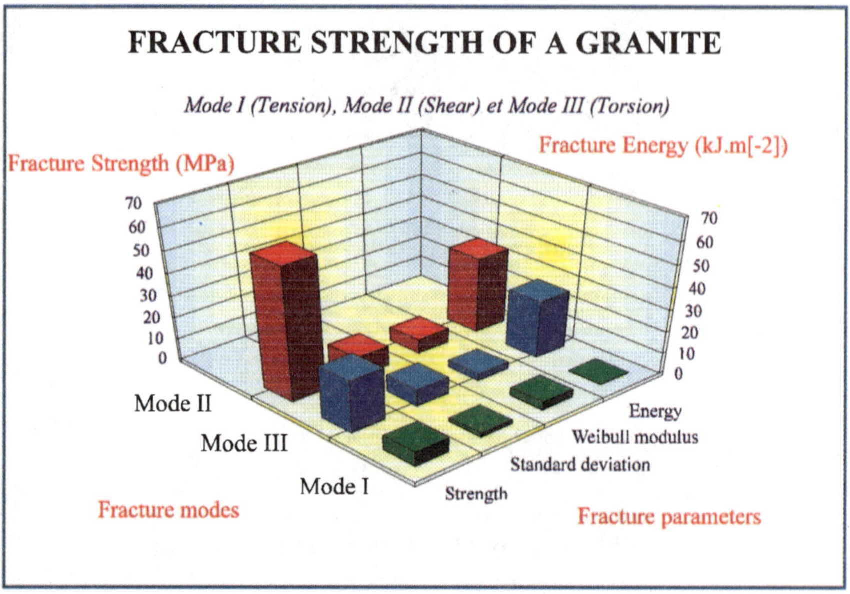

Figure 7: Mechanical characteristics of the tested granite.

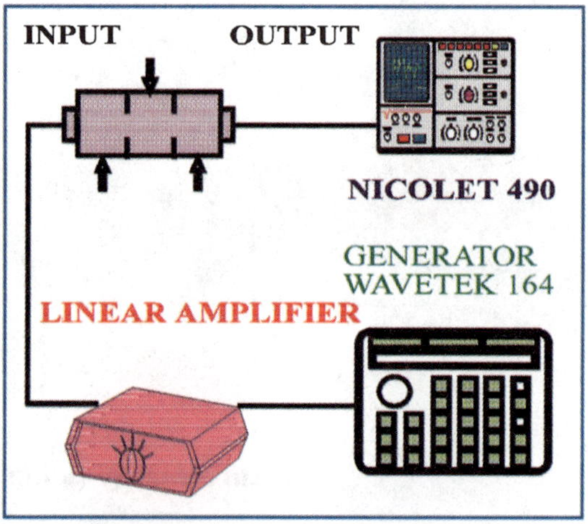

Figure 8: Testing arrangement
for microcracking detection in granite.

7 Detection of microcracking in granite

In most rocks, both the acoustic velocity and the attenuation vary greatly. According to Koltonski and Malecki [12] (1958), granite, for instance, depending on its grain structure can reach values from 1.7 to 5.0 $km.s^{-1}$. The nondestructive testing (NDT), most extensively used at present, is probably the ultrasonic technique. Several researchers (Jones [11] 1952 for instance) used sonic and ultrasonic method to study the fracture of concrete and reported that cracking noises occurred at about 25 % to 75 % of ultimate load. It was recognized that these significant changes in the property of concrete and rocks are probably caused by microcracking, resulting in the increase of volume under compression and the increase of Poisson ratio.

An input-output non parametric procedure (Liu and Vinh [17] 1991, Thomas [28] 1995), based on ultrasonic pulse propagation and using a non-linear analyser for data reduction, has been chosen to portray the non-linear behavior of rock (Luong [24] 1995). The pulse transmission method, using video scan piezoelectric transducer V150 (250 kHz), has been applied in conjunction with a pulser-receiver Panametrics 5052PR that provides high-energy broad band performance. The pulser section produces an electrical pulse to excite a piezoelectric transducer, which emits an ultrasonic pulse. The pulse travels through the specimen that is subjected to a given static shearing, to a second transducer acting as a receiver. The transducer converts the pulse into an electric signal that is then amplified and conditioned by the receiver section and made available for non-linear analysis.

The arrangement, used for the detection of microcracking in granite subjected to direct shear loading, is described in Figure 8. Signal records indicate that at 42 % of mode II shear strength, the high order kernels are still negligible. They become very significant at 70 %. They evidenced: (1) the occurrence of microcracking that modifies the longitudinal part (the faster) of the pulse signal, and (2) the effects of micro cracking that affect its transversal part (the slower).

A simple quantitative parameter on which a criterion could be applied is necessary to characterize the material response. Data reduction based on Volterra series required identification of Volterra kernels through a multi-amplitude procedure, giving access to energy of different terms of the series. Comparison of these energies provides a quantitative evaluation of the significance of non-linear phenomena (Luong [24] 1995). The non-linear ratio $\rho(a)$ is defined as follows:

$$\rho(a) = \frac{\text{non-linear energy}}{\text{total energy}} \tag{3}$$

It expresses the energetic part of non-linear components in the response of the material when subject to an impulse excitation of intensity a. In order to compare two different measurements, it is interesting to use an other parameter called non-linear contrast (Thomas [28] 1995).

$$\Delta_B^A = \int_{a_{min}}^{a_{max}} \left|\delta_B^A(a)\right| da \quad \text{where} \quad \delta_B^A(a) = \rho_B(a) - \rho_A(a) \tag{4}$$

8 X-ray tomodensitometry of granite

Originally used for medical analysis (Hounsfield [9] 1973), the X-ray computerized tomography (CT) technique has been recently used for engineering materials (Vinegar [29] 1986, Latière and Mazerolle [16] 1987, Withjack [31] 1987, Colleta et al. [4] 1991). As for conventional radiography, the basic principle of the CT technique is the attenuation of X radiations through materials. This attenuation is mainly dependent on material density, effective atomic number, and thickness. In computerized tomography, a planar fan of rays penetrates the object, and 1024 detectors measure attenuation. Source and detector are attached to a rotor and move on a circular track around the object.

Attenuation is determined at a rate of 150 measurements per second on each of the 1024 detectors during a 360° rotation, which lasts for 3.4, 6.8 or 13.6 s. During the rotation the X-ray fan investigates a slice of the object that may be 1.5 mm, 3 mm or 8 mm thick. The large number of measurements (about 10^6 for a 6.8 s time of acquisition) enables attenuation values to be computed for a 512 × 512 matrix of elements of volume (voxels) in the investigated slice. The computed values produce a 512 × 512 pixel image, which represents a cross section of the object.

The scanned specimen is translated horizontally through the investigation plane by a motor-driven bed with an accuracy of 1 mm. Thus it is possible to obtain serial cross sections of the model. Attenuation values, which are given in Hounsfield units (HU) for 130 kV energy, are compatible with the range of the medical scanner. One of the most significant advantages of computed tomography is its sensitivity to very small differences in density (Figure 9). The internal cracks are visualized. Voids and inclusions can be identified and localized. Figures 10 and 11 demonstrate the great interest of coupling these two non-destructive techniques for the detection of the occurrence of micro cracking in granite.

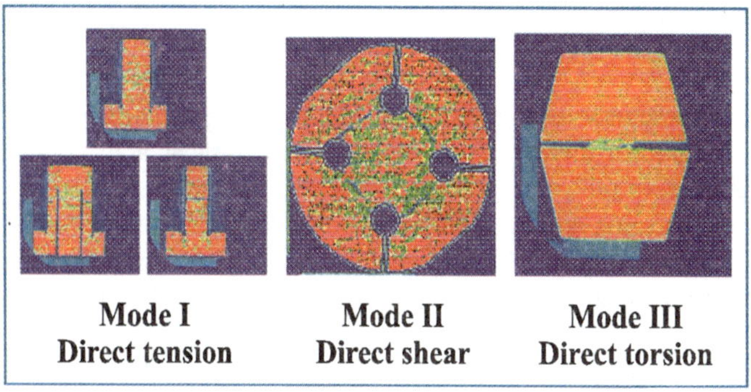

Figure 9: Tomography of a granite under fundamental modes of fracture.

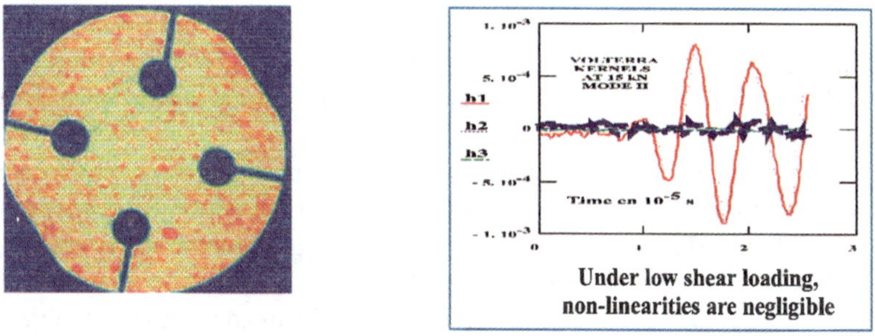

Figure 10: Coupled non-destructive techniques used for the detection of microcracking in granite.

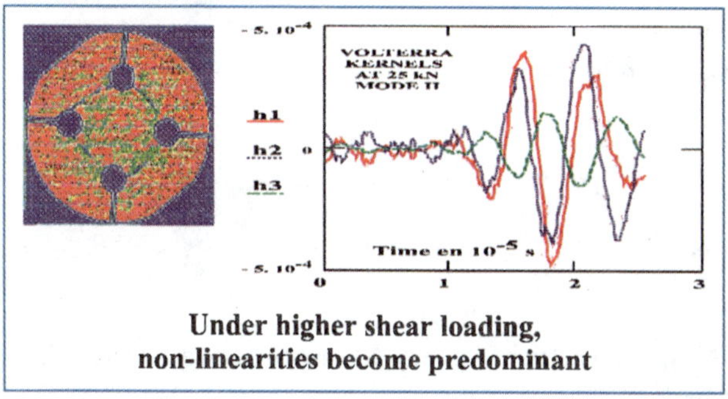

Figure 11: X-ray tomography validating the non-linear detection of microcracking in granite.

9 Concluding remarks

The versatility of these new testing devices allows a ready evaluation of the behavior of granite during mode I direct tension, mode II direct shear and mode III direct torsion loading.

The proposed testing arrangements for both direct shear and torsion shear strength measurements simplify the loading equipment by the use of a uniaxial test machine with the combined compression, shear and torsion loading frames described in the paper.

The test set-up allows the normal stresses to be monitored and measured during the experiments under varying shear stresses. The resulting responses are shear stress versus slip and shear stress versus normal stress, which are of course very useful in the interpretation of conventional pull-out tests. They are practical and reliable. They also facilitate the testing procedure when the materials are subject to severe and hostile environmental conditions. They can be used as routine tests more readily achieved than on the conventional triaxial tests in the cases of low mean stress level. In addition they demonstrate the significance of fracture modes on the mechanical behavior of granite.

This ultrasonic technique, using ultrasound propagation characteristics in non-linear cracked medium, provides a ready mean to analyze the non-linear response of granite. This method relies on the fact that during the process of micro cracking, rock material is locally unstable and hence obstructs the propagation of ultrasound pulse. The non-linear analyser reveals to be very useful for the detection of the micro cracking process announcing the occurrence of propagating damage. It can be used to monitor nondestructively and continuously the whole fatigue damage process of rock infrastructure so that the damage mechanisms can be quantitatively interpreted.

X-ray computerized tomodensitometry is also of great interest for studying cracking occurrence of rock specimens in mechanical tests. Serial cross-sections provide useful data for morphological analysis and three-dimensional descriptions of fractures.

Acknowledgments

The support of ANDRA (French Agency for Radioactive Waste Management) is gratefully appreciated.

References

1. B.I.G. Barr et al., *Int. J. Cem. Comp. Light. Concr.*, **4**(4), 221 (1982).
2. N. Barton, *Int. J. Rock Mech. Mining Sc.*, **13**, 255 (1976).
3. H.D. Bui, *C.R. Acad. Sci. Paris*, **235**, 521 (1982).
4. B. Colletta et al., *Geology*, **19**, 1063 (1991).
5. G. Everling, *Int. J. Rock Mech. Mining Sc.*, **1**, 145 (1964).
6. W.L. Fourney, *Comprehensive Rock Engineering*, **4**(2), 39 (1983).
7. I. Hawkes and M. Mellor, *Engineering Geology*, **4**(3), 177 (1970).
8. A. Hillerborg, *Fract. Mech.: Appl. Concrete,* ACI **SP-118**, 157 (1989).
9. G.N. Hounsfield, *British J. Radiology*, **46**, 1016 (1973).
10. Y.S. Jenq and S.P. Shah, *Report to RILEM,* **Committee 89-FMT**, (1989).
11. R. Jones, *British J. Applied Physics,*V, **3**(7), 229 (1952).
12. W. Koltonski and I. Malecki, *Acoustica,* **8**, 307 (1958).
13. P. Krpan and M.P. Collins, *J. Struct. Div.,* **107**(6), 1107 (1981).
14. D. Krsmanovic et al., *Proc. 1st Cong. Int. Soc. Rock Mech.*, **1**, 537 (1966).
15. R.D. Lama and V.S. Vutukuri, *Handbook on mechanical properties of rock*, Trans Tech Pub., **1**, (1978).
16. H.J. Latière and F. Mazerolle, *Engineering Fract. Mech.*, **27**(4), 413 (1987).
17. H. Liu and T. Vinh, *Mechanical Systems Signals Processing*, **5**(1), 61 (1991).
18. M.P. Luong, *Fracture Concrete & Rock*, Elsevier Appl. Sc., 18 (1989).
19. M.P. Luong, *Micromechanics of failure of quasi-brittle materials*, Elsevier Appl. Sc., 275 (1990a).
20. M.P. Luong, *Engineering Fracture Mechanics*, **35**(1/2/3), 127 (1990b).
21. M.P. Luong, *Fracture Processes in Concrete, Rock and Ceramics*, E & FN SPON, 727 (1991a).
22. M.P. Luong, *Mech. Behaviour Materials*, Pergamon Press, **4**, 141 (1991b).
23. M.P. Luong, *J. Nuclear Engineering and Design*, **133**, 83 (1992).
24. M.P. Luong, *Rock Foundation*, Balkema , 157 (1995).
25. M.P. Luong and N. Hoteit, *Environmental and Safety Concerns in Underground Construction*, Balkema , 525 (1997).
26. S. Melin, *Int. J. Fracture*, **30**, 103 (1986).
27. A. Roberts, *Geotechnology*, Pergamon Press, **103** (1977).
28. J.J. Thomas, *Doctorate Ecole Polytechnique*, (1995).
29. H.J. Vinegar, *J. Petroleum Technology*, 257 (1986).
30. Y. Wang and S. Backer, *Int. J. Cem. Comp. Light. Concr.*, **11**(1), 11 (1989).
31. E.M. Withjack, *Society Petroleum Engineers*, SPE **16951**, 183 (1987).
32. H. Yoshikawa et al., *J. Struct. Div.*, ASCE **115**(4), 771 (1989).

BEHAVIOR OF COAL CAUSED BY WATER OR CARBON DIOXIDE

J. GUSTKIEWICZ
Strata Mechanics Research Institute Polish Academy of Sciences
Reymonta 27, 30-059 Kraków, Poland
e-mail: gustkiew@aga.img-pan.krakow.pl

Y. ORENGO
Lirigm: Géologie et Mécanique, Université Joseph Fourier, Grenoble
Maison de Géosciences - 1381 rue de la Piscine, Domaine Universitaire
38-400 Saint Martin d'Hères
e-mail: orengo@ujf-grenoble.fr

After the pores have been evacuated from the atmosphere in vacuum conditions, the hard coal was saturated either with water at the atmospheric pressure or with carbon dioxide at pressure. During saturation the behavior of coal was being observed together with the special effects: swelling, stress caused by swelling in the conditions of rigid constrains imposed upon the samples, and the acoustic emission.

1. Preface

Knowledge of the chemical structure of coal as well as relevant physical model are of vital importance in understanding of its behavior. This knowledge is inevitable for studying the interaction between pore fluids (water and carbon dioxide) and the coal structure and, consequently, its influence on mechanical properties of the coal. The paper deals with the mechanical behavior of hard coal during saturation with water and carbon dioxide.

2. Coal and fluids

From the chemical point of view, hard coal can be described as an organic polymer of complicated and only partly recognized structure.

Several concepts have been developed in order to explain the coal structure. According to Shinn[1] the coal matter consists of independent organic macromolecules which may include from several hundreds to several thousands carbon atoms.

A single macromolecule usualy comprises several tens of groups of the condensed aromatic rings, some tens of alicyclic nuclei, alkylic chains and numerous fragments of aliphatic chains. Moreover, nitrogen- and sulphur-bonding structures are present. Hard coal contains also significant amounts of oxygen bonded in various functional groups: hydroxyl, carboxyl, metoxyl, carbonyl and ether.

Due to the bonding of condensed aromatic rings, macromolecules coalesce to form sheets (so called "lamellae") varying in diameter (few to some tens of nanometers) and in thickness (from some tenths to one nanometer). Lamellae are bonded with oxygen bridges, aliphatic chains and hydroxyl groups. As coalification proceeds, the van der Waals forces bring the lamellae together, forming crystallites similar to those known from the graphite structure.

The existence of functional groups causes local polarity of an essentially electrically neutral coal matter.

Structural elements-crystallites, individual lamellae and chains are separated with submicropores. These pore spaces comparable in size with the dimensions of fluid molecules penetrating the coal structure are responsible for thermally-activated absorption of fluids driven by diffusion.

Lamellae and chains separated by submicro- and/or micropores form structure resembling the compliant spatial network with the embedded larger crystallites and mineral aggregates. It seems that there is a range of force values which cause bending and straightening of chains or flattening of lamellae although spacings between adjacent carbon atoms practically do not change in comparison with the size changes of the individual structural elements.

Such deformations are directly facilitated by the presence of pores. Increasing forces may generate the stretching strains in the carbon chains and, consequently, their breakdown. Friction of the neighboring structural elements during deformation may give rise to energy dissipation. Plastic deformations can be expected if adjacent lamellae slide over each other crystallites, and, may be also in minerals.

Fluids penetrate and migrate through the coal structure as a result of various processes: filtration, several types of diffusions and probably also capillary forces. Several modes of pore space filling by fluids may be distinguished (the infilling of pore spaces due to the sorption potential, capillary condensation, mono- and polimolecular adsorption). Behavior and mechanical properties of coals are controlled by pore pressure of free fluid, temperature as well as by adsorbed and absorbed fluids.

Water and carbon dioxide may be adsorbed and absorbed on the coal structure. Influence of these processes on behavior of coal are discussed in the following paper.

Due to polarity of its molecules, water is adsorbed on those parts of coal structure which contain polar functional groups. This sorption mechanism is driven by hydrogen bonds. Carbon dioxide is physically sorbed on the apolar parts of coal structure by fluctuation forces. Volume of adsorbed fluid decreases with the increasing temperature.

Absorption is a diffusional filling of submicropores and thus, is thermally activated. Carbon dioxide has relatively low diffusion activation energy which, together with small size of its molecules, facilitates its penetration into pores unaccessible for larger molecules. In the case of water energy of diffusion into the interspaces is amplified by additional forces resulting from the polar interaction between H_2O molecules and functional groups.

3. Mechanical behavior of coal during sorption of water or carbon dioxide.

Attention will be focused on a laboratory coal sample. Such a sample is immersed in the fluid at the initial moment of the experiment and its whole external surface remains in contact with the fluid up to the end of the test. As the experiment proceeds the fluid penetrates into the sample and fills the pore spaces.

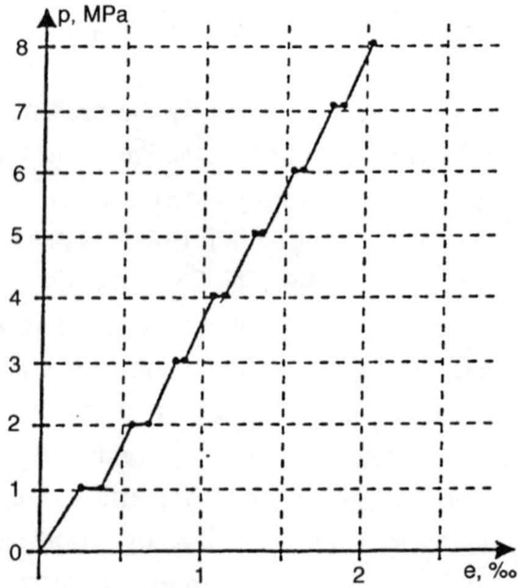

Fig. 1. Volumetric strain versus nitrogen pressure

Instantaneous load applied at the beginning of the experiment may impede the relative movements of structural elements due to friction along their contacts. Penetrating fluid separates the structural elements and reacts as the lubricant. As a result, retarded strain is produced, as illustrated in Fig. 1. Plot segments parallel to strain axis correspond to 30-minutes-long intervals after each 1 MPa nitrogen pressure increment. This time spans appeared to be sufficient to full equilibration of nitrogen penetration into the sample. Deformations produced are compressional during both the pressure jumps and the following periods of constant pressure. As nitrogen sorption on coal is insignificant at room temperature, its effect on coal structure can be regarded as mechanical.

Both water and carbon dioxide are strong coal sorbates. Carbon dioxide is readily absorbed because of its low diffusing activative energy but can also be adsorbed. Water dipoles are, in turn, adsorbed on the active centers of coal structure. Such centers facilitate also penetration of water molecules into the submicropores. As compared with nitrogen, both H_2O and CO_2 fill the coal structure in much larger volumes and their filling times are much longer.

Penetration of sorbates causes the swelling of coal. Basing on a brief review of available literature, Czaplinski[2] concluded the lack of definite solution of the dilemma whether swelling is an effect of adsorption or absorption. The volumes of sorbed water as a function of time along with the corresponding volumetric strain are exemplified in Fig. 2. Water has been sorbed under atmospheric pressure. In each experiment pore spaces were evacuated prior to saturation.

Maximum swelling was found to be somewhat lower in the second pore-filling cycle than in the first one but volume of sorbed water was slightly larger in the second cycle. Similar dependence between the volume of sorbed water and the sorption

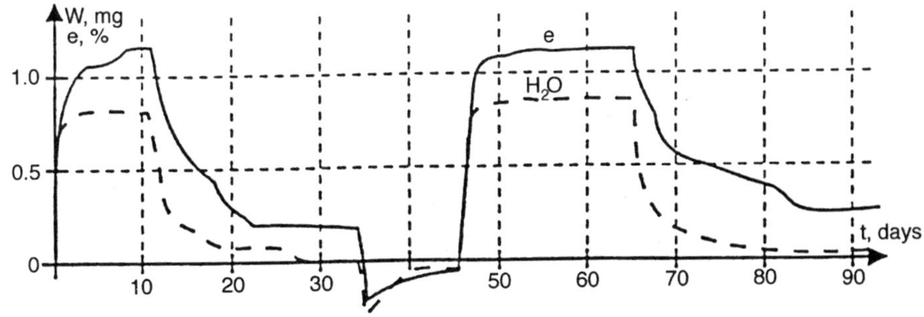

Fig. 2. Amount of sorbed water and swelling-induced strain versus time. Two sorption cycles separated by drying of sample at atmospheric conditions.

cycles has been reported by Lason and Zyla[3] who explained such a relationship by means of insignificant changes in pore structure caused by swelling-induced cracks. Appearance of microcracks is suggested by acoustic emission recorded during the sorption process. Acoustic activity has been measured as number of pulses per

Table 1.

Cycle	Emission time (hours)	Maximum acoustic activity (number of pulses/g·min)
1	36	55
2	20	45
3	11	25
4	3	1.5

minute and per mass unit. Changes of acoustic activity in time for the three from total four succeeding sorption cycles of the same sample (dried and degassed before each cycle) are illustrated in Fig. 3. Smooth curves enveloped the plots which expressed the acoustic activity versus time for second sorption cycle is presented in Fig. 4. Attention should be payed to a very distinct reduction of both the time and the acoustic emission after each cycle (see also Tab. 1). The coal reveals specific memory of its contact with water.

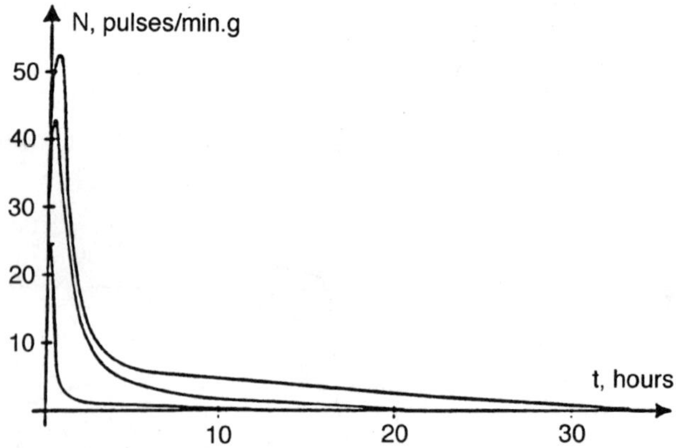

Fig. 3. Acoustic activity versus time for three succeeding sorption cycles

Acoustic emission appears also during the sorption of carbon dioxide as proved by Pawlowski and Gutkiewicz[4], Majewska[5] and by unpublished authors data.

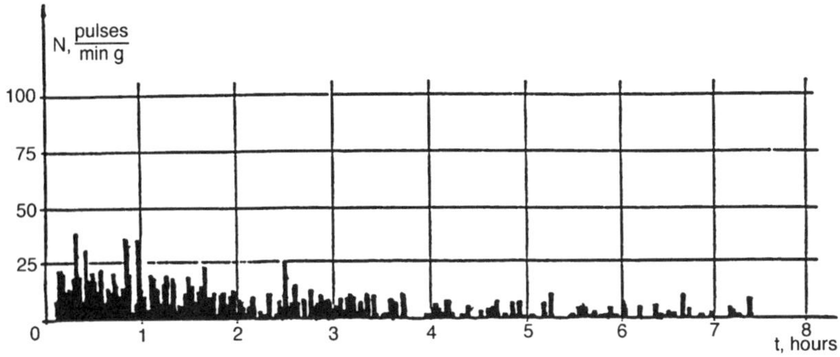

Fig. 4. Acoustic activity versus time for second sorption cycle.

Proceeding and intensity of the CO_2-induced emission are similar to the effects caused by the water sorbate. Each sorption cycle causes maximum emission immediately after the beginning of experiments and then decrease is observed. Acoustic activity of coal saturated with CO_2 under pressure 1.5 MPa is illustrated in Fig. 5.

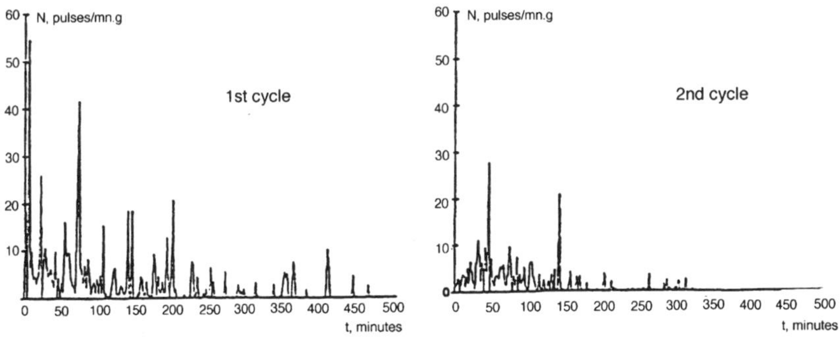

Fig. 5. Acoustic activity versus time for first and second sorption cycles for CO_2 pressure 1.5 MPa

Both curves imply that for constant pressure both emission time and maximum intensity decrease in the succeeding cycles. The increasing pressure increases time of emission and its maximum intensity.

Acoustic emission may result from breaking of atomic bonds during penetration of fluid into the intermolecular spaces or from shearing along the contacts between fluid-saturated, swelling, and fluid-free, yet non-expanding parts of the coal structure.

With the increasing pressure the amount of fluid accumulated in the pore spaces increases. Consequently, the swelling-derived strain also increases. However, as both the confining and the pore pressures are equal, the rise of pressure should result in contraction of the sample. Such a process has been observed in inorganic rocks and in coals saturated with non-sorbing fluids (Fig. 1). It is suggested that during compression the coal sample filled with sorbing fluid is subjected to the two contrasting processes: swelling which seems to be controlled by the limited sorption capacity of the coal and contraction which happened due to the mechanical action of the pressured fluid. For the pressure values close to zero swelling predominates whereas with the increasing pressure the contraction prevails (see Fig. 6).

Other example of a combined effect produced by the two contrasting deformation types is illustrated in Fig. 7. Coal sample has been subjected to the stepwise pressure increase in CO_2 atmosphere. After each jump the pressure has been kept constant for one hour and the corresponding strains have been recorded.

The experiment proved that each pressure jump has been accompanied by instantaneous strain (expressed by plot segment terminated by circles) followed by the retarded one. Initially, the strain was always compenssional (positive) and resulted from penetration of gas into the coal structure (as in the case of nitrogen, see above). Subsequently, it changed into the tensional (negative) strain which reflects the dominance of sorption-driven swelling of the structure.

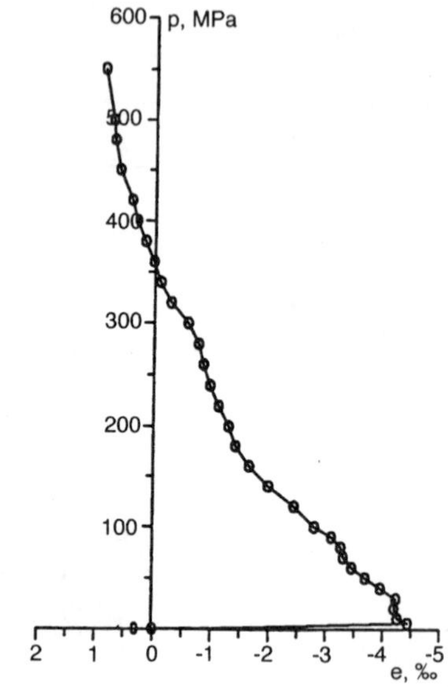

Fig. 6. Volumetric strain due to swelling and compression of coal versus CO_2 pressure

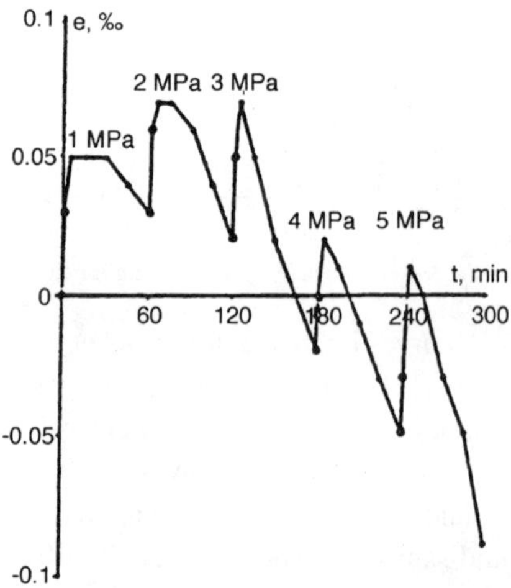

Fig. 7. Volumetric strain versus time for various values of CO_2 pressure

Sorption of carbon dioxide or water causes swelling of the sample. This process can be limited or eliminated by rigid constrains imposed on sample in a given direction, which induces stress (so called "sorption stress"). The stress changes in time with increasing volume of sorbed fluid. An example is shown in Fig. 8. In this exper-

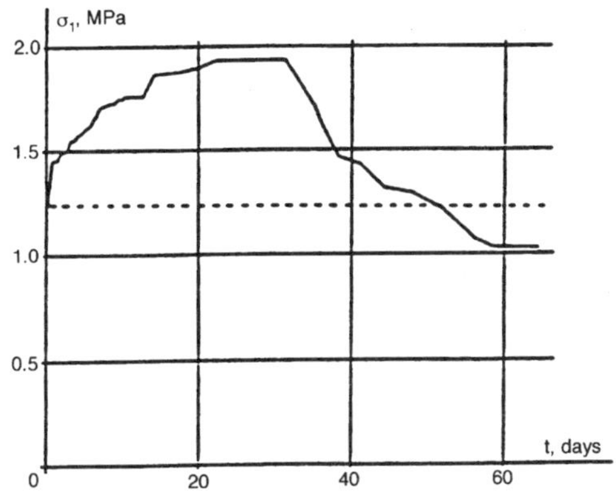

Fig. 8. Sorption stress versus time during water sorption and drying

iment coal sample placed in a rigid frame has been subjected to pore evacuation, then saturated with water and loaded with initial axial stress 1.25 MPa. As a result an increasing stress was observed, followed by its decrease below the initial value during drying of the sample. Such a stress pattern suggests that the sample has been effected by the two processes: swelling, which has generated sorption stress, but also stress relaxation proceeding as a rheological feature. The latter is evidenced by the stress decrease below the initial value.

It seems therefore, that the resultant stress depends on the dominance of one of above mentioned processes which, in turn, seems to be controlled by the initial stress value. Such an interpretation is supported by Fig. 9, which shows stress changes in time for a water-sorbing sandstone sample. For the lowest initial load graph undoubtly reflects the swelling whereas for the highest initial load the relaxation predominates. Successive changes of the initial load allowed to select a particular value for which relaxation and sorption stresses are balanced. Both types of stress appear at the initial load 0.75 MPa.

Similarly to water sorption under the atmospheric pressure, experiments has been carried with carbon dioxide as a sorbate. Czaplinski and Gustkiewicz[6] loaded an evacuated sample with succeeding uniaxial stresses in vacuum and then intro-

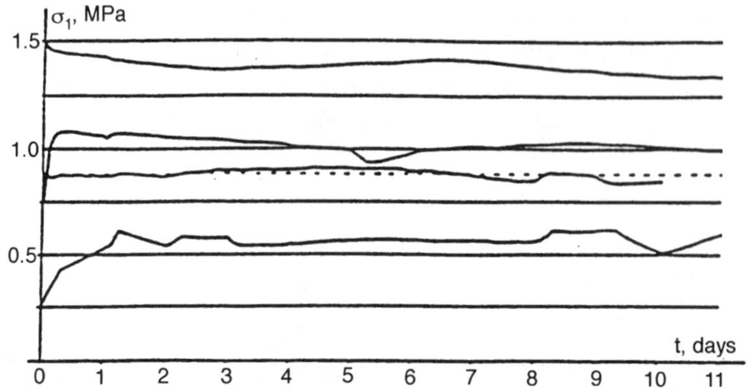

Fig. 9. Sorption stress versus time during water sorption under various initial loads. Sandstone.

duced CO_2 under pressure. Resulting stress has been recorded during 72 hours until the system reached sorption equilibrium. Experiment has been repeated for several CO_2 pressure values. Sorption temperature was always constant whereas initial stress in vacuum and CO_2 pressure were variable.

Attention has been focused on sorption stress obtained in an equilibrium state reached after 72 hours. These values were expressed in relation to the initial stress applied in vacuum and to the sorption pressure (i.e. pressure under which the sorption equilibrium has been attained at thermostat-controlled constant temperature 298 K).

Fig. 10 presents the sorption stress itself as a function of sorption pressure at given initial stress as a parameter and sorption pressure as a function of initial stress

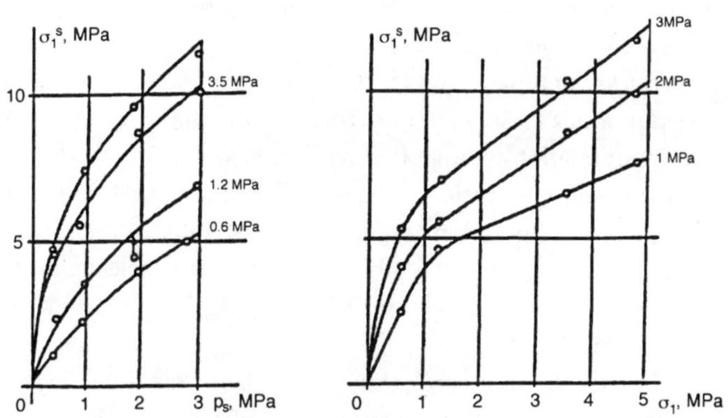

Fig. 10. Sorption stress versus sorption pressure with initial stress as a parameter, and sorption stress versus initial stress with sorption pressure as a parameter.

with sorption pressure as a parameter. Increase of the sorption stress with the sorption pressure at the constant initial stress seems to originate from migration of CO_2 into the coal structure and resulting swelling. Mass of penetrating gas increases with the rise of sorption pressure. Consequently, the rise of sorption stress with the initial stress under constant sorption pressure can be an effect of increasing packing density of sample - a response of structure forced by the initial stress. Particularly evident is the closure of cracks which corresponds to the initial 2 MPa stress interval, treated as an independent variable. It must be emphasized that uniaxial compression of studied sample with the load up to several MPa (in order to avoid permanent deformations) proved that 2 MPa load causes the closure of cracks transversal to the stress direction. The measured remaining stress after each sorption-desorption cycle decreases below the initial value of stress. It indicates the presence of stress relaxation.

Swelling of coal is accompanied by decrease of its permeability in time, despite the type of sorbate (water or carbon dioxide). It is confirmed by the results of experiments carried out by Dyrga et al[7]. Significant decrease of filtration coefficient can be the effect of the experiment conditions which impede the expansion of sample transversally to the filtration direction. As a result, the pressure closed the cracks and reduced other open spaces.

References

1. J. H. Shinn, Fuel 63: 1187 (1984)

2. A. Czapliński, Scien. Bull. of the Stanisław Staszic Academy of Mining and Metallurgy, 1215, Chemistry Bull. 9: 39 (1988)

3. M. Lasoń and M. Żyła, Scien. Bull. of the Stanisław Staszic Academy of Mining and Metallurgy, 366, Mining Bull. 46: 21 (1973)

4. Z. Pawłowski and P. Gutkiewicz, Strata as Multiphase Medium, Vol. I, Wydawnictwo AGH: 195 (1990)

5. Z. Majewska, Strata as Multiphase Medium, Vol. I, Wydawnictwo AGH: 221

6. A. Czapliński and J. Gustkiewicz, Strata as Multiphase Medium, Vol. II, Wydawnictwo AGH: 455 (1990)

7. L. Dyrga et. al., Arch. Min. Sci. 37: 313 (1992)

COUPLED EFFECTS BETWEEN MECHANICAL BEHAVIOR AND MASS TRANSFER PHENOMENA IN ROCK SALT

Ph. COSENZA*#, M. GHOREYCHI*

* *Groupement pour l'étude des Structures Souterraines de Stockage (G.3S), Ecole Polytechnique, 91128 Palaiseau Cedex, France.*

Département de Géophysique Appliquée, Université Paris VI, 4 Place Jussieu, 75005 Paris, France
E-mail: cosenza@ccr.jussieu.fr

The results given in this paper deal with experimental and theoretical studies performed on rock salt permeability, in the framework of hydrocarbon or radioactive waste storage. The results of a field experiment carried out in a salt layer in the Mines de Potasse d'Alsace is in agreement with those of the experiments performed in other salt formations : rock salt is permeable to gas and to brine, at least around underground facilities. Its permeability is about 10^{-21} m^2. This permeation is due to the damage resulting from excavation or tectonics. Considering rock salt as a porous, permeable and reactive medium, a theoretical framework is used to discuss dissolution/crystallization phenomena leading to a change in transport properties. The results of calculations are in agreement with those of laboratory tests, and show that a mechano-chemical coupled effect can explain the extremely low permeability of rock salt at a geological time scale. Nevertheless, the results of calculations on underground structures show that a very low permeability which may not be measurable using existing techniques, modifies in a significant way the long term evolution of an underground storage.

1 Introduction

Salt formations are considered as suitable repository media since they are assumed to be impervious to fluids : rock salt has been selected in many countries for underground storage of hydrocarbons, industrial and radioactive wastes.

Long term behavior of such facilities depends strongly on both rheological properties and mass transfer properties : salt solubility in water and rock salt permeability. Very often, studies dealing with long term safety of storage include two separate approaches : 1) hydrochemical stability studies and 2) mechanical stability studies.

The main objective of this research was to reexamine the hypotheses of these uncoupled approaches for long term safety assessment of the storages in rock salt.

This research is mainly based on a field experiment performed at the Mines de Potasse d'Alsace (M.D.P.A., France) which required preliminary theoretical and experimental studies for the design, performance and interpretation of the test. The experiment is presented in the first part of the paper.

In the second part, a theoretical macroscopic framework is presented, considering the rock salt as a deformable porous medium in which a chemical reaction between the fluids and the matrix is taken into account.

The third part of the paper is a contribution to the safety analysis of underground disposals in saline formations. In this framework, the so-called "healing" process leading to the closure of the discontinuities (microcracks and grain boundaries) is discussed. Then, a preliminary study of the effect of an extremely low permeability on the long term evolution of a storage is undertaken with the help of a model based on a simple geometry and *in situ* experimental results.

2 In Situ Rock Salt Permeability Measurement

A review of hydro-mechanical behavior of rock salt shows that two main mechanical phenomena are involved in permeability evolution : (a) deviatoric loading leads to a microfracturing and dilatation of grain boundaries, (b) hydrostatic loading induces an instantaneous elastic closure of the existing microcracks and a continuous "healing" of the discontinuities (microcracks and grains boundaries).

The first phenomenon explains why the zone surrounding an underground storage in salt has reduced hydrogeological and mechanical properties (Stormont[1]). But little is known if the second phenomenon can restablish the initial permeability.

In order to study the conditions in which rock salt is really impervious, an *in situ* test experiment was performed in the Amelie mine belonging to the Mines de Potasse d'Alsace (M.D.P.A., France) (Cosenza et *al.*[2]). A vertical borehole, 20 m long, was drilled from an excavated depression to access to the S1 bed, having a low insolubles content. The selected salt bed is approximately 1 m thick and is located 16 m away from the gallery floor. This distance was assumed to be sufficient to put the tested zone beyond the area disturbed by mining.

A preliminary study on thermo-hydro-chemico-mechanical processes in salt showed that the measurements can be disturbed by various effects : dissolution-crystallization associated with brine percolation, thermal expansion and creep (Cosenza and Ghoreychi[3]). To control them, the following recommendations were made and adopted : (a) tests with constant pressure steps, (b) thermal regulation of the test. Morever, a special set-up was designed and the principal components (packer, resin sealant, injection system...) were tested in the laboratory prior to the field test.

The experimental results show that rock salt is permeable to gas and to brine even at some distance from the underground facilities. The low flow rates of brine that were measured are relatively well represented using Darcy's law. The obtained values of intrinsic permeability and initial pore pressure are respectively equal to 2.10^{-21} m^2 and 1 MPa. The part of experiment with injection of nitrogen, shows the following results :

- after percolation of brine within the rock mass, the capillary pressure makes the gas flow rate lower by one order of magnitude.
- fluid migration in salt is not controlled by Darcy's law. The

Klinkenberg effect and the presence of an initial capillary pressure may play a important role.

3 Theoretical Framework of Coupled Effects in Salt

The effect of an extremely low permeability on the long term evolution of a storage must be discussed since disposal of industrial or nuclear waste in salt caverns is under study in many industrialized countries. A preliminary study of such an effect is proposed with the help of a theoretical framework based on *in situ* experimental results.

The developed framework is based on the macroscopic description of the coupled phenomena obeying to the principles of Thermodynamics of Open Continuum Media (Biot[4]; Coussy[5]).

This approach allows to set up a general constitutive laws for the reactive porous media, following three steps :

1st Step : An Equivalent Continuum Medium is defined using the Representative Elementary Volume.

2nd Step : Each element of the Equivalent Continuum Medium is an elementary system, the thermodynamical equilibrium of which is defined by the following state variables: $\underline{\underline{\varepsilon}}$: elastic strain tensor; T : temperature; m : liquid mass supply and ξ : mass exchange between solid phase and liquid phase normalized by the initial volume.

3rd Step : The fundamental state equations linking the state variables are obtained by writing the first and second principles of thermodynamics. For instance, the mechanical state equation is written as follows :

$$\underline{\underline{\varepsilon}} = \frac{1+\nu_o}{E_o}\left(\underline{\underline{\sigma}} - \sigma_o \underline{\underline{1}}\right) - \frac{\nu_o}{E_o} tr\left(\underline{\underline{\sigma}} - \sigma_o \underline{\underline{1}}\right)\underline{\underline{1}} + \frac{b}{K_o}\frac{P-P_o}{3}\underline{\underline{1}} + \alpha_o^t (T-T_o)\underline{\underline{1}} + \alpha_o^c \xi \underline{\underline{1}} \quad (1)$$

$\underline{\underline{\varepsilon}}$: elastic strain tensor; E_o : drained Young's modulus; ν_o : drained Poisson's coefficient; $\underline{\underline{\sigma}}$: total stress tensor; σ_o : initial stress; b : Biot's coefficient ; K_o : drained bulk modulus; P : pore pressure; P_o : initial pore pressure; T: temperature; T_o : initial temperature; α_o^t : thermal drained linear expansion.

A chemical strain associated with the chemical reaction is involved :

$$\underline{\underline{\varepsilon}}^{ch} = \alpha_o^c \, \xi \, \underline{\underline{1}} \tag{2}$$

α_o^c : chemical drained linear expansion (obtained if P=P$_O$).

Using diffusion laws (Darcy's law; Fourier's law, Fick's law) and the different balance equations (brine mass, salt mass in solution and energy), a system of 4 equations (3 scalar and 1 vectorial) with 4 unknowns ($\underline{u}$: displacement; P: pore pressure; T : temperature; C : salt concentration) is obtained :

Hydraulic equation :

$$\frac{1}{M}\dot{P} + b\dot{\varepsilon} - \frac{(\rho_o^{fl})^2 L}{M\,T_o}\dot{T} + (-1 + \frac{\alpha_o^c K_o}{M\,b})\phi_o \frac{S}{V} K_c(C_s - C)$$
$$= Div(\frac{k}{\eta}\underline{gradP} + \frac{K_c^p}{\rho_o^{fl}}\underline{gradC} + \frac{K_t^p}{\rho_o^{fl}}\underline{gradT}) \tag{3}$$

Solute mass equation :

$$\phi_o(1 + C_o\alpha_{fl}^c)\dot{C} + C_o\,b\dot{\varepsilon} + C_o(\frac{b-\phi_o}{K_o})\dot{P} + C_o(\phi_o\alpha_{fl}^t - \alpha_o^t)\dot{T}$$
$$= Div(D\underline{gradC} - \underbrace{C\underline{U}}_{\text{convective term}} + K_t^c\,\underline{gradT}\,) + (1 + C_o b\alpha_o^c)\phi_o\frac{S}{V}K_c(C_s - C) \tag{4}$$

Thermal equation :

$$(m_o\,C_\varepsilon + \frac{\rho_o^{fl^2} L^2}{MT_o})\dot{T} + (\alpha KT - L\rho_o^{fl} b)\,\dot{\varepsilon} - \frac{\rho_o^{fl} L}{M}\dot{P} + \frac{\Re}{T_o}\phi_o\frac{S}{V}K_c(C_s - C)$$
$$= Div(\lambda\underline{gradT} + K_c^t\underline{gradC} + K_c^t\underline{gradT} - \rho_o^{fl}\,c^{f1}\frac{k}{\eta}\underline{gradP}) \tag{5}$$

Navier (mechanical) equations :

$$(K_o + \frac{4}{3}\mu)\underline{grad}(Div\underline{\dot{u}}) - \mu\,curl(curl\,\underline{\dot{u}}) - b\underline{grad\dot{P}}$$
$$-3\alpha_o^t K_o\underline{grad\dot{T}} - 3\alpha_o^c K_o\underline{grad}(\phi_o\frac{S}{V}K_c\left(C_s - C\right)) + \dot{m}\underline{F} = 0 \tag{6}$$

These equations can provide a basis for a finite element (or finite difference) computer program to handle coupled problems in the framework of safety analysis of storage.

To discuss the conditions for which one or several processes can be uncoupled, a monodimensional evolution was considered and the transport dimensionless equations were written. The dimensionless analysis show that the coefficients related to the kinetics of the chemical reaction impose a chemical equilibrium ($C=C_S$).

For rock salt, the terms related to the pore pressure (convection terms) have no effect on the salt mass transfer and the heat transfer due to the very low permeability of the medium.

Nomenclature :

α_o^t : thermal drained linear expansion;
α_o^c : chemical drained linear expansion;
α_{fl}^c : brine chemical linear expansion;
α^t : thermal undrained linear expansion;
α^c : chemical undrained linear expansion;
b : Biot's coefficient ;
C_S-C : volumetric undersaturation ;
c^{fl} : brine heat capacity;
$\underline{\underline{\varepsilon}}$: elastic strain tensor;
ε : trace of tensor $\underline{\underline{\varepsilon}}$;
ϕ : porosity;
η : dynamic viscosity ;
k : intrinsic permeability ;
K_o : drained bulk modulus ;
D : diffusion coefficient of salt in brine;
Div : divergence operator;
$\underline{\text{grad}}$: gradient operator;
K_c : dissolution rate;
K_b^a : coupling coefficient of the term $\underline{\text{grad}}$b related to conduction law on state variable a;
C_ε : heat capacity at constant volumetric strain;
λ : thermal conductivity ;
L : specific latent heat per unit value of fluid mass supply;
μ : shear modulus;
m : brine mass supply;
M : Biot's modulus;
P : pore pressure;
ρ^{fl} : brine density;
$\Re$: undrained reaction enthalpy;
S/V : ratio of surface aera of the mineral divided by volume of the solution;
σ_o : initial stress;
T : temperature;
curl : curl operator;

Subscript o indicates the initial state.

4 Contribution to the Safety Analysis of Underground Disposals in Saline Formations

4.1 Permeability Evolution of Rock Salt under Mechano-Chemical Stresses

The previous theoretical framework is used to discuss two phenomena involved in geochemical interactions leading to the changes in transport properties : (1) dissolution/crystallisation in microcracks due to brine solubility changes (2) local chemical potential gradient induced by local normal stress gradient.

In the first case, a simple permeability model, based on isotropic distribution of microcracks is used. Microcracks are considered to be the planes characterized by three microstructural parameters : average microcrack spacing, two mean characteristic lengths (length and half-aperture).

Brine solubility evolution induced by diffusion of pore pressure or diffusion of temperature is supposed to change the mean half-aperture.

On the basis of the previous theoretical study, the following system of equations with pore pressure P and temperature T as the unknowns, are written :

$$\left(\frac{1}{M}+\frac{b^2}{K+4/3\mu}\right)\frac{\partial P}{\partial t}+\left(-3\alpha_m+\frac{3\,aK}{K+4/3\mu}\right)\frac{\partial T}{\partial t}+\frac{\partial \phi^{chim}}{\partial t}=Div\left[\frac{k\left(\phi^{chim}\right)}{\eta}\underline{gradP}\right] \tag{7}$$

$$(\rho C)\frac{\partial T}{\partial t}=Div\left[\lambda_T\,\underline{gradT}\right] \tag{8}$$

M : Biot's modulus; b: Biot's coefficient; K : undrained bulk modulus; μ : shear modulus; α : thermal undrained expansion; α_m : thermal expansion of the mixture; k : intrinsic permeability; η : dynamic viscosity; ρ : salt density; C heat capacity; λ_T : thermal conductivity;

This system was completed by the laws governing the induced chemical porosity change ϕ^{ch} and the induced chemical permeability change k^{chim}. These laws are obtained through the permeability model, and by using an experimental equation linking salt solubility and brine state variables (temperature and pressure).

The results of coupled numerical calculations on a cylindrical cavity show that the dissolution-crystallization effect does not change the order of magnitude of the initial transport properties and remains much smaller than the influence of thermal loading and that of effective pressure (figure 1).

In the second case, a local chemical potential gradient induced by local normal stress gradient is supposed to produce a local mass transfer from grain boundaries towards the microcracks, leading to "healing" of the latter. At a microscale, the driving force is a chemical potential difference related to normal stress as follows (Olivella *et al.*[6]) :

$$\delta\mu_s=\phi(\sigma-P)V_m \tag{9}$$

ϕ : porosity; σ : applied hydrostatic pressure; P : pore pressure; V_m : molar volume.

The results of the calculations made by using the permeability model mentioned above are in agreement with those of Peach[7] (figure 2), and suggest that mechano-chemical coupled effect associated with viscoplastic deformation of grains can explain the extremely low permeability of rock salt at a geological time scale.

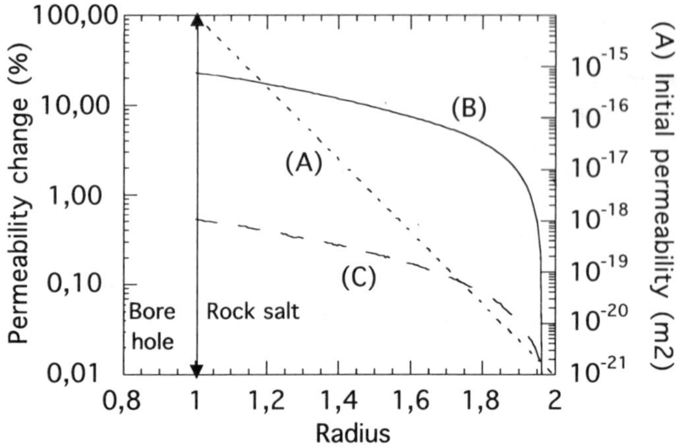

Figure 1. Permeability changes due to cristallization (C) and thermo-mechanical. loading (B) (both normalized by initial permeability) versus normalized distance from borehole wall Right : logarithm of initial permeability (A), close to the experimental data obtained at the W.I.P.P. site (Stormont.[1]). Boundary conditions are $P_0=P_1=12$ MPa; $T_1=10$°C at borehole wall (R=1, if R is a normalized distance from borehole wall); $T_0=50$ °C (initial temperature and temperature at the boundary of disturbed zone R=2).

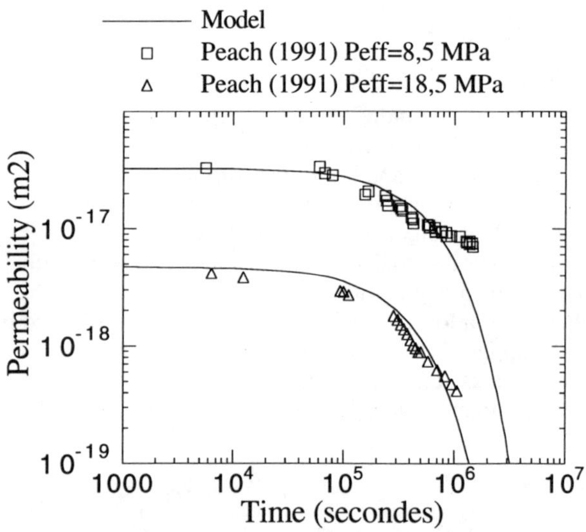

Figure 2. Comparison between experimental data and model . Peff=σ-Pf=8,5 MPa; confining pressure σ =10 MPa and pore pressure Pf =1,5 MPa. Peff=σ-Pf=18,5 MPa; confining pressure σ =20 MPa and pore pressure Pf =1,5 MPa.

4.2 Effect of Very Low Permeability of Salt on the Long Term Evolution of a Storage

To study the effect of very low permeability of salt on the long-term evolution of a storage, a simple calculation was made on a spherical cavern in a steady state evolution. Rock salt is considered as a poro-viscoplastic medium obeying Norton-Hoff's law based on a steady state creep constitutive model.

In these conditions, equilibrium pressure P_1 in the cavern satisfies the following equations :
- fluid mass balance in the cavern ;
- pore pressure diffusion in the rock mass;

From these equations, we obtained a non linear equation with P_1 as the unknown :
(if permeability k=constant in the rock mass)

$$\frac{1}{2}\eta_s(T_1)\left(\frac{3}{n}\right)^n\left(\frac{\sigma_o - P_1}{2\sigma_n}\right)^n = \frac{k}{\eta}\frac{P_1 - P_o}{r_o} \tag{10}$$

$\eta_s(T)$: salt fluidity, function of temperature ; n : exponent of viscoplastic criterion ; σ_o : initial hydrostatic stress; σ_n=1MPa; k : intrinsic permeability; P_o : initial pore pressure; r_o : cavern radius.
or
(If k=k(r) in the Disturbed zone)

$$\frac{1}{2}\eta_s(T_1)\left(\frac{3}{\mathrm{n}}\right)^{\mathrm{n}}\left(\frac{\sigma_o - P_1}{2\sigma_n}\right)^n = \frac{1}{\eta}\frac{P_1 - P_o}{r_o^2}\frac{1}{\int_{r_o}^{r_e}\frac{dr}{k(r)\,r^2}} \tag{11}$$

$r_e - r_o$: Disturbed zone thick.

Regarding the experimental results obtained at the WIPP site (Stormont[1]), the following equation $k(r) = k(r_o)\left(\frac{r}{r_o}\right)^\alpha$ (with $\alpha \approx -20$) can be considered for the permeability expression.

The results show that long term evolution of the storage is sensitive to the permeability, even very small non measurable (smaller than 10^{-22} m^2) (figure 3).

On the other hand :

- The higher the permeability value, the more the equilibrium pressure P_1 tends towards the initial pore pressure P_o. Reciprocally, the lower the permeability, the more the equilibrium pressure P_1 tends towards the initial lithostatic pressure σ_0.

• The equilibrium pressure P_1 is more sensitive to permeability in the small caverns (small radius and volume) at a shallow depth.

• For a given permeability value, the higher the exponent n (of the creep law) and salt fluidity η_s, the higher the equilibrium pressure P_1.

• In presence of a disturbed zone, the equilibrium pressure P_1 tends towards the initial pore pressure P_0, when the permeability at the cavern wall k_0 is high (at constant external radius r_e) or when the extent of the disturbed zone is significant (i.e. the parameter r_e increases at constant k_0).

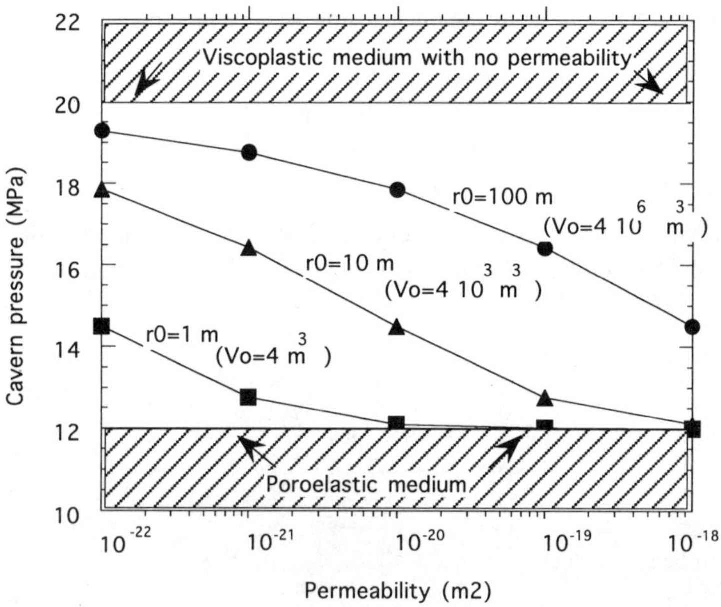

Figure 3. Fluid pressure in the cavern at a steady state for different values of permeability (Depth : 1000 m; Initial pore pressure : 12 MPa; Geostatic pressure : 20 MPa; r_0 : cavern radius).

5 Conclusion

On the basis of the experimental results and the theoretical analyses, it can be concluded that :

. The performed field experiment confirms that rock salt is permeable to gas and to brine, even relatively far (16 m) from underground openings.

. The measurements provided by the tests performed with brine were interpreted in a satisfactory way by a model based on Darcy's law (intrinsic permeability value of 2. 10^{-21} m^2).

. The proposed theoretical framework based on thermodynamics of open and reactive porous media allows to take into account coupled effects occurring in underground facilities in rock salt.

. The elaborated model shows that "healing" process for a damaged salt is partially controlled by mass transfer from grain boundaries to microcracks. This can explain the extremely low permeability of rock salt at a geological time-scale.

. A very low permeability which may not be measurable using present technology (permeability lower than 10^{-22} m^2) has a significant influence on the long term evolution of a disposal in a saline formation.

Acknowledgements

This research was performed under a contract with European Commission to which took part the Agence Nationale pour la gestion des Déchets Radioactifs (A.N.D.R.A.). The authors express their gratitude to these organizations.

References

1. J.C. Stormont, In situ gas permeability measurements to delineate damage in rock salt, *Int. J. Rock Mech. Min. Sci. & Geomech. Abstr.*, **34**, 1055-1064 , (1997).
2. Ph. Cosenza, M. Ghoreychi, B. Bazargan-Sabet, In situ gas and brine permeability measurements in salt, (Proc. of the 4th Conference on Mechanical Behavior of Salt, M. Aubertin, H Hardy Eds.; 3-16, Trans Tech Publications, Clausthal, 1996).
3. Ph. Cosenza, M. Ghoreychi, Coupling between mechanical behavior and transfer phenomena in salt (Proc. of the 3rd Conference on Mechanical Behavior of Salt, M. Ghoreychi, P. Bérest, H Hardy, M. Langer. Eds., 286-307, Trans Tech Publications, Clausthal, 1993).
4. M. Biot, General theory of three-dimensional consolidation, *J. Applied Physics*, **12**, 155-164 (1941).
5. O. Coussy, Mechanics of porous media, (John, Wiley & Sons, New York, 1995).
6. S. Olivella., A. Gens E. E. Alonso, J. Carrera, Constitutive modelling of porous salt agregates, (Numerical models in Geomechanics, Pande et Pietruszcak Eds, 179-189, Balkema, Rotterdam, 1992).
7. C. Peach, Influence of deformation on the fluid transport properties of salt rocks, *Geologica Ultraiectina*, **77**, Netherland (1991).

A PROGRAMME OF UNDERGROUND INSTRUMENTATION AND LABORATORY TESTS, TO UNDERSTAND THE ROCK MASS BEHAVIOUR ON THE MERENSKY REEF, BUSHVELD COMPLEX

B.P. WATSON

CSIR, Mining Technology, P.O. Box 91230, Auckland Park, 2006, Republic of South Africa

E-mail: bwatson@csir.co.za

A comprehensive underground instrumentation programme and geomechanical laboratory testing was carried out to understand the rock mass behaviour around pillar supported tabular stopes in the platinum mines of the Bushveld Complex. Very little inelastic movement was measured in the hangingwall but a large amount was measured in the footwall. A borehole camera survey, performed after mining had been completed, showed numerous open joints and fractures below the footwall contact. Observations made near the "crush" pillars indicated that the pillars were punching into the footwall. The uniaxial compressive tests showed that the pillar material was weaker than either the footwall or the hangingwall and that the footwall material was the strongest, indicating that punching into the footwall should not occur. However, the geomechanical behaviour observed in the laboratory tests indicated that the footwall material could be more susceptible to blast damage or to being split apart by the dilating pillar than either the reef or hangingwall. The point load tests indicated that the tensile strength of the footwall, in a horizontal direction, was relatively low which could have enabled extension fractures to develop in a high horizontal stress environment.

1 Introduction

An instrumentation programme and geomechanical tests were carried out to gain a fundamental understanding of the rock mass behaviour around pillar supported tabular stopes in the platinum mines of the Bushveld Complex.

A typical geological sequence above the Merensky Reef is shown in Figure 1. The stratigraphy consists of various igneous rock types that grade into one another. The only definite stratigraphic discontinuity is at the base of the Bastard Reef, which occurs 10 m – 30 m above the Merensky Reef. The dip of the strata is usually about 15^0.

There were two major joint sets in the hangingwall of the panel that was instrumented, as shown in Figure 2, the more persistent being on dip, the A set; the "B" set, being younger, often terminated on the "A" set. Both sets were almost vertically inclined.

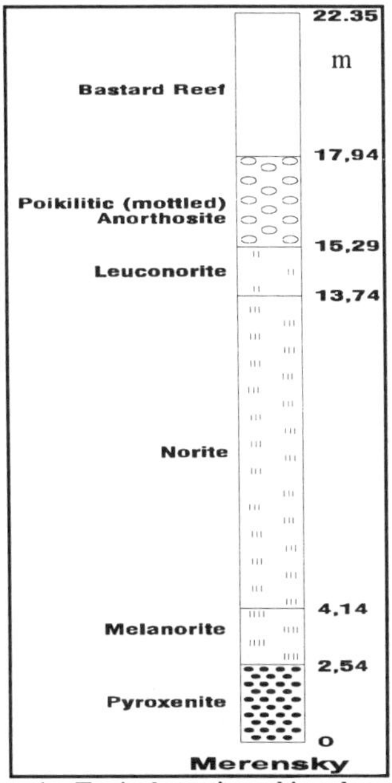

Figure 1 – Typical stratigraphic column above the Merensky Reef

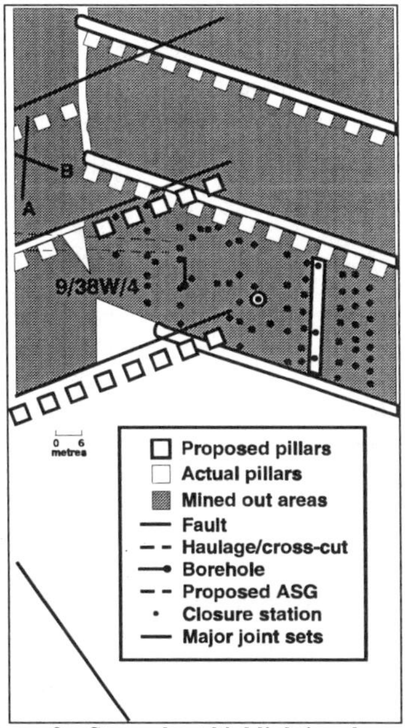

Figure 2 – Stope sheet highlighting the instrumentation panel

The instrumentation site had a typical layout for Merensky Reef mining, i.e. a centre raise from which 35 m panels are mined on breast, on either side of the raise. The individual panels were supported by 4 m x 4 m "crush pillars" (with 2 m ventilation holings) below each strike gully and 150 mm diameter timber props within the panel. The stope in which the instrumentation panel was located, consisted of five panels on each side of the centre raise and was separated from the upper stopes by 6 m wide barrier pillars, and from adjacent stopes by very narrow, variable width ventilation pillars on dip. The down dip limit of the stope was an abutment in which no stoping had taken place. The instrumented panel itself was against this abutment which could have influenced the results slightly.

The depth of the site was approximately 500 m, which indicated that the vertical virgin stress should have been about 15 MPa. However, stress measurements performed near to the site indicated that the k-ratio was close to two, with a vertical field stress component of about 23 MPa.

1.1 Instrumentation method and layout

Two 50 m long boreholes were drilled from the haulage below the Merensky Reef, to intersect the Bastard Reef, ahead of the mining face. Figure 2 shows the position of the holes relative to the instrumentation panel. An extensometer was installed in one of the holes to detect movement around the Merensky Reef and the other was used as an inspection hole.

The type and number of instruments installed as part of the investigation are shown in Table 1 and the positions of the borehole and closure-meters are marked in Figure 2.

Table 1 – Instrumentation programme

TYPE	INSTALLED	TYPE	INSTALLED
Extensometer	2	Geotechnical logging	1
Closure-meter	71	Timber prop load-cells	1
Crack-meter	3	Joint survey	1
Borehole camera survey	2	Geomechanical testing	475

Most of the closure-meters were installed less than 0,5 m behind the stope face. The first line of closure-meter stations was installed when the face was about 50 m from the borehole/reef intersection. Initially the closure-meters were installed on a 4 m grid pattern, but the cleaning equipment destroyed many of the stations before readings could be taken. This resulted in the scattered layout shown in Figure 2

Uniaxial and triaxial compressive tests were performed on core from the inspection hole to establish the geomechanical properties of the pillars and the immediate footwall and hangingwall of the Merensky Reef. Radial and axial point loads were performed at 0,1 m intervals along the core to determine if any strength abnormalities existed, and to calculate the tensile strengths and anisotropy ratios.

2 Results

2.1 Closure/ride-meter stations

A relatively high closure rate occurred within 6 - 8 m of the face and then reduced asymptotically with distance from the face as shown in a typical graph, Figure 3 (the exact locations of these closure stations have been highlighted in Figure 2). Closure also continued for periods greater than 30 days when no blasting took place, indicating creep behaviour.

In all cases more closure was measured at the top of the panel than the bottom and generally systematically increased with distance from the abutment.

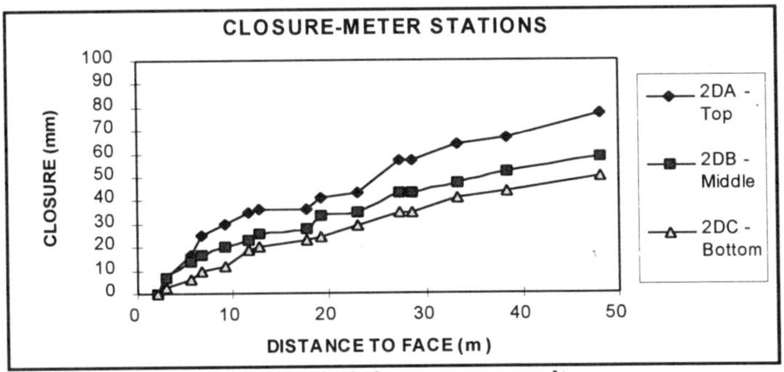

Figure 3 – Typical closure-meter result

2.2 Extensometers

Figures 4 and 5 show rock mass movements that occurred behind the face i.e. after the extensometer had been exposed in the panel. The positions of the reference points relative to the collars of the holes are shown in Figures 6 and 7. During the first metre of face advance, the footwall movement was equal to the measured stope closure and no hangingwall movement occurred. This trend did not persist and, with further face advances, the final combined hangingwall and footwall dilation constituted about half the measured closure.

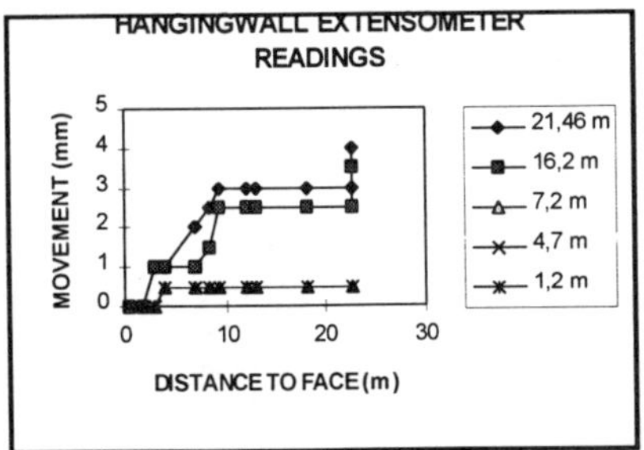

Figure 4 – Hangingwall extensometer results

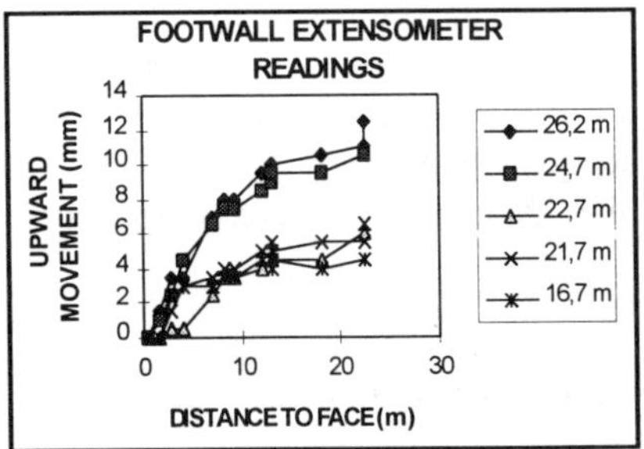

Figure 5 – Footwall extensometer results

Most of the measured hangingwall movement (Figure 4) occurred between 7,2 m and 16,2 m above the panel, but the inelastic component was small. The footwall movement occurred in two zones, between 0 m and 1,6 m (which was not measured being above the 26,2 m anchor) and between 3,1 m and 5,1 m (relates to 24,7 m – 22,7 m in Figures 5 and 6) below the panel. The dilation in the upper 1,6 m zone of the footwall was estimated by subtracting the elastic closure and the extensometer results from the measured closure.

2.3 Crack-meters

Joints of various directions and dips were monitored in the hangingwall, near the top, middle and bottom of the panel. No differential shear movement or dilation was detected.

2.4 Borehole camera survey

A survey was performed in the hangingwall and footwall inspection holes to observe the rock conditions where the extensometer had recorded movements. The section highlighted near the top of Figure 6 shows the region in the footwall anorthosite where a large number of open discontinuities were seen. This zone extended to about 5,5 m below the reef contact. The parting with the largest aperture was at 5,41 m, and it was the only discontinuity showing evidence of minor shearing. However, the horizontal component of shear was less than 1 mm. Many more, near-horizontal discontinuities were observed near the reef intersection than were in the original borehole core, indicating a fracture zone.

The dashed lines in Figure 7 show the positions of open joints in the hangingwall, most of which occurred on steeply dipping joints with the exception of the Bastard Reef contact. All movements were small and some of the inelastic component probably occurred ahead of the face. There was no evidence of very high horizontal stresses in the borehole.

2.5 Observations made in the panel

Loose rock was barred out of the footwall, to a depth of about 100 mm, next to a crush pillar that uncovered a major fracture zone, filled with rock powder. Generally, the immediate footwall appeared highly fractured whereas no mining induced damage was visible in the hangingwall.

Crush pillars in panels higher up in the stope showed a much higher degree of spalling and there was evidence of much greater closure within the panels.

2.6 Geotechnical log

The core from the visual observation hole was logged geologically and geotechnically. The aim was to identify any weak zones which could cause hangingwall or footwall movements, see Figures 6 and 7.

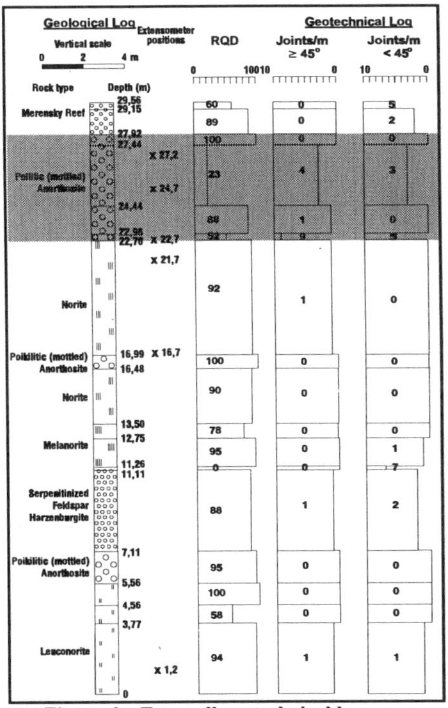

Figure 6 – Footwall geotechnical log

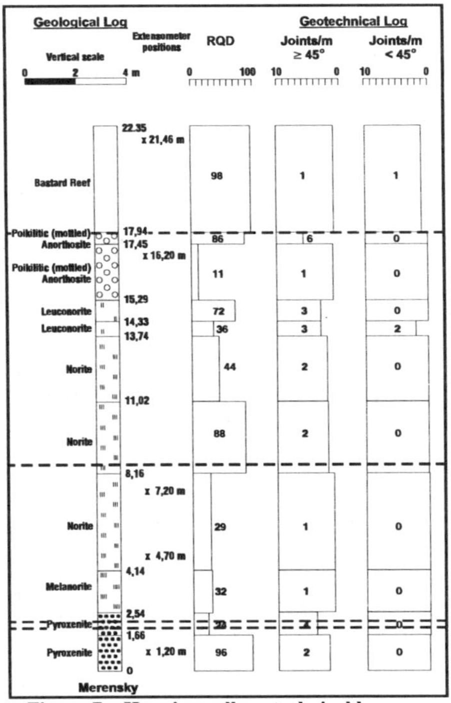

Figure 7 – Hangingwall geotechnical log

2.7 Geomechanical testing

A summary of the average uniaxial, triaxial and point load test results is shown in Table 2. The strength parameters were derived from straight line regression analyses, relating the axial stress at failure to the confining stress. The stress-strain curves for the triaxial tests of the footwall, reef and hangingwall are shown in Figure 8. It should be noted that, while inelastic radial dilation initiated at lower strains in the reef and pyroxenite, the radial dilation was greater in the anorthosite for any given strain. Similarly, the anorthosite had a larger radial strain, than the pyroxenite, for a given level of stress.

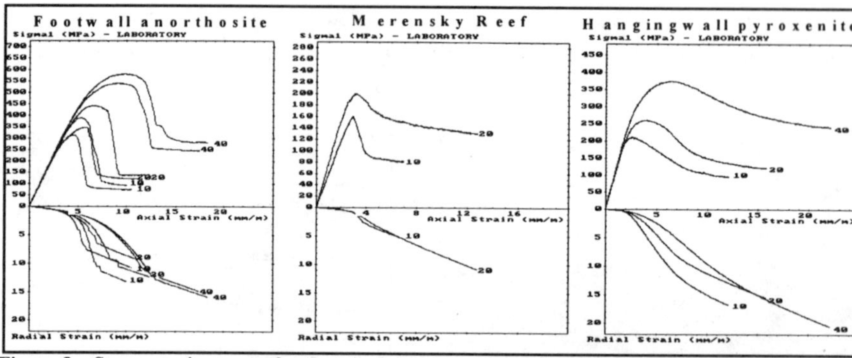

Figure 8 – Stress-strain curves for the triaxial tests performed on the footwall, reef and hangingwall rocks

Table 2 – Summary of laboratory derived rock properties for the footwall, reef and hangingwall materials

Rock Type	Thickness of the lithological unit (m)	UCS (MPa)	Cohesion at Failure (MPa)	E (GPa)	ν	Axial tensile strength derived from point load (MPa)	Radial tensile strength derived from point load (MPa)
Footwall Anorthosite	5,16	227	41,1	90	0,22	13,2	12,6
Merensky Reef	1,64	123	31,4	74	0,18	10,3	6,5
Hangingwall Pyroxenite	2,54	130	26,7	115	0,15	9,6	13,3

The equation used to calculate point load strengths was taken from C. Butenuth [1]

2.8 Three dimensional elastic modelling

MINSIM-W[2] was used to model the elastic movements in the stratigraphic interval covered by the extensometer hole (Figure 9) to compare the extensometer results against pure elastic movements. The elastic constants used in the model were Young's modulus 70 MPa and Poisson's ratio 0,25. A worst case scenario was adopted for the model, assuming that the crush pillars were not providing any significant support resistance, i.e. if any greater movement was measured by the extensometer than predicted by this model, then inelastic movements could be considered to have occurred.

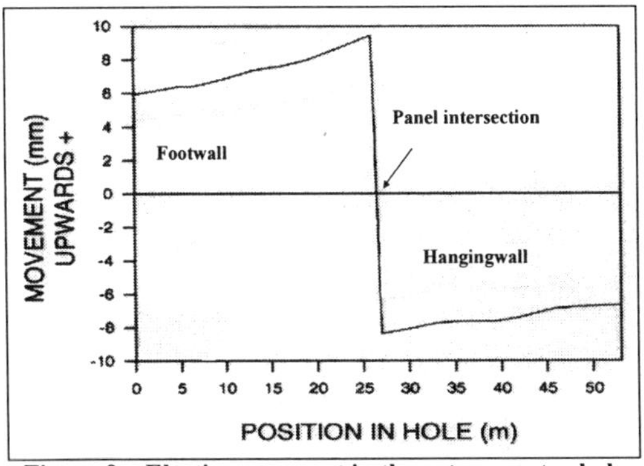

Figure 9 – Elastic movement in the extensometer hole

Figure 9 shows that the elastic component in the footwall interval was less than 3 mm and in the hangingwall section was less than 2,5 mm. The maximum elastic convergence in the panel would have been about 17,5 mm compared to measurements of 45 mm.

3 Discussion

A comparison between the results of the instrumentation and the elastic model shows that a large degree of inelastic movement took place in the footwall. The observations and instrumentation suggest that pillar punching was probably the mechanism. More evidence of a yielding system was indicated by the "stepped" profile of the closure curves (Figure 3). It was apparent that blasting in the panel above, and the cutting of pillars in particular, resulted in these "steps". Very little inelastic movement was detected by the hangingwall extensometer, which was confirmed by the crack-meters and borehole camera observations.

Most of the footwall dilation appears to have occurred in two zones. The upper zone, which was not measured by the extensometer, appears to have been fractured due to mining (the observed open discontinuities were many more than in the original borehole core), whereas the deeper zone could have been as a result of movement along existing discontinuities. However, the horizontal component of movement was small and only observed on one discontinuity in the borehole camera survey. Figures 6 and 7 indicate that there were a few more adversely orientated joints in the footwall than in the hangingwall, which could account for the movements observed in the lower zone. However, there was a competent 0,5 m beam in the immediate footwall that became highly fractured, and crushed at the edge of the pillars, as mining progressed. The large number of shallow dipping fractures in the upper 0,5 m is not explained by the bending beam or buckling theory.

About 21 mm of vertical displacement was calculated to have occurred in the crushed zone 0 m – 1,6 m below the footwall. This relatively large movement was confirmed by the observations made in the borehole camera survey.

The instrumented panel was against the abutment and it appears to have influenced the panel behaviour by causing a decrease in closure towards the bottom of the panel. Physical observations made of the pillars showed that they had not failed. This was confirmed by the calculated, relatively low average pillar stress, of about 75 MPa, acting on the pillars. This confirms the hypothesis that the pillars were punching into the footwall rather than crushing. Observations made in other panels showed that more closure appeared to occur in panels where "crush" pillars were located at both extremities of the panel. To better understand the mechanisms involved with this phenomenon, geomechanical tests were performed.

The footwall anorthosite, when tested normal to stratification, had a higher compressive strength than either the hangingwall or the reef, indicating that, if no discontinuities were present, footwall punching should not have occurred. The reef was the weakest material with the lowest "effect of confinement", thus being weaker than the other rock types at all levels of confinement tested, confirming that pillar punching should not have occurred. However, the comparatively high residual strength observed after failure at the 20 MPa confinement level indicated that the reef rock was approaching the brittle / ductile transition at this relatively low confinement (see Figure 8). Any confinement at or above the brittle / ductile transition level causes plastic flow to occur and the possibility of brittle failure is excluded. It is likely that at high vertical field stresses, relatively high confinement levels are generated within a pillar. A comparison of the results shown in Figure 8 shows that the footwall is by far the most brittle of the three materials. The brittle/ductile transition of the hangingwall material appears to be slightly higher than that of the reef but is probably still at a relatively low confinement (see Figure 8). If compressive horizontal stresses occur above the pillars, as is likely, this could prevent brittle failure. Also brittle stress fracturing and blast damage would probably be less than in a more brittle rock, such as the immediate footwall. For the same reason the hangingwall would also be less susceptible to being torn apart, by a dilating pillar or face, than the footwall at relatively high confinement levels.

In addition, the relatively high Poisson's ratio of the anorthosite could have caused a higher horizontal stress to develop in the footwall than the hangingwall. A high horizontal stress combined with the low tensile strength shown in the point load tests could account for the high degree of horizontal extension fracturing observed in the borehole camera survey.

4 Conclusions

Measurements and observations demonstrated that the "crush" pillars punched into the footwall, causing footwall damage to occur in two main zones, resulting in footwall heave. The deeper zone appears to have had movement, with a small horizontal component, along pre-existing discontinuities, while the shallower zone appears to have been crushed by the pillars and face. Relatively high horizontal stresses could have developed as a result of the geomechanical properties of the footwall material relative to the other rock types, which may have caused horizontal stress fracturing. The footwall anorthosite was much more brittle than the reef or hangingwall pyroxenite, implying that blast fracturing could develop more readily in the footwall than in either the reef or hangingwall. Dilation of the pillars or face could also have been part of the mechanism by tearing the footwall apart.

Acknowledgements

The work involved in this paper was sponsored by SIMRAC and they are thanked for allowing publication. The management of the mine involved are thanked for allowing the work to be carried out on their mine, for their co-operation during the project and for giving permission to publish the results. The following people are thanked for their valuable contributions in terms of discussions and insights: Mr A Jager, Mr G York and Mr K Kabeya.

References

1. C Butenuth (Rock Mechanics and Rock Engineering 1997)
2. MINSIM-W (boundary element code for calculating elastic stress and displacements 1997)

EXPERIMENTAL STUDY ON PHYSICAL MECHANICAL CHARACTERS OF QUASI-SANDSTONES

LIXING HUANG
Institute of Rock and Soil Mechanics, Wuhan
The Chinese Academy of Sciences, Wuhan 430071,China

This paper describes the unique characteristics of the rockmass surrounding the Yutan hydropower station. The unique characters has been studied by using laboratory and ***in-situ*** physical mechanical test of various parameters. The testing results provide a reliable scientific foundation to the design of the engineering project .

1 Background

The Yutan hydropower station in Dayong city is located in Xianhe Valley of the Lishui River's middle reaches. The mountains along the banks in the dam site are mostly over 800 meters in height with an average slope of 75 degrees. A general land form is formulated and can be described as "high in the west, north and south, as well as low in the east".

The rock strata emerged at the dam site belong to Banxi Group, Crefaceous System and Lacuna Carboniferous System. The Banxi Group consist of slate and epi-metamorphic sandstone. Beyond two km from the dam banks, the strata consist of limestone.

The unitary mono-declined strata in the dam site has a median-thick to thick attitude with a strike of 25-45° in general, inclining towards the upper reaches, and a dip of 25~35° .The faults are of small scale containing rather developed joints. The most developed joint set of 310~330°/NE(SW) 70~80° stretches 20~30m and even more along the strike with some joints deeply cutting the rock mass. The joint set of 30~40° /SE (opposite against the bedding) is also quite developed and it has a great cutting depth of 10~20 meters limited within the bedding. There are no large fault and active structure in the region but the strata are rich in underground water.

The lithological composition of the base rock in the dam site is quartz sandstone of Yuntaiguan Group of Miocene Epoch of Devonian Period, with quartz silt stone and silt shale (locally) in them. However, the base rock has mechanical properties obviously different to those of other similar rock mass, the differentiation for example is mainly that the elastic wave velocity of the former is 20~25% lower than that of the latter no matter the compressive strength of the base rock mass is normal. For this reason, the regional geologists call the base rock as "quasi-sandstone"

To clear the above described unique phenomenon, a series of lab and *in-situ* experimental studies on physical and mechanical properties of both rock specimens and rock mass have been carried out. In testing, the possible measuring errors caused by the factors such like measuring technique were fully removed.

2 Test methods and techniques

The research work include two parts: *in-situ* measuring and lab testing. The measuring items are mainly the velocity of elastic waves of rock mass, *in-situ* elastic module of the rock mass, wave velocity and strength of rocks.

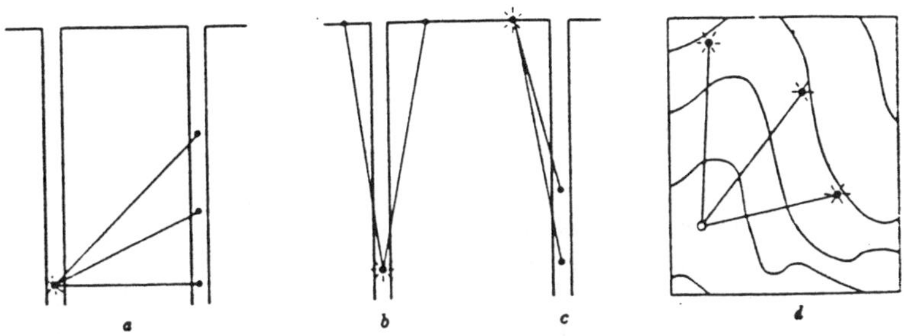

Fig.1 The schematic diagram of measuring elastic wave velocity

2.1 *In-situ Measurement*

According to the condition of the measuring site, four techniques has been used to measure the elastic wave velocity of the country base rock mass in the dam site, which are the methods of cross-hole(Fig.1a),upper-hole (Fig.1b),down-hole (Fig.1c) and fan-shaped emission (Fig.1d), and the characters of the country rock were assessed according to the testing results.

A dilatometer (Goodman jack) was used to determine the elastic module of the rock mass(Fig. 2).

Fig 2 The schematic diagram of dilatometer (Goodman Jack) measuring

2.2 Laboratory tests

Various rock specimens were collected from bore hole cores for laboratory testing. The testing items can be described as follows; varying law of acoustic wave velocity of specimens under uniaxial compression, composition of specimens by using x-ray diffraction meter, electronic microscope observation of specimen's structure and texture, void ratio distribution of specimens determined by using a mercury pressure meter, study on the stress history of the rock mass based on Kaiser effect by using AE technique (Fig3),determination of variation of specimen's acoustic wave and

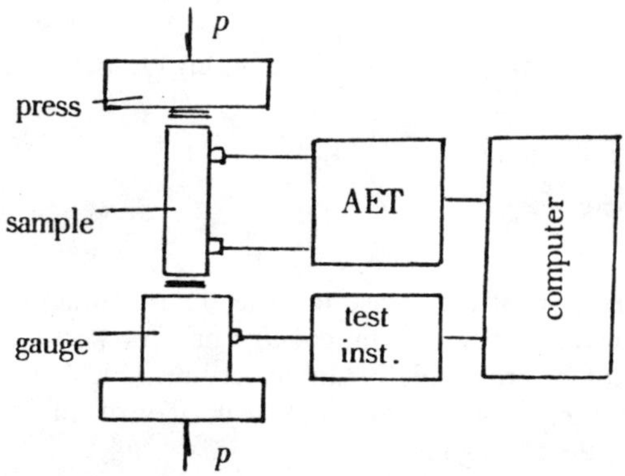

Fig 3 The schematic diagram of acoustic emission tests

effect of stratification on wave velocities under dry and saturated conditions, as well as determination of deformation and strength by using MTS testing machine.

3 Measuring results and analyses

3.1 *In-situ Measuring Results*

The layout of measurement is shown in Fig. 4.

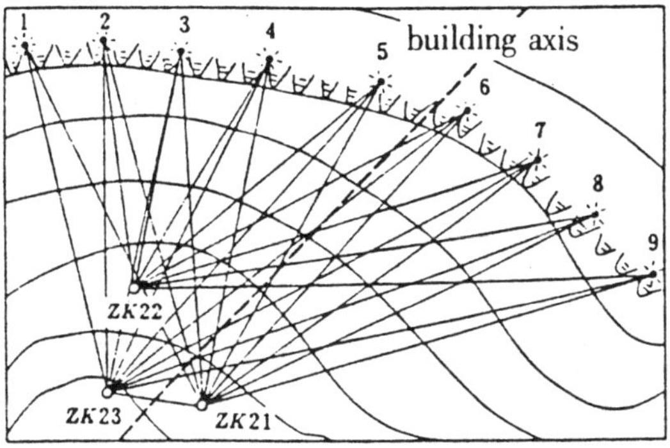

Fig. 4 Layout of monitoring points for comprehensive measuring

3.1.1 Cross-hole measuring

The cross-hole measuring was carried out in three bore holes. One hole was used to emit vibration and others to receive signals, altogether has been fired 29 blasting, resulting in 53 sets of measuring data corresponding to 23 spatial rays. Within the depth range of 42~60 meters while the underground chambers are located in the elevation of 210~230 meters, the measured average longitudinal wave velocity ordinary ranges from 2700 m/s to 3280 m/s but with a velocity lower than 2500 m/s at some measuring points.

The analysis on the information of bore holes indicates that the lower wave velocity occasionally happened because of the fractured zone in the rock mass.

3.1.2 Upper-hole measuring

6 measuring points have been arranged to receive the signals on the ground surface along the periphery with a diameter of two meters from the mouth of bore hole No.23 and 6 blasting have been fired to give 24 sets effective data. The blasting at the depth of 36~64m in the bore hole gave wave velocities ranging between 1921~2329 m/s

3.1.3 Down-hole measuring

3 blasting have been fired on the ground surface along the periphery with a definite diameter from the mouth of bore hole No.22 and signals have been received at the depths of 36~46 meters in the bore hole. The received 6 effective sets of measuring information show that the wave velocity ranged between 2307~2697 m/s

3.1.4 Fan-shaped emission measuring

Use a bore hole as a central measuring point to receive the signals. The special environment of a highway made a fan-shaped arrangement of blasting points. 23 blasting have been fired to produce 96 sets of effective data. The average velocity of

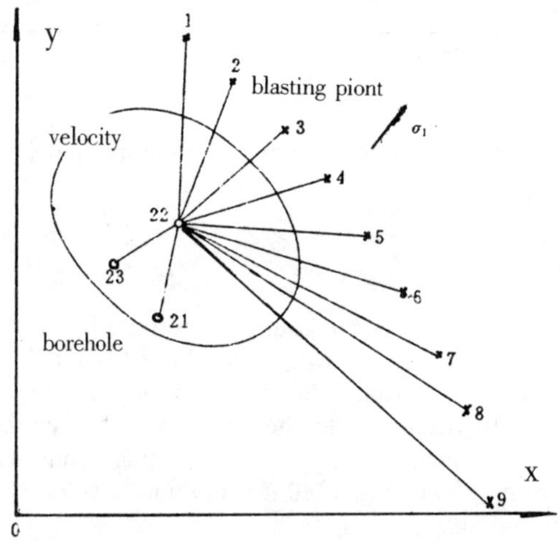

Fig. 5 The fan-shaped AE measuring

elastic wave propagation in the rock mass at the measuring points with an elevation of 220 m fell into the range of 2200~3300 m/s as shown in Fig.5

3.1.5 *In-situ* determination of rock mass elastic module

The determination of elastic module of the rock mass was performed in bore hole No.22. The measured deformation module of the rock mass along the direction of 270° towards North at the depth of 46m (an elevation of 220 m) is 1.93 GPa and the curve of pressure versus displacement is shown in Fig.6.

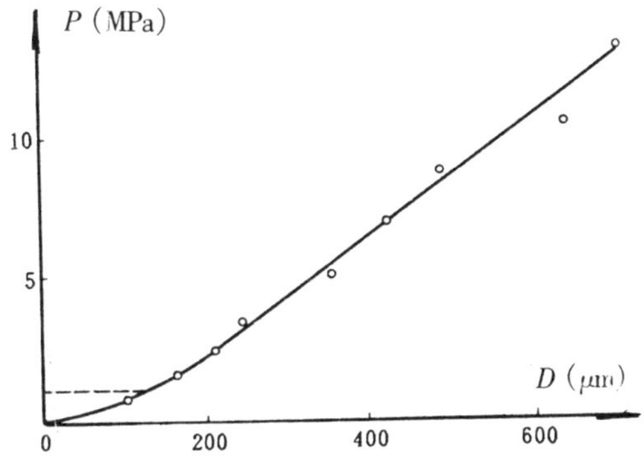

Fig. 6 The testing curve of pressure versus displacement by using a dilatometer (Goodman jack)

3.2 *Laboratory Testing Results*

3.2.1 Determination of elastic wave velocity of rock specimens

Tests were conducted on 26 rock samples, among which seven were tested to measure the variation of acoustic wave under pressure with the stress range of 0~8 MPa ,meanwhile, the elastic wave velocity of these samples have been measured after being saturated with water for 24 hours. The testing results have shown that these samples have a lower acoustic wave velocity and the increment of pressure(0~8 MPa) has a great effect on the acoustic wave velocity(Fig. 7). Compared with dry samples, the saturated ones have a 20~40% decrement of wave velocity (from 2600~3500 m/s to 1350~2270 m/s)

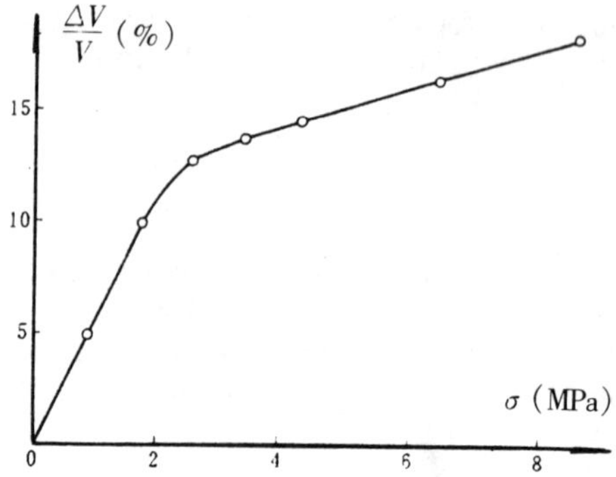

Fig.7 The relation curve of uniaxial stress-wave velocity

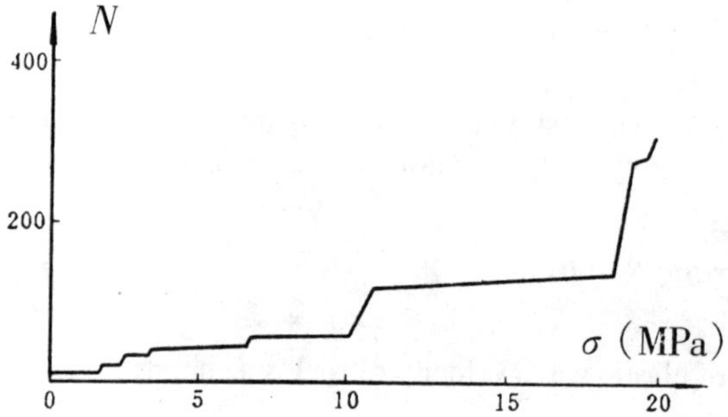

Fig.8 The testing result by Kaiser effect

3.2.2 Diffraction of x-ray

X-ray diffraction test were conducted on three specimens,collected from bore holes No.21,22 and 23 respectively. The testing results gave the lithological composition of the specimens as follows:quartz -90%, clay minerals - less than 5 %,hematite - less than 5% and quite low content of carbanate cements.

3.2.3 Electronic microscope observation

Three typical specimens overcored from the above mentioned bore holes were observed under electronic microscope. The testing results have shown that the specimens have a very low content of cements. The cements are loose in structure with obvious micro tensile cracks and the grains of uniform size are cemented in a contact mode.

3.2.4 Determination of void ratio

A mercury pressure meter was used to measure the void ratio of the typical samples collected from the bore holes. Finally, a void ratio of 8~12 has been obtained.

3.2.5 Acoustic emission measuring

Four specimens from the three bore holes were tested and a typical testing curve was obtained and shown in Fig. 8,from which it can be seen that the maximum principal stress of these samples in history was 11~18 MPa and there has an obvious turning point of AE activity at stress level of 2 MPa indicating a change in stress history.

3.2.6 Uniaxial compressive tests by using MTS testing machine

Three groups of rock samples were tested to understand their strength property and deformation ability. The tested specimens have a uniaxial compressive strength of over 200 MPa with exception of some individual ones with a strength of up to 280 MPa.

4 Conclusions

(1) Mineral components of the quasi-sandstone under consideration are: Quartz 90%,clay minerals - <5% , Hematite - <5% and quite few carbonate cements. The rock still belong to quartz sandstone according to the mineral composition classification criterion.

(2)The electronic microscope observation has shown that the rocks have a very low content of cements and the cements are loose in structure, present a contact cementing mode and contain obvious tensile cracks, which are the main reason causing the low wave velocity of both samples from rock or rock mass.

(3)The *in-situ* measured average elastic wave velocity falls between 2200~3300 m/s and the wave velocity of samples measured in lab falls between 2500~4000 m/s, about 20% more than those results obtained from rock mass. The *in-situ* measured elastic module is 1.9 GPa and the uniaxial strength of intact rock samples is over 200 MPa ,all falling into the normal range.

(4)The average acoustic wave velocity of the rock samples is about 3000 m/s. The wave velocity increases with the increment of pressure, the increasing rate is high up to 25 %. As compared with the dry condition, the saturated samples have a 20~40 decrement of wave velocity, which indicates that underground water play a very important role in design of engineering.

(5)Under a pressure of 2 MPa, Kaiser effect presents an obvious activity step of acoustic emission ,but *in-situ* rock mass is less effected by geo-stress.

(6)The analyses on the comprehensive testing results show that the phenomenon of low wave velocity in rocks and rock mass is normal because of its loose inner structure with a low content of cements of the quasi-sandstone.

(7)The high strength exists due to the quartz and its homogeneity and the low wave velocity is caused because of the rock structure and underground water. Above conclusions of research work bring great benefits to the design and excavation of engineering project.

Acknowledgements

Author heartily expresses his thanks to his colleagues in this research work, they are:Chunhe Yang,Jie Zhang, Wulin Wang,Yaping Wu, Nanyan Yu, Meiying Zhang and Zheliang Kang, of the Institute of Rock & Soil Mech. Wuhan, China

References

1 Sun Guangzhong, Book of Structure Mechanics of Rockmass, Publishing House of Science,Beijing,(1988)
2 Tao Zheyu, Book of Practical Cases of Rock Mechanic Application to Hydralic Engineering Projects in Foreign Countries and Empirical Data, Publishing House of Water Conservancy and Power,Beijing,(1976)

Ultrasonic and Micro-Seismic Measurements

DAMAGE PROCESS OF SEDIMENTARY ROCKS: ADVANCED PROCESSING OF ULTRASONIC WAVES

J.F. COUVREUR, J.F. THIMUS
Civil Engineering Department, Université catholique de Louvain,
Place du Levant,1, 1348 Louvain-la-Neuve, Belgium
E-mail: jeanfrancois.couvreur@fina.be, thimus@gc.ucl.ac.be

A. VERVOORT
Civil Engineering Department, Mining Unit, Katholieke Universiteit Leuven,
W. de Croylaan, 2, 3001 Leuven, Belgium
E-mail: Andre.Vervoort@bwk.kuleuven.ac.be

M.S. KING
Department of Earth Resources Engineering, Imperial College of Science,
Technology and Medicine, Royal School of Mines,
Price Consort Road, London SW7 2BP, United Kingdom
E-mail: m.s.king@ic.ac.uk

It is shown that appropriate processing of attenuation of P and S ultrasonic wave-trains results in more and better information on internal damage in rocks than conventional methods, such as for instance strain measurement and ultrasonic velocities. A number of uniaxial, triaxial and hydrostatic tests have been conducted on a chalky limestone under dry conditions. Based on the results of these tests, an interpretation of the ultrasonic parameters is validated in terms of the closure-opening-creation of cracks. This is followed by the proposal of a conceptual model. In a diagram plotting stress deviator versus confining pressure, the existence of correlations is revealed between deformation and damage fields and the variation of ultrasonic parameters (velocity, quality factor, energy). In particular, a detailed study of the quality factor results in detection of the critical failure onset in rocks. Energy is shown to be a new parameter to determine attenuation easily for practical purposes. The authors hope that the schemes proposed will be applied in the near future on other rock materials and will eventually become commonly used to determine the successive steps occurring in the damage process.

1 Introduction

The presence of existing as well as induced cracks and fractures significantly affects the mechanical behaviour of rocks. In the field of rock mechanics and petroleum engineering, it is therefore important to detect the existence of such weakness zones. Indeed, in many cases failure onset and growth must be controlled: borehole stability, underground constructions such as tunnels, mines and storage caverns, hazardous waste disposal, oil recovery from fractured reservoirs and geothermal energy extraction.

It is therefore of primary importance to describe the successive steps occurring in the damage process of rocks. The research presented here investigates if appropriate processing of ultrasonic attenuation of P and S waves results in more and better information than conventional methods (such as strain measurements and ultrasonic velocities). In particular, the research aims at detecting the critical failure onset of a sample under loading conditions.

2 Scope of the research

It is well known that ultrasonic wave **velocities** are influenced by a number of parameters, as for example density, water content, temperature and clay content [2]. However, the key factors for ultrasonic velocities related to rock failure are porosity and microcracking. Laboratory measurements of ultrasonic velocities in rock specimens as a function of the confining pressure have indicated that the presence of microcracks significantly influences the velocities [2,7,8,10]. If the load increases, velocities increase also because of the closure of existing cracks (*crack-like* or *soft porosity*), but it does not occur significantly if the specimen is only characterised by porosity in the form of well-rounded pores (*pore porosity*). Therefore ultrasonic anisotropy can result from the application of an anisotropic state of stress and crack closure in a preferential direction. As soon as all cracks are closed, velocities tend to stabilise. If the load is still increased and starts causing damage, velocities are influenced by the occurrence of cracks. The velocities are function of the direction of wave propagation and polarisation (in case of shear waves) and of the induced state of stresses [2,8].

Moreover, not only velocities can be obtained from the signal, but also **attenuation**. Modification of the waveform during propagation through the material is a measure of the nonelastic components of deformation [2]. Attenuation may therefore be more sensitive than velocity to defects, but it is more complex to analyse [8]. Attenuation is influenced by the same parameters as velocities but the effect is larger [2,6,8]. Attenuation should lead to a better understanding of the state and damage of the rock, but non-elastic interactions between waves and particles complicate the understanding of attenuation.

It can be concluded, therefore, that velocities of wave propagation appear to reflect the degree of microcracking and porosity of rocks. The velocities indeed increase with stress at low values of stress, but this effect becomes negligible at high values of stress comparable with the strength of rock. It is therefore unlikely to provide a mean of measuring failure approach, when such measurements would be of most interest. The question is whether a measure of attenuation is more likely than velocities to provide such a mean. This research aims to determine appropriate schemes employing ultrasonic attenuation to investigate successive steps in the damage process of rocks and to detect their failure onset.

3 Rock material tested

The rock tested was a chalky limestone from Caen (France), which is a highly porous and homogeneous rock. In table 1, the physical properties based on the ISRM recommendations[5] are summarized, while in figure 1 a thin section (1.a) and a CT X-rays scan (1.b) are presented. This rock type was selected because it is homogeneous, weak, highly dilatant and not clayey. Such properties are definitely favourable in the framework of the research.

Table 1: Physical and mechanical properties of the Caen limestone: γ: volumic weight, γ_s: specific weight, n: absolute porosity, n_a: apparent porosity, w_{sat}: water content in saturated state, c: cohesion, ϕ: friction angle, D_f: crack density, σ_r: unconfined compressive strength (UCS), σ_t: splitting tensile strength, V_p and V_s: P and S wave velocity, Q_p and Q_s: quality factor of the P and S waves, E: static and dynamic Young's modulus, ν: static and dynamic Poisson's ratio.

γ [kN/m^3]	γ_s [kN/m^3]	n [%]	n_a [%]	w_{sat} [%]	c [MPa]	ϕ [°C]	D_f [%]	σ_r [MPa]
18.6	26.4	29.6	25	15.6	7.7	17.2	19	12.6
σ_t [MPa]	V_p [km/s]	V_s [km/s]	Q_p	Q_s	E_{st} [GPa]	E_{dyn} [GPa]	ν_{st}	ν_{dyn}
1.9	2.61	1.58	94	309	5.5	11.3	0.33	0.22

A number of uniaxial, triaxial and hydrostatic tests were conducted under dry conditions. The experimental set-up continuously monitored the applied load, the axial and radial strains and the signals of ultrasonic waves. A coupling agent was used to improve the wave transmissibility through the interface between the transducers and the rocky specimen: an exhaustive study has determined lead foil as the best choice[3]. Ultrasonic signals are processed in three parameters: velocity, quality factor and an original one, the energy of the entire signal[4].

4 Results and discussion

4.1 *Mechanical behaviour*

The brittle failure of rocks under uniaxial compression has been investigated by many researchers. They have shown that the stress σ - strain ϵ curves could generally be divided in 4 phases, following the model of Bieniawski[1]

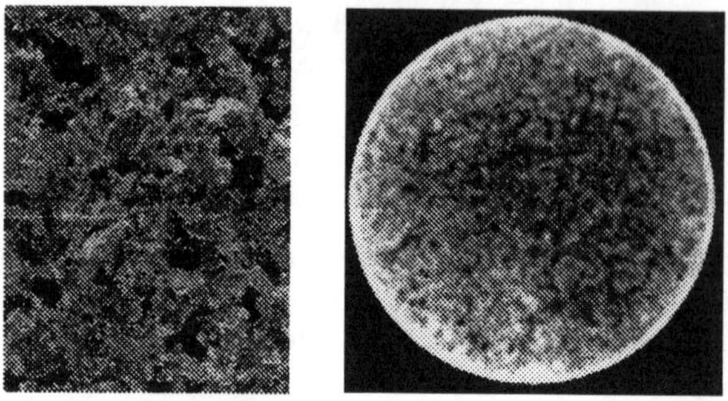

Figure 1: a. Thin section (2 × 3 mm) and b. CT X-rays scan (diameter: 50 mm) of Caen limestone.

(figure 2). The stress ranges relative to the strength σ_r are indicative: **0 to 10 %** σ_r **:** setting of the sample and the closure of existing cracks explain the slight upwards concavity; this is a non linear elastic region; **10 to 50 %** σ_r **:** linear elastic deformation; **50 to 80 %** σ_r **:** stable propagation of cracking; **80 to 100 %** σ_r **:** unstable propagation of cracking (dilatancy), possibly with rupture of grains, leading to macroscopic failure[11].

4.2 Velocities

During both uniaxial and triaxial compression tests, the behaviour of the velocities is in agreement with the common principles described in the second section. The velocities increase with stress at low values of stress, but this effect becomes negligible at higher values of stress, velocities therefore stabilise. At stresses close to rupture (at about 80-85 % σ_{1r}, i.e. approximately when dilatancy is appeared), V_s decreases. V_p is observed to decrease just before failure (about 90-95 % σ_{1r}), possibly due to coalescence. In conclusions, velocities are indeed sensitive to the presence of existing microcracks but they indicate the failure process, only at the start of dilatancy (unstable cracking). The early beginning of cracking (phase 3 of Bieniawski's model: stable cracking) cannot be determined by the velocities.

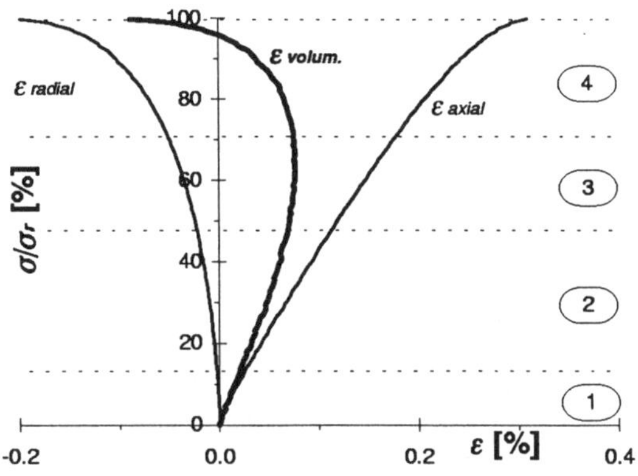

Figure 2: Failure mechanisms of Caen limestone under uniaxial compression conditions after Bieniawski's model.

4.3 Attenuation: quality factors and energies

While it presents some uncertainties during the crack closure and elastic phases, energy is suitable to determine dilatancy and crack coalescence. The quality factor is found well suited to distinguish the main steps in the damage process of the limestone: crack closure, elastic region, stable cracking and possibly crack coalescence. The quality factor Q_s indicates early onset of failure, well ahead of the occurrence of dilatancy, while Q_p appears to detect crack coalescence. To observe dilatancy, the quality factors are not so helpful, but energies and velocities are complementary to the quality factors as indicators of dilatancy.

4.4 The conceptual model

Based on the results of the tests (more detailed in [4]), an interpretation of the ultrasonic parameters is validated in terms of closure-opening-occurrence of cracks, followed by the proposal of a conceptual model. Such an original and synthetic display is inspired by the critical state concepts of soil mechanics [9]. In a diagram plotting stress deviator ($\Delta\sigma$) versus confining pressure (σ_3) in figures 3.a to .d, the existence of correlations is revealed between deformation and damage fields and the variations of the ultrasonic parameters (velocity, quality factor, energy).

From the results of each triaxial test, key points are displayed on figure 3.a: maximal stress or strength (-), dilatancy onset (black rhombus) and end of closure of existing cracks (x) as measured on stress–strain diagrams. Hydrostatic tests showed the stress level at which pore collapse appears for the dry Caen limestone (black circle). For reasons of clarity, results from uniaxial compression tests are not shown on the diagram, but they are not significantly different from those of the triaxial tests. Based on these stress-strain measurements, a deformational map (figure 3.a) is conceptualised to summarize the different damage mechanisms: crack closure, elastic region, stable cracking phase, dilatancy and pore collapse. For a stress deviator equal to zero, i.e. during hydrostatic compression, an "infinite" compaction of the material takes place: in the limit the mechanical and acoustic properties of the solid material itself (calcite in this case) will be reached. By following the vertical axis ($\sigma_3 =$ 0 MPa), the behaviour of a uniaxial compression test is analysed and the 4 phases of Bieniawski's model are obtained again (cf. figure 2).

Figures 3.b to .d present information provided by ultrasonic parameters selected to better understand the mechanisms of the damage process of the limestone studied: velocity V, quality factor Q and energy E for P and S waves propagating axially, i.e. parallel to the loading axis. In particular, the quality factor of the S wave (black triangles) shows the onset of critical failure, i.e. the start of the initially stable cracking. No other parameter gives an indication of this start. As with strain measurements, velocities confirm that they are sensitive to dilatancy which appears later in the damage process. In the case of crack closure, probably related to the geological in situ state of stress, all parameters indicate the end of this phase. Finally, energy is able to distinguish dilatancy and perhaps coalescence, but this has to be confirmed by further experiments. Energy has the advantage that it can be processed in real time (knowledge of the arrival time is not needed, nor is FFT) and it is efficient even in the case of a low signal-to-noise ratio.

A summary of this discussion is provided in table 2. Two findings are estimated to be new and original for rock mechanics and geophysics: 1. quality factor of S waves is able to determine the early onset of critical failure, whereas other ultrasonic parameters or more common measurements are not sensitive enough. 2. The energy of the entire signal is much easier than the other parameters, such as velocity or quality factor, to process in real time and under adverse recording conditions; and energy can distinguish the onset of dilatancy and perhaps crack coalescence.

Table 2: Information given by several ultrasonic parameters to highlight the successive steps occurring in the damage process of the dry Caen limestone ($\star$ means that the corresponding parameter is able to highlight the corresponding behaviour phase).

behaviour phases	V_p	V_s	Q_p	Q_s	E_p	E_s
end of crack closure	$\star$	$\star$	$\star$	$\star$		
beginning of stable cracking				$\star$		
dilatancy onset		$\star$			$\star$	
coalescence appearance	$\star$	$\star$		$\star$		$\star$

5 Conclusions

The results obtained so far show that a detailed processing of the P and S wave signals improves the understanding of the damage process. In particular, a detailed study of the quality factor results in the detection of the critical failure onset in rocks. Energy is shown to be a new parameter to determine attenuation easily for practical purposes. The authors hope that the presented schemes will be applied in the near future on other rock materials and will eventually be used commonly to determine the successive steps occurring in the damage process.

Acknowledgments

The authors are keen to acknowledge the experimental support of E. Bouchonville, B. Galopin & B. Sine at Université catholique de Louvain, and of R. Marsden & J. Dennis at Imperial College. They also thank Prof. G. Marchal for permission to use the CT-scanner at the Gasthuisberg University Hospital in Leuven. The first author was funded by the F.D.S. (Fonds de Développement Scientifique) of the Université catholique de Louvain. His stay at Imperial College was made possible thanks to the A. De Potter's grant from the Belgian Royal Academy for Sciences.

References

1. Z.T. Bieniawski. *Int. J. Rock. Mech. Min. Sci. & Geomech. Abstr.*, 4(4):395–430, 1967.
2. T. Bourbié, O. Coussy, and B. Zinszner. *Acoustics of porous media.* Technip, 1987.
3. J.F. Couvreur and J.F. Thimus. *Int. J. Rock. Mech. Min. Sci. & Geomech. Abstr.*, 33(4):417–424, 1996.
4. J.F. Couvreur. *PhD thesis.* Université catholique de Louvain, 1997.

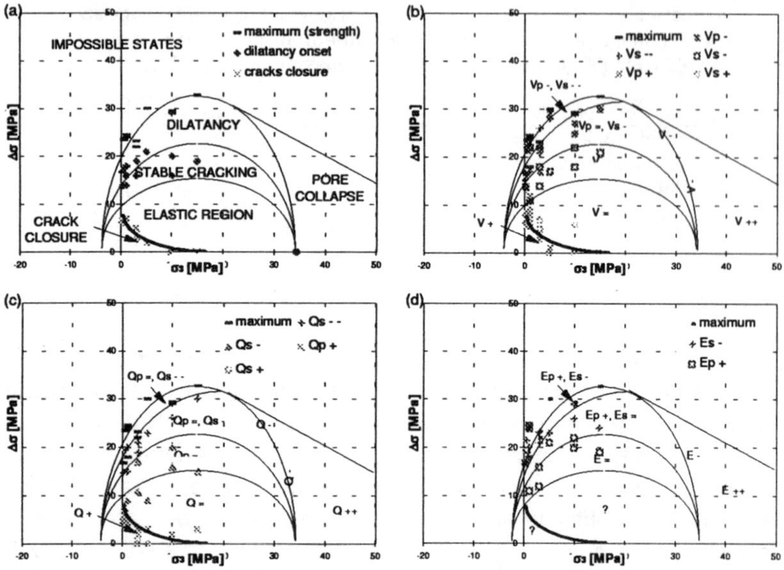

Figure 3: Damage mechanisms of Caen limestone described in a diagram with stress deviator versus confining pressure: a. strain and damage fields, b. ultrasonic wave velocities map, c. quality factors map (wave attenuation), d. energies map (attenuation of entire signal).

5. J.A. Franklin *et coll.* In E.T. Brown (ed), ***Rock characterization testing and monitoring***, chapter 4, pages 79–94. Pergamon Press, 1981.
6. D.H. Johnston and M.N. Toksöz. *J. Geophys. Res.*, 85:925–936, 1980.
7. M.S. King. *Geophysics*, 31(1):50–73, 1966.
8. D.A. Lockner, J.B. Walsh, and J.D. Byerlee. *J. Geophys. Res.*, 82(33): 5374–5378, 1977.
9. R.J. Marsden, B. Wu, J.A. Hudson, and J.S. Archer. In V. Maury and D. Fourmaintraux (eds), ***Rock at great depth***, pages 753–760, Pau, Balkema, 1989.
10. A. Nur and W. Murphy. In Brulin and Hsieh (eds), *4th int. conf. on continuum models of discrete systems*, pages 311–327, Stockholm, 1981.
11. M.S. Paterson. *Experimental rock deformation: the brittle field.* Springer-Verlag, 1978.

MICROSEISMIC MONITORING OF LONGWALL CAVING PROCESSES AT GORDONSTONE MINE, AUSTRALIA

X. LUO, P. HATHERLY
Division of Exploration and Mining, CSIRO, PO Box 883, Brisbane, Qld 4069, Australia
E-mail: x.luo@dem.csiro.au, p.hatherly@dem.csiro.au

B. MCKAVANAGH
Central Queensland University, Rockhampton Old 4702, Australia
E-mail: b.mckavanagh@cqu.edu.au

In this paper we present results from a detailed study on microseismic monitoring of longwall caving processes at Gordonstone Mine, Australia. The objective of the study was to investigate whether fracturing induced by mining activities extends up into aquifers above the German Creek Coal Seam. A seismic system comprising 27 triaxial geophones was installed above a longwall panel. 629 events were located. The microseismicity shows that the fracturing occurred mainly in the roof within the panel and directly correlated with the mining production. Significant rock failure was found near the working face. The intensity of the failure decreases rapidly with increasing distance from the coal seam. It is evident that the fracturing occurred 0-70 m ahead of the longwall face at seam level and formed a symmetric arched zone passing obliquely at about 50° back over the face. The fracturing in the zone is dominated by reverse faulting with a fault plane angle of about 50° sweeping back to the face. The maximum height of the detected breakage reaches 125 m above the coal seam being mined and this is 50 m below the base of the aquifers. Very few events were located in the mined area behind the face. Results from this work have opened up new understanding of longwall caving processes.

1 Introduction

Longwall mining commenced in the Bowen Basin of central Queensland, Australia in 1986 and undergone rapid growth. Raw coal production by longwall mining in central Queensland was more than 15.5 mt in the year 96-97, an increase of about 40% than on the previous year. The number of longwall coal mines in this region is set to more than double in the next five years. However the longwall operations in central Queensland do not always go smoothly. Production delays can occur due to unpredictable geomechanical problems.

Longwall mining results in a large volume of coal being taken out from underground, changing the stress distribution and causing the roof to cave and the floor to heave behind the working face. In the caving process the strata around the working area are in an unstable condition. Managing this is one of the major concerns in longwall mining.

Longwall caving is controlled by factors such as the dimensions of the longwall panel, the conditions of the roof and floor rocks, the thickness of the coal to be mined, geological structures, the size of pillars and the mining speed. Because longwall mining in central Queensland is relatively new, knowledge and experiences of caving processes are still limited. Microseismic monitoring has been successfully used in metalliferous mines for the determination of unstable regions (for example Young[1] and Mendeki[2]). This technique can dynamically map fracturing locations and mechanisms and so it could improve understanding of longwall caving processes.

Gordonstone Mine was the largest underground coal mine in the Bowen Basin. The German Creek Seam(GC) of 3 m thickness was being mined by the retreat method about 240 m below the ground surface. Above the seam is 170 m of Permian formations with a further 70 m of Tertiary volcanic and unconsolidated sediments containing aquifers. There are thin coal seams, Corvus(CO), Tieri 2(T2), Tieri 1(T1) and Aquila(AQ) above the German Creek Seam. Typical roof strata consist of interbeded sandstone and siltstone. The immediate roof is considered to be weak with its UCS values ranging 5-15 MPa and to have a high moisture content (Kelly et al.[3,4]). The regional stress is about 12 MPa and in a direction approximately parallel to the longwall panels. To establish the height of longwall caving and the potential for fractures to breach the Tertiary boundary, a microseismic monitoring system was installed (Hatherly et al.[5,6]).

2 Field Monitoring

The microseismic system deployed at Gordonstone Mine comprised 27 triaxial geophones. The longwall panel (LW103) which was monitored was 250 m wide and about 1700 m long. In order to accurately map the height of fracturing three boreholes in a triangular arrangement and spaced at about 350 m were drilled from the ground surface down to the floor of the German Creek Seam. In each of the boreholes nine triaxial geophones were grouted at different depths with eight below the Tertiary boundary at spacings ranging from 10 to 30 m. The other triaxial geophone was installed in the Tertiary layer 15 m below the ground surface. Fracturing which occurs in the roof between the coal seam and ground surface and within or near the network was expected to be detected using this configuration.

The field monitoring started on August 8, 1994 when the longwall face was 300 m away from the microseismic network, and terminated on November 15 when the face approached the end of the panel. During this period the longwall face advanced about 700 m and more than 530,000 m^3 coal was taken out from underground. Approximately 1,900 microseismic events were detected within this period. However the best data set was obtained from Sep 16 to Oct 29 when the seismic system was in full operation and the mining was close to the geophones. During this period 1,200 events were recorded.

However, not all of the 1,200 events can be located. Nearly half of the events were weak and contaminated by background noise. As the first arrival times for these could not be comfortably determined, these events could not be reliably located and therefore were rejected from the data processing. 629 events with clear compressional and shear wave arrival times were processed in this study. Figure 1 shows the locations determined for these events.

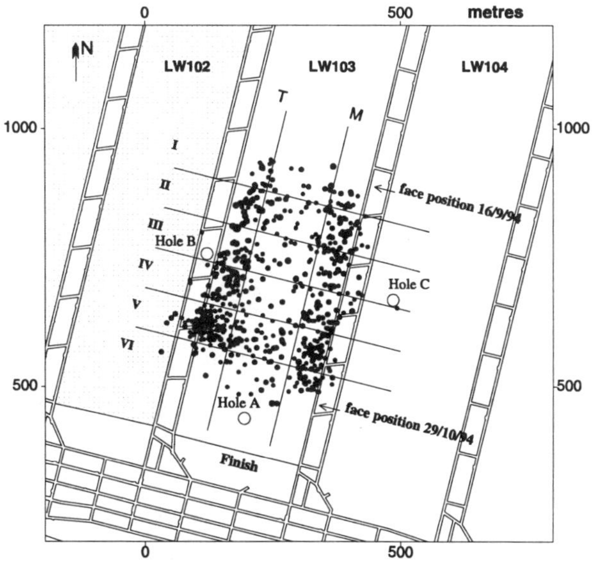

Figure 1: Locations of the three geophone boreholes, A, B and C and seismicity from 16/9/94 to 29/10/94 in plan view of longwall panel 103 (LW103) at Gordonstone Mine. LW102 had been mined and LW104 was under development. I, II, III, IV, V and VI label the regions for the vertical cross sections shown in Figure 2. Lines T and M define the boundaries of the tailgate and maingate regions for the vertical seismic sections shown in Figure 4. The finish line is where the LW103 mining was terminated.

3 Results

3.1 Patterns of seismicity

As shown in Figure 1, except for a few events, nearly all of the events are located within the longwall panel being mined. There is a symmetrical concentration of events to the sides of the panel (the tailgate and maingate) with a noticeable decrease in the centre. Figure 2 shows the events projected on vertical cross sections looking northeast along the longwall panel. Except for section VI, the event distributions in the roof are similar to each other, suggesting that the geomechnical response of roof strata to the mining progress is consistent along the longwall panel. Figure 3 shows the vertical cross section for all the seismic events. As with the plan locations in Figure 1, the rock failure defined by the zones of seismicity is symmetrical. The failure above the tailgate and maingate are confined by envelopes of the events at angles of about 16° from the vertical. Most of the events are located in a region extending from 20 m in the floor to 125 m in the roof.

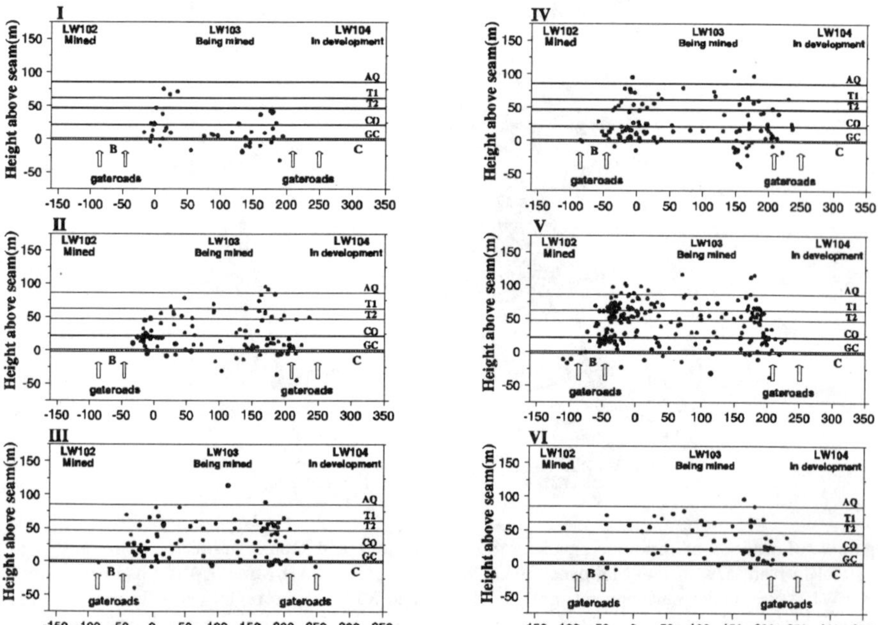

Figure 2: Seismic events for cross-sectional regions, I, II, III, IV, V and VI as shown in Figure 1, looking northeast along LW103. GC denotes the German Creek Seam. AO, T1, T2 and CO are thin coal seams. B and C represent the locations of boreholes B and C.

Looking northwest side on to the longwall panel, the microseismicity in the tailgate, central and maingate regions are shown in vertical cross sections in Figure 4. It is noticeable that more events were recorded as the face passed within the network. This is due to the proximity of these events. Apart from this, there is no significant variation in seismicity on the three sections, suggesting again that the geomechanical processes are constant along the longwall panel.

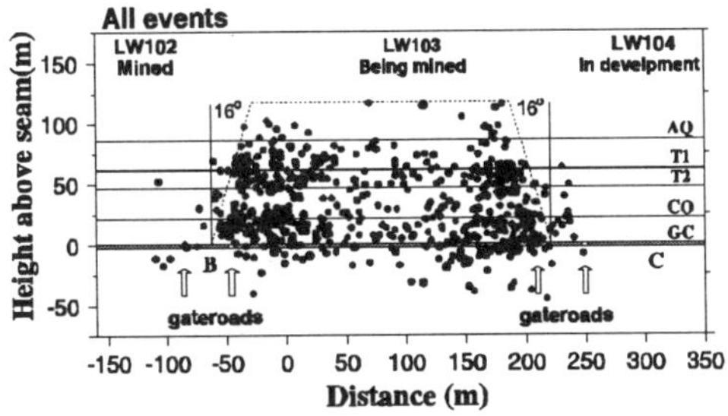

Figure 3: Projection of all events on a vertical section looking northeast. A failure zone in the roof is defined by the dashed lines.

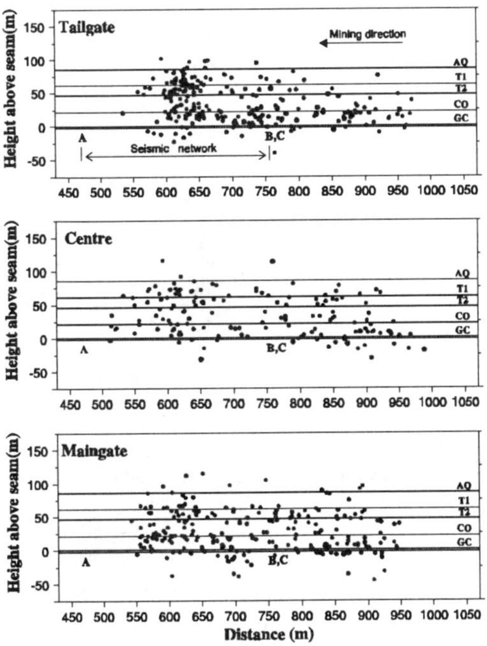

Figure 4: Vertical cross sections (looking northwest) of events near the tailgate (west of line T in Figure 1), in the central region between T and M, and near the maingate (east of line M). A, B and C are the locations of the boreholes.

Figure 5 shows the total and representative daily seismicity on vertical sections side on to the longwall panel. It shows clearly that the location of the fracturing is controlled by the progress of the face. When all the seismicity is normalised to a fixed longwall face, it is apparent that the fracturing zone is 0-70 m ahead of the face at the seam level. It passes obliquely at about 50 degrees from horizontal back over the face.

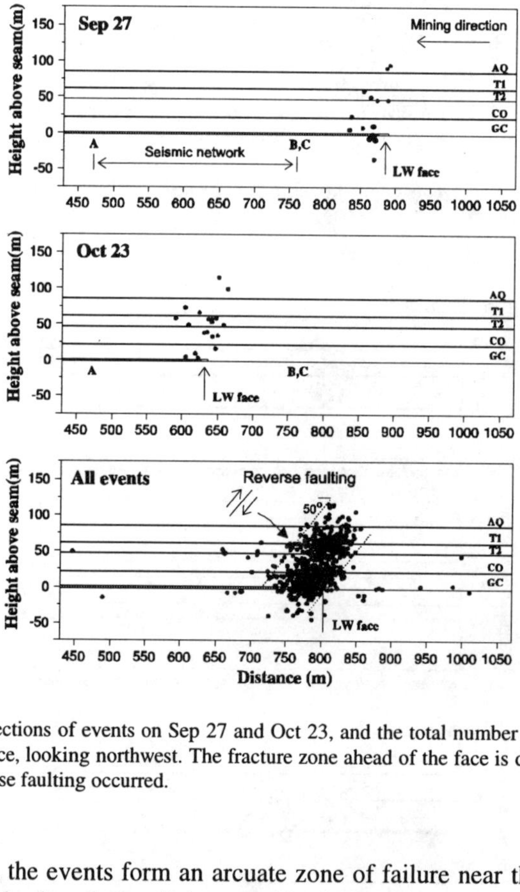

Figure 5: Vertical sections of events on Sep 27 and Oct 23, and the total number of events normalised to a fixed longwall face, looking northwest. The fracture zone ahead of the face is defined by the dashed lines. Failure by reverse faulting occurred.

In plan view, the events form an arcuate zone of failure near the longwall face. Figure 6 shows the locations of the events relative to a fixed face position for different height intervals in the roof and floor. With increasing height the arcuate zone extends back over the face.

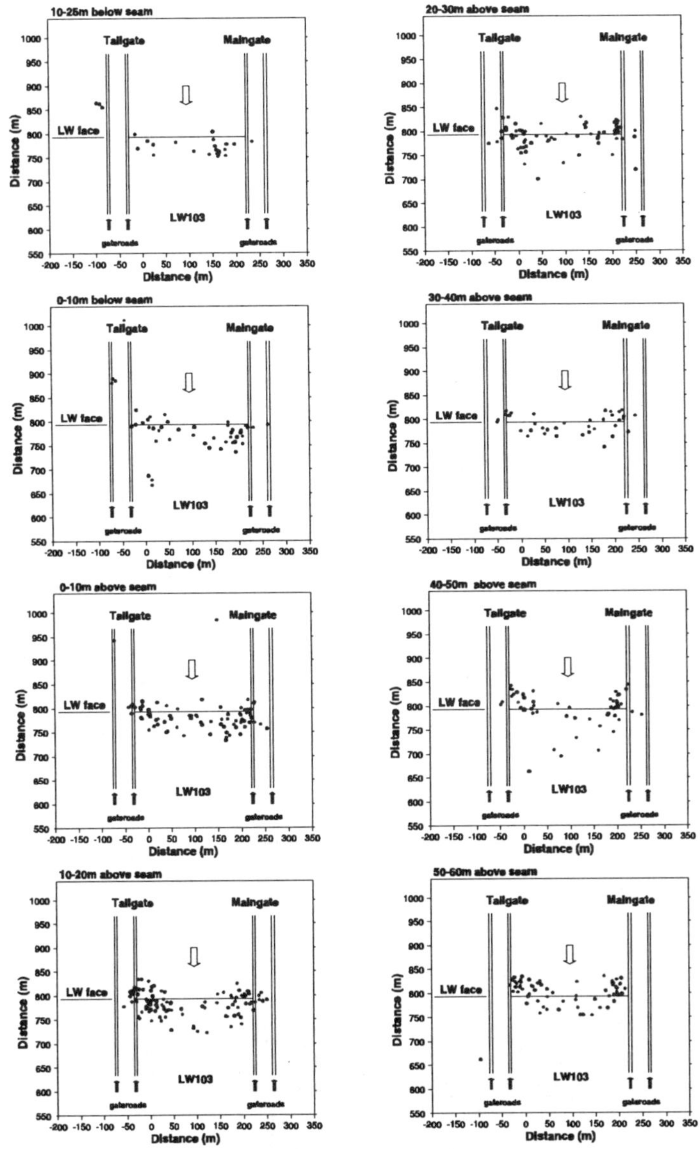

Figure 6: Plan view of event locations at different height intervals relative to a fixed longwall face position. The longwall retreat direction is marked by the arrow.

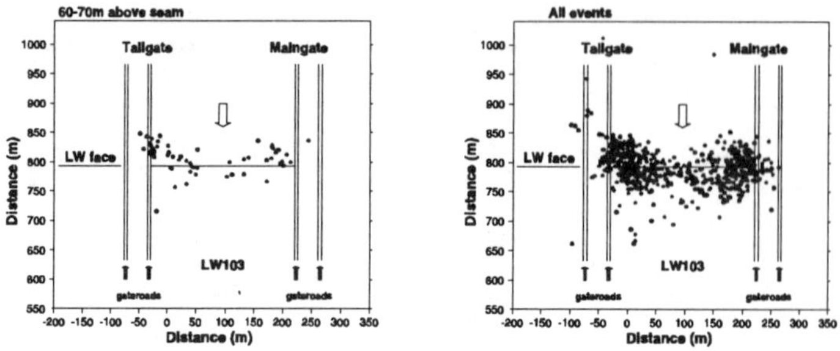

Figure 6: Continue

3.2 *Source mechanisms*

Analysis of source mechanisms was carried out using a moment tensor solution to selected events located ahead and above the face. The mechanisms are dominated by shear failure in the form of reverse faulting. The angle of the fault plane was determined to be about 50° from the horizontal. This is consistent with the dip angle of the fracture zone observed ahead of the longwall face shown in Figure 5.

3.3 *Correlation with mining*

The occurrence of microseismic activity correlated closely with mining activity. As shown in Figure 7 there is a direct correlation between mine production, represented

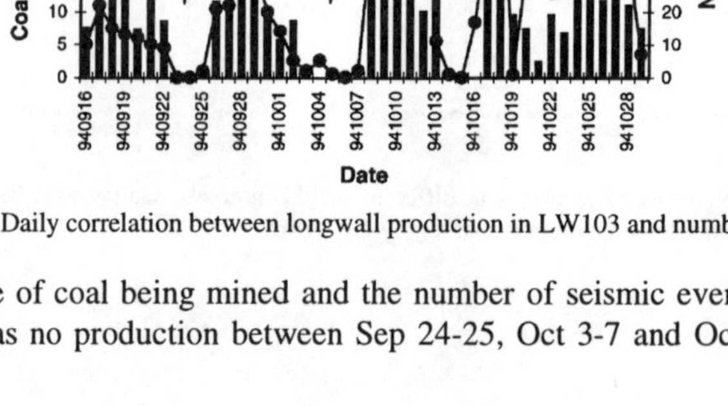

Figure 7: Daily correlation between longwall production in LW103 and number of events.

by the volume of coal being mined and the number of seismic events recorded per day. There was no production between Sep 24-25, Oct 3-7 and Oct 14-16 and the seismicity quickly cessed. The seismic activity resumed after commencement of mining. The longwall face entered the network on Oct 10 and the number of events increased from about 12 to 30 per day.

The microseismicity also shows cyclic characteristics. Figure 8 shows the number of events per hour over the period from 16/9/94 to 19/10/94. The cycle for active seismicity is about 3-5 hours. This pattern seems to correlate to mine shift changes and crib times. At Gordonstone, a shift change was made every 8 hours at 7.30 am, 3.30 pm and 11.30 pm, and crib time (rest time) was for about 30 minutes half way through a shift. At each of these times there was a temporary halt to the production. This seems to be reflected in the general pattern of seismicity.

The cyclic seismicity is another demonstration that rock failure is closely correlated with mining production. Apparently the rock failure reaches its maximum level about 2 hours after the start of mining.

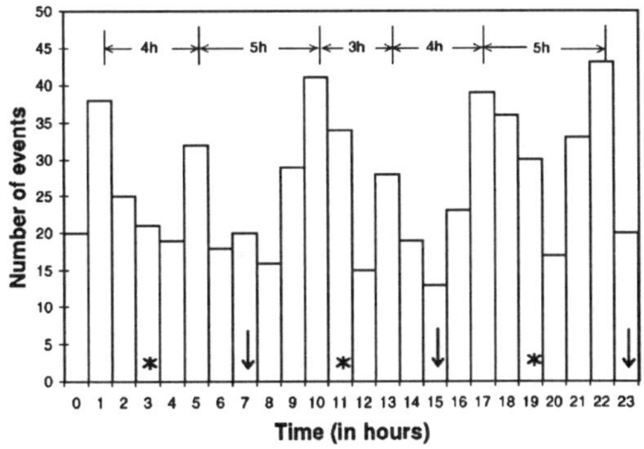

Figure 8: Hourly seismicity of a day showing cyclic seismic activity. The mine shift change times are marked by arrows and the crib time is marked by * .

3.4 *Vertical distribution of events*

The number of events plotted within 5 m height intervals is shown in Figure 9. Approximately 80% of the events are located above the seam. They are concentrated in two zones separated by an aseismic zone 35 m above the seam. This zone is also evident in Figures 3 and 5. The greatest number of events occurred 20 m above the seam, indicating that more fractures were induced in the immediate roof than elsewhere.

The intensity of the seismic energy correlates with the source process and size. An event with greater magnitude can be considered to be associated with a source of larger rupture length. Therefore events of different magnitudes at the same height would indicate fracturing of different intensity. Figure 10 shows the released seismic energy at different height intervals above the seam, in terms of differential magnitude. It was obtained by subtracting the average magnitude at a depth interval from the average magnitude of all events. Clearly the intensity of the fracturing is greatest near the working face and becomes weaker with increasing distance from the seam. At a height of about 100 m above the seam the intensity was very low. This height is still 75 m below the Tertiary boundary. Therefore no breaching into the Tertiary aquifers due to mining induced fracturing is expected.

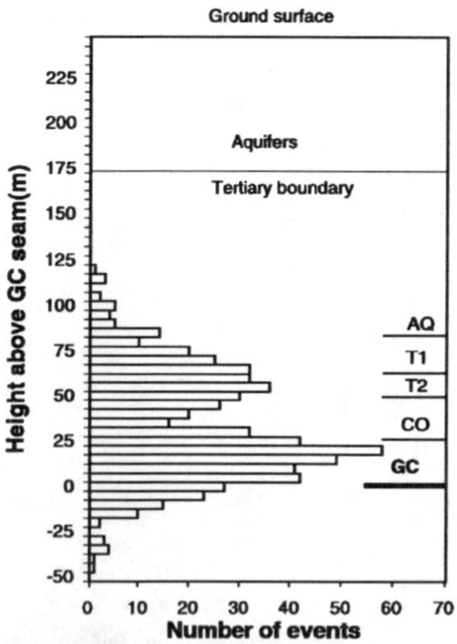

Figure 9: Number of events in 5 m height intervals

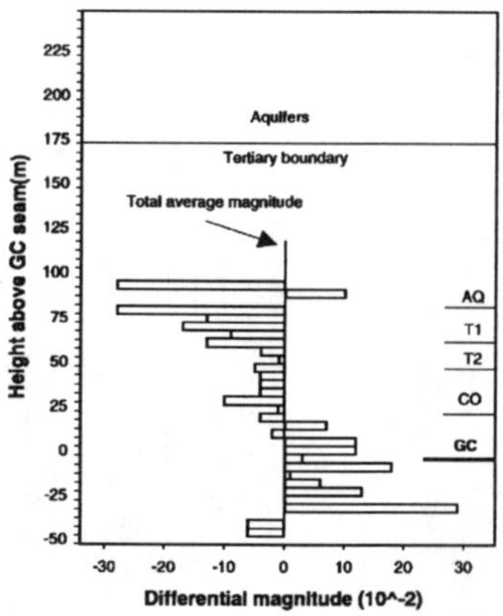

Figure 10: Differential magnitude in 5 m height intervals.

4 Discussion

4.1 Rock failure ahead of the face

Results from some previous longwall studies suggest that abutment stresses should only about 10-30 m in front of a longwall face (Thomas[7], Wilson[8]) and that its maximum value should be near the face. However, the microseismicity observed at LW103 clearly shows that rock failure occurred up to 70 m ahead of the longwall face. This suggests that the induced stress was built up and exceeded the strength of rock mass over that distance.

This result is supported by water pressure monitoring carried out for the same panel during the seismic monitoring period. Three piezometers were installed in a borehole in the center of the seismic network and at heights of 20 m, 55 m and 100 m above the seam. The lower two piezometers showed that changes in pore pressure first occurred at a distance of 160 m from the face, becoming more pronounced at about 80 m shortly before they ceased to function. The changes in pore pressure and the failure of the piezometers both imply failure at significant distances ahead of the face. Rock failure of 30-90 m ahead of a retreating longwall face was also observed from a microseismic experiment at the Foidel Creek Mine of Colorado, USA (Wilson and Cox[9]).

4.2 Why few events are located in the caving area

That there are very few events located in the caving area behind the face was a surprise to us. Studies (Christiaens[10], Zhu et al.[11]) have shown that tensile failure should occur in the roof near the face. The caving which occurs behind the face should result in tensile fracturing development in the roof. Kusznir et al.[12] have reported that events with collapse source mechanisms were determined in the caving area of a longwall in UK but indicated the events are very weak in energy.

It could be that the caving is a two stage process. In the first stage, the shear failure occurs ahead of the face as a result of the regional compressive stress and the abutment stress. These were the events located in this study. In the second stage the tensile failure could be along the pre-existing shear failures after the longwall passes by. The seismic events would therefore be very weak.

As a matter of fact, nearly half of the events recorded between Sep 16 and Oct 29 were not processed due to their weakness in energy and indistinct arrivals. It could be that some of these are associated with tensile failures. This conjecture will remain until an efficient approach to extract locations and focal mechanisms from this remaining data set is developed.

4.3 Aseismic zone

The aseismic zone 35 m above the German Creek Seam suggests that less fracturing was developed within this height range than in the two active seismic zones above and below (Figures 3, 5 and 9). There are no significant anomalies from geophysical logs at the same depth range. However there is nothing distinctive in the geological logs in this zone. It could be that it marks the transition between the caved zone below where the piezometer indicated a loss of pore pressure and a fracture zone above where the piezometer showed an increase in pore pressure but this still does not explain the reason for the existence of this zone. Perhaps slow creep along bedding planes occurred rather than brittle failure.

5 Conclusions

The microseismicity observed at Gordonstone was directly associated with longwall mining activity. It shows that rock failures occur mainly in the roof within the longwall panel being mined and with symmetric concentrations near the maingate and tailgate. More and stronger rock failures occur in the immediate roof and floor and they decrease rapidly with increasing distance from the working face. The maximum height of fracturing induced by the mining activity is 125 m above the working face, approximately 50 m below the Tertiary boundary. Water ingress from aquifers in the Tertiary sediments due to fracturing induced by longwall mining is unlikely.

The locations of the microseismic events show that the fracturing occurs 0-70 m ahead of the longwall face at seam level and forms a symmetric arched zone which passes back over the face at an angle of about 50°. The fracturing in the zone is dominated by reverse faulting with a fault plane angle of about 50°. An aseismic zone with less rock breakage is shown 35 m above the seam. Very few events were located in the mined area behind the face.

A seismic monitoring system comprising vertical geophone arrays is essential to accurately determine the height of rock failure. However, this method is relatively expensive due to the cost of drilling. Deployment of seismic detectors on the ground surface or in the mine workings will reduce the cost but result in a loss of vertical resolution. Results from this work have opened up new understanding of longwall caving processes and proved that microseismic monitoring is a very useful technique for studies of longwall geomechanics.

Acknowledgment

We are grateful to Mike Kelly and Prof S.S. Peng for their comments on this work. We would also like to thank Robert Dixon, Michael Berry and Zak Jecny for installation of the seismic system and transferring of seismic data sets. This project was funded by CSIRO Exploration & Mining, BHP Australia Coal, ACARP and Gordonstone Mine. We would like to thank these organisations for their supports.

References

1. Rockbursts and Seismicity in Mines, ed. P.R. Young (Balkema, Rotterdam, 1993).

2. Seismic monitoring in Mines, ed. A.J. Mendecki (Chapman and Hall, London, 1997).
3. M. Kelly, B. Lawrence and D. Devey, Roof Control at Gordonstone leads to Australian records, The Australian Institute of Mining and Metallurgy, Bulletin **6** (1994).
4. M. Kelly, W. Gale, P. Hatherly, R. Balusu and X. Luo, New understanding of longwall caving processes: results of microseismic monitoring and initial computational modeling at Gordonstone Mine, In *Proc. Symp. Geology in Longwall Mining*, p. 153-161, eds. G.H. McNally and C.R Ward (Springwood NSW Conference Publications 1996).
5. P. Hatherly, X. Luo, B. McKanvanagh, R. Dixon and D. Devey, Seismic monitoring for mapping longwall caving at Gordonstone Mine, In *Proc. Bowen Basin Symp.*, Geological Society of Australia, p. 117-126 (1995).
6. P. Hatherly, X. Luo, B. McKanvanagh and R. Dixon, Seismic monitoring of ground caving process associated with longwall mining of coal, In *Proc. 4th Int. Symp. on Rockbursts and Seismicity in Mines*, p. 121-124, eds. S. J. Gibowicz and S. Lasocki (Balkema, Rotterdam 1997).
7. An Introduction to Mining, ed. L.J. Thomas (Hicks Smith & Sons, Sydney, 1973).
8. A. Wilson, Stress and stability in coal ribsids and pillars, In *Proc. 1st Ann. Conf. on Ground Control in Mining*, ed. S.S. Peng (West Virginia Univ. 1981).
9. P.E. Wilson and R. M. Cox, Integrated monitoring and analysis techniques for evaluating longwall panel ground behaviour, In *Proc. 35th US Symp. on Rock Mechanics*, p. 587-592, eds. Daemem and Schultz (Balkema, Rotterdam, 1995).
10. P. Christiaens, Proposed calculation model for powered supports on a longwall face, In *Proc. 7th Int. Conf. on Strata Control*, p. 139-155, Liege, Belgium (1982).
11. D. Zhu, M. Qian and S.S. Peng, A study of displacement field of main roof in longwall mining and its application, In *Proc. 30th US Symp. on Rock Mechanics*, p. 149-156, ed. A.W. Khair (Balkema, Rotterdam, 1989).
12. N.J. Kusznir, I.W. Farmer, D.P. Ashwin, L.G. Bradley and N.H. Al-Saigh, Observations and mechanics of seismicity associated with coal mining in North Staffordshire, England, In *Proc. 21st US Symp. on Rock Mechanics*, p. 632-640, ed. D.A. Summers (University of Missouri USA 1980).

THE LABORATORY STUDY OF DEFORMING BEHAVIOR AND RELATED ACOUSTIC EMISSION OF DIFFERENT ROCKS

PETR KONEČNÝ
Institute of Geonics, Academy of Sciences of the Czech Republic
Bulharská 1426, 708 00 Ostrava, Czech Republic
E-mail: konecpe@ugn.cas.cz

The deformation of rock samples during uniaxial loading is a complicated process which differs from the pure elastic and/or elasto-plastic behaviour. The deviation of the real course of stress strain diagram from the theoretical one is, among others, determined by the nature of rock material, by its complicated structure and composition. The course of deformation and accompanying induced acoustic emission during uniaxial loading were studied. Three main phases were usually distinguished during the whole deformation process: the initiative one, in which the adjustment of rock specimen to the loading frame occurred, the middle one, preliminary with elastic deformation manifestation and the final one, in which more or less rapid development of cracks resulting in the fracture was observed. The comparison of the volumetric deformation of the sample with its acoustic emission enable to distinguish different behaviour of particular types of rock material.

1 Introduction

The qualification of the rock deformation and failure processes is one of the most important preconditions for geomechanical applications. The process is usually studied in laboratory conditions using the deformation and strength measurements on rock specimens. The common methods of such measurements are based on the determination of axial and lateral deformations during uniaxial loading of the rock sample and the arrangement of tests, as well as the calculations of characteristic data are standardised.

With respect to the complicated rock structure, attempts are still made to learn this process in detail using acoustic emission measurements. There are many publications dealing with this topic, which cannot be analyzed in this paper. Let me mention therefore only one of the first publications in Czech by Šimáně [4], who developed the method for acoustic measurements with respect to the rock material predisposition to rock bursts.

Sedimentary rocks from the Czech part of the Upper Silesian Coal Basin have different deformational behaviour depending on their mineralogical components and structure. Some of them manifest themselves as quite plastic material, the others as a very brittle one. To find out how is this behaviour influenced by the microcrack generation during the loading, special measurements of acoustic emission using a newly developed set of apparatus were realised (Knejzlík [2]).

The deformation of rock sample and the acoustic emission were registered simultaneously. The loading of rock samples was produced by mechanical screw press type with computerised control system which allowed the regulation of deformation velocity in a wide range.

2 Work hypothesis

The work hypothesis was based on the assumption, that the deformation process in a rock specimen is crucially influenced by the crack generation and that in the rock material the cracks arise at different levels of loading. So it is expected, that in the rocks with plastic behaviour the cracks are induced already at the very beginning of their loading. On the other hand the brittle rocks are characterised by the crack generation mainly during the final stage of load, before the peak strength (i.e. the compressive strength) is reached.

According to this hypothesis the following stages are usually observed during uniaxial loading of rock specimens (after Bock [1]):

-stage 1 (section OA on the deformation diagram Fig. 1), with a decreasing axial deformation and small lateral deformation. At this stage the adjustment of rock specimen to the loading frame occur. During this stage the first cracks are formed and their amount depends mainly on the quality of rock specimen preparation, namely on the flatness and parallelism of contact surfaces. Therefore this stage depends mainly on the quality of the sample preparation, not on the mechanical properties of rock itself.

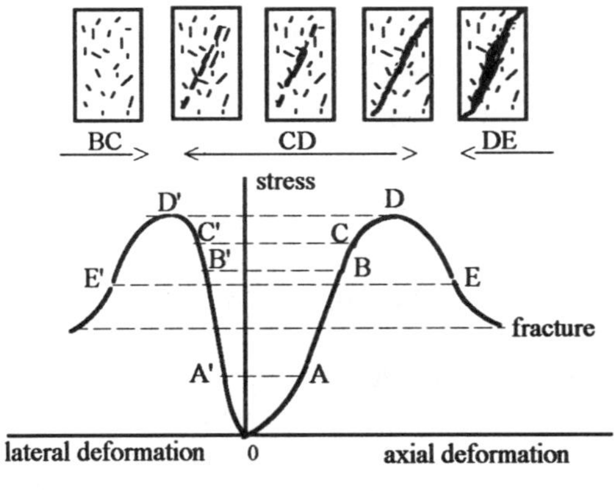

Figure 1: Scheme of deformation process and crack development during uniaxial loading of rock specimen (after Bock [1]).

- stage 2 (section AB) with the linear course of deformation diagram. Elastic behaviour of rock material is typical for this stage, no crack development is expected,

- stage 3 (section BC) where the increase of deformations is recognisable. In this part the cracks in the central portion of the loaded specimen are formed. These cracks are preliminary oriented in the direction of maximum shear or tensile stress. The stress strain diagram is no more linear, both, the axial and lateral deformations have an increasing trend,

- stage 4 (section CD) reaches the compressive strength. The increased number of cracks is quick at this stage. In the vicinity of the compressive strength the cracks join and they form the areas of fracture surfaces. The growth of axial deformation is accelerated and the lateral deformation, at the last stage in the vicinity of the maximum stress, is usually so large, that the volume of specimen starts to increase,

- stage 5 (section DE) is the post peak stage. The specimen has still some load - carrying capacity and the deformations occur mostly on the fracture surfaces, which were created during stage 4. The amount of cracks remains practically the same, almost no new cracks are generated. Behind point E the break down of specimen takes place.

The length of the above described stages is different, depending on the mechanical properties of the rock. It is very important for geomechanical practice, particularly for the assessment of the elasticity of rock material and its capability to store elastic energy, to distinguish particular stages of the deformation process. The aim of the experiments was to inquire whether the examined rocks behaved according to the above described hypothesis and whether the method used for measuring of acoustic emission made it possible to obtain some new information about the behaviour of rock during uniaxial loading for the appreciation of deformation behaviour of rock.

3 Laboratory equipment and procedure of measurement

The used set of apparatus was developed specially for this kind of measurements (Knejzlík [3]).

The main parts and the parameters of the device are:

- loading frame, which consist of the screw press type ZWICK 1494 with computerised control system which allowed the regulation of deformation or loading velocity in a wide range. Maximum loading force is 600 kN, stiffness of mechanical part is 4,3x108 $MPa.m^{-1}$,

- deformation are measured with the help of a chamber with strain gauge type sensors. The axial deformation is measured by 2 gauges situated symmetrically to the axis of the cylindrical specimen. The range of this measurement is 2 mm and the accuracy better than 1,5 %. Lateral deformations are indicated in 4 symmetrical points in the lateral area of the cylindrical specimen in the half of its height. The range is 1 mm and the accuracy also better than 1,5 % (Knejzlík, Konečný, Rambouský [2]). All measured values are taken off with the frequency up to 5 Hz and registered in the data acquisition system featured by data storage on the PC,

- the acoustic emission was measured with the newly developed laboratory device (Knejzlík, [3]). As the acoustic sensors the gauges of type S 9223, produced by the Physical Acoustic Corporation, USA were used. They allowed to measure in the frequency range from 5 kHz up to 300 kHz. Only one gauge was cemented by means of paraffine wax on the surface of the cylindrical specimen in the middle of its height,

The specimens were cut from the core sample with diameter 48 mm, the height of them was 96 mm. During the uniaxial loading of rock specimen with constant velocity of deformation about 0,5 mm/min the axial and lateral deformation and the parameters of acoustic emission, i.e. the maximum amplitude, first wave onset, time of arrival, time of phenomena duration and cumulated number of events was measured.

The analysis of data set, obtained from the measurement with the developed device and configuration of gages showed, that the primary indubitable information is, that an impulse came into existence, which implied that a crack in the rock specimen was created. The arrangement with one acoustic sensor only did not allow to calculate the position of the crack.

A very rough estimation of the crack energy was possible from the maximal amplitude of the first registered half wave. The result is influenced by the distance between the crack position and its orientation to the gage position, but it can be considered, that the higher is the maximal amplitude of the whole set of registered events, the higher is the acoustic energy and the larger are the induced cracks.

4 Results and their interpretation

More then 50 specimens prepared from different rock types were measured and evaluated up to now. Typical courses, prepared from measured data and evaluated using the Microsoft Excel are plotted on the Fig. 2a, b to 5a, b. The results of measurements on the selected samples are plotted on two types of diagrams:

The first one displays the axial, lateral and volumetric deformation depending on uniaxial stress. The volumetric deformation was calculated from the well known equation

$$\varepsilon_V = \varepsilon_L - 2 \times \varepsilon_Q \tag{1}$$

where ε_V is the volumetric strain, ε_L.axial strain and ε_Q lateral strain.

The acoustic emission is represented by points; the abscissa of each point corresponds always to the stress of the crack induction in MPa and the ordinate to the amplitude of the first half of the acoustic wave of the particular event (Fig. 2 to 5, labelled by a). It is necessary to remark, that these amplitudes were plotted only in the voltage units, because the device was not yet calibrated in absolute values of displacement.

The second type demonstrates the course of the development of the volumetric deformation and cumulative number of events, also depending on the applied stress (Fig. 2 to 5, labelled by b). For the representation reasons the volumetric strain was expressed in the scale 100 display unit equals 10^{-3} (i.e. per mile) and the stress was expressed in MPa, as usually.

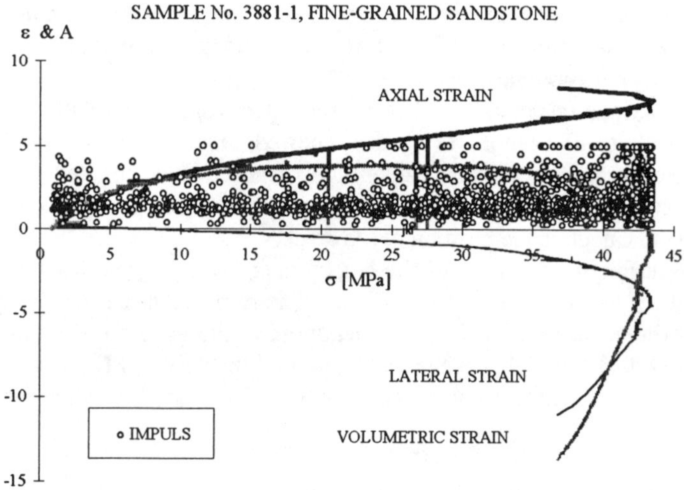

Figure 2a: Deformations and acoustic emission depending on stress. Weak type of fine grained rock material. Stress value σ is in MPa, axial and lateral deformation ε in 10^{-3} (per miles), amplitude of impulses A is in volts.

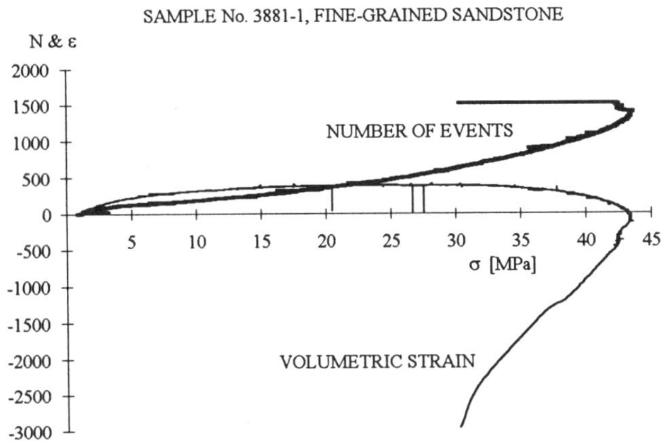

Figure 2b: Cumulative number of events N and volumetric strain ε depending on stress. Weak type of fine grained rock material. The volumetric strain is represented in the scale 100 display unit equals 10^{-3} (1 per mile).

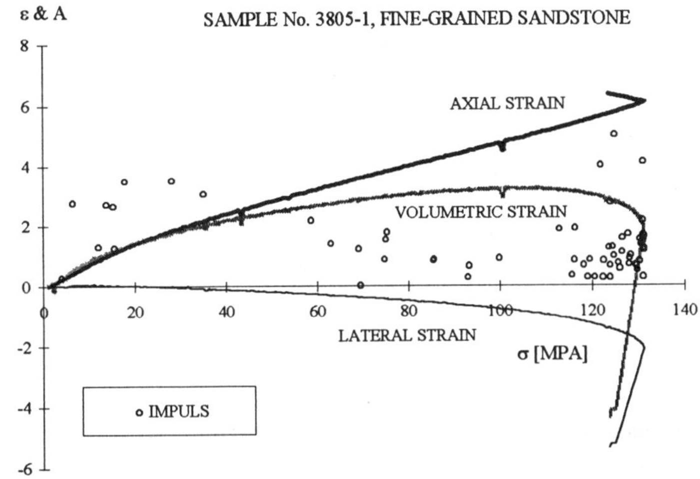

Figure 3a: Deformations and acoustic emission depending on stress. Hard type of fine grained rock material. Stress value σ is in MPa, axial and lateral deformation ε in 10^{-3} (per miles), amplitude of impulses A is in volts.

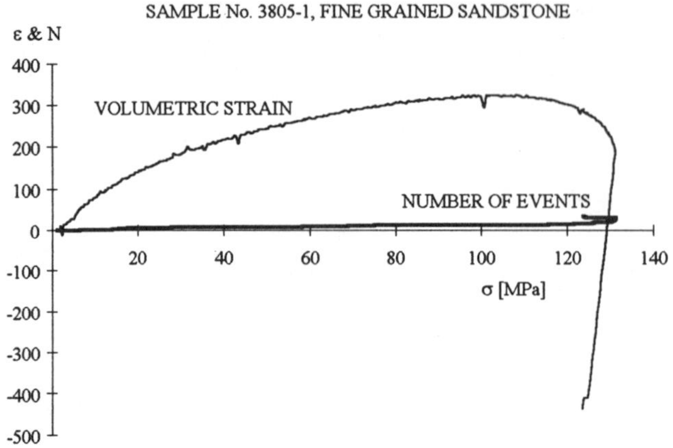

Figure 3b: Cumulative number of events N and volumetric strain ε depending on stress. Hard type of fine grained rock material. The volumetric strain is represented in the scale 100 display unit equals 10^{-3} (1 per mile).

An example of entirely different development of the acoustic emission was observed on the samples No. 3881-1 (Fig. 2a, 2b) and 3805-1 (Fig. 3a, 3b). From the structure point of view both samples were described as medium grained sandstone, but the second one was much more stronger that the first one (compressive strength approximately 43 MPa of the first and 130 MPa of the second one). The deformation of the weaker material was attended by very frequent acoustic emission during the whole loading process while the acoustic emission from the stronger material was much less frequent and limited above all on the final stage of loading. Though the deformations were of the same order, the course of volumetric deformation was different, too: the weaker rock material showed its expansion already in the middle stage of loading while the stronger material manifested the dilatation in the final stage of loading.

The acoustic emission from the third sample - the medium weak to strong, medium grained sandstone with compressive strength 51 MPa approximately, was of different character. The gradual acceleration of acoustic emission nearly from the very beginning of loading was observed (Fig. 4a, 4b) and the dilatation of material started also very soon.

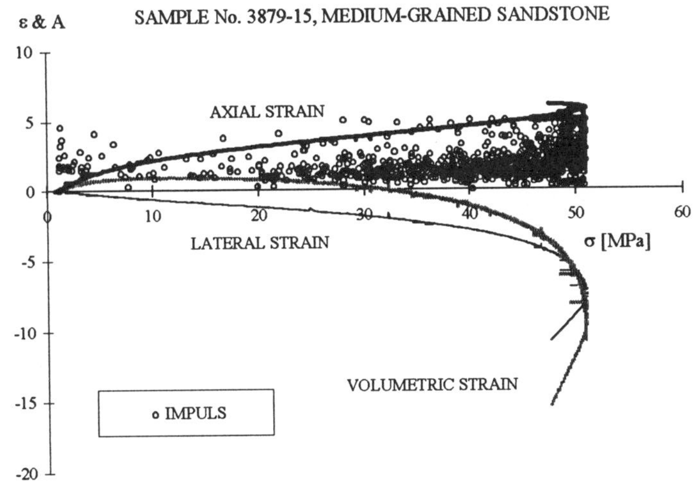

Figure 4a: Deformations and acoustic emission depending on stress. Medium grained rock material. Stress value σ is in MPa, axial and lateral deformation ε in 10^{-3} (per miles), amplitude of impulses A is in volts.

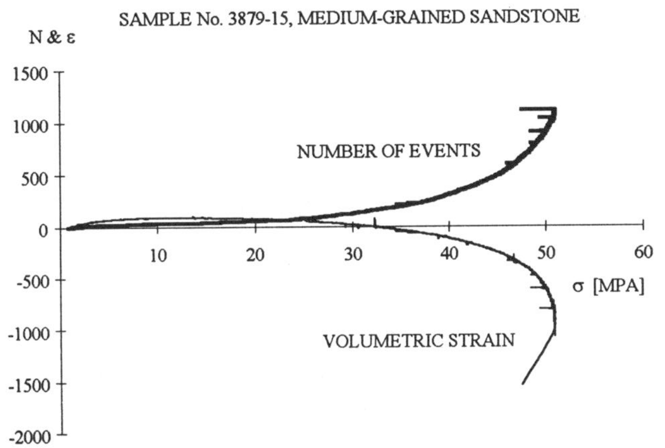

Figure 4b: Cumulative number of events N and volumetric strain ε depending on stress. Medium grained rock material. The volumetric strain is represented in the scale 100 display unit equals 10^{-3} (1 per mile).

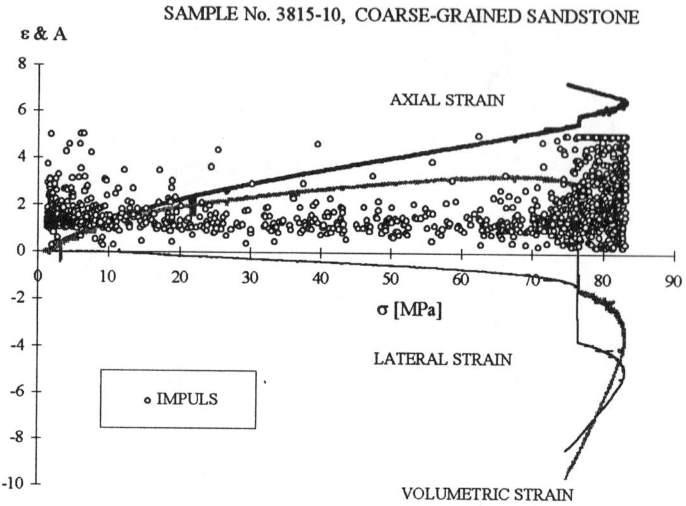

Figure 5a: Deformations and acoustic emission depending on stress. Course grained rock material. Stress value σ is in MPa, axial and lateral deformation ε in 10^{-3} (per miles), amplitude of impulses A is in volts.

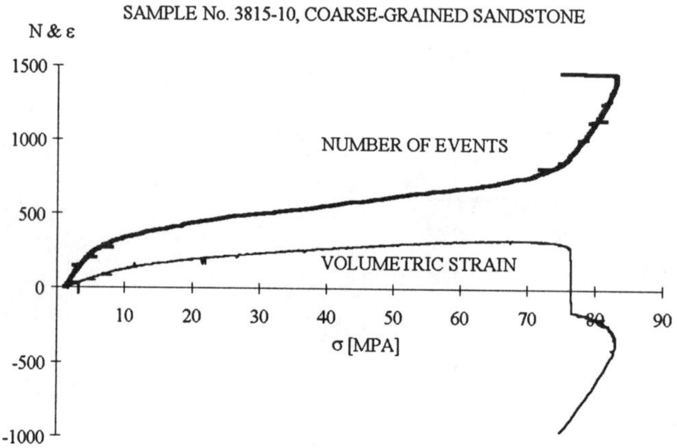

Figure 5b: Cumulative number of events N and volumetric strain ε depending on stress. Coarse grained rock material. The volumetric strain is represented in the scale 100 display unit equals 10^{-3} (1 per mile).

Finally the forth sample, which was a strong, coarse grained sandstone (compressive strength about 84 MPa), represented the behaviour with three stages of acoustic emission. Relatively high emission in the very beginning of the loading, lower emission in the middle part and accelerating emission in the final part of loading. Such behaviour was associated with pertinent deformations. Let us remark to this, that the sudden drop on the deformation curves might be caused by the arising of a spacious crack.

5 Conclusions

The measurements of acoustic emission, made on rock specimens prepared from sedimentary carboniferous rock from the Czech part of Upper Silesian Coal Basin proved, that in compliance with the working hypothesis, stages with different share of the crack induction, related to the whole deformation process during uniaxial loading of rock samples, can be distinguished. The extent of different stages depends on the strength and structure of the investigated rock type. The comparison of the cumulative number of events with the development of volumetric strain facilitates the good understanding of the character of deformation and crack induction processes.

Acknowledgements:

This paper has been elaborated at financial support of Grant Agency of the Czech Republic, Reg. Grant No. 105/96/1531 and the Project No. RA K10 12601

References

1. H. Bock: An introduction to rock mechanics. *Mir*, Moscow (1983).
2. J. Knejzlík, P. Konečný Jr., Z. Rambouský: Measurement of the stress - strain behaviour and acoustic emission of rock in the uniaxial strain. *33 Conf. of Experimental Stress Analysis*, Třešť, pp 129-133 (1995).
3. J. Knejzlík: The equipment for measurement of acoustic emission during laboratory tests. *Int. Symp. on Mine Planning and Equipment Selection*, Ostrava (1977).
4. J. Šimáně: Qualitative determination of the rock burst parameters of rocks. *Acta Montana 55*, UGG ČSAV Prague (1980).

STUDY ON STABILITY OF SHAFT WALL BY ACOUSTIC EMISSION MONITORING SYSTEM

SHULIN LI, KESENG YU, YUFA SANG

Changsha Institute of Mining Research, 236 Lushan South road, Changsha, Hunan 410012, P. R. China

E – mail: cimrod@public.cs.hn.cn

Monitoring of stability of the shaft wall by acoustic emission technique can overcome the disadvantages caused by traditional measurement of displacement and stress, and can realize the continuous and long-term monitor. This paper introduces the geological and engineering background of new auxiliary shaft at Fankou Lead-zinc Mine, AE monitoring system and its results. Based on these results, the stability of the shaft wall is analysed and assessed in using grey-cusp catastrophe coupling model and coefficient completeness of rockmass.

1 Introduction

Traditional method of analysing the stability of shaft wall is based on the measurement and stress on shaft wall. But it has a lot of disadvantages, such as heavy amount of measuring work, discontinuous measurement, affecting normal production, and difficulty in long-term measurement. In addition, there is no good theory on data analysis and data processing to assess the stability of shaft wall. However, acoustic emission monitoring system can overcome these disadvantages, realize AE monitoring all day long under the complicated stress state, and find out fracturing state of rockmass around shaft wall. Moreover, as it was connected with terminal computer directly, it is very convenient to analyse and process data. Unfortunately, there has been no existing theoretical analysing method which can be referred to for shaft wall stability by AE parameters so far.

There are few reports on rockmass stability analysis based on AE parameter now in China. The analysing method for rockmass stability by AE parameters is commonly based on in-site monitoring and empirical determination, and a critical value of fracture must be determined. However, destroying tests are not permitted in many rockmass engineerings to obtain this critical value. Because of complexity of rockmass projects, the value obtained from destroying tests has no universal significance. So, it is necessary to look for general analysing method theoretically. This paper introduces the AE monitoring system, its application at Fan kou mine and its monitoring results, and tries to explore theoretically analysing method of stability of shaft wall by AE parameters.

2 Geological and Engineering Background of New Auxiliary Shaft

New auxiliary shaft is one of the main production shafts at Fankou Lead-zinc Mine. It serves for miners, material, waste rock lifting and ventilation below-200level, and it is also a main shaft which should be stretched down for deep mining. It has a depth of over 600 meters, from the surface of the ground(+ 128.7 meters) to the present depth (− 475.85 meters), with a net diameter of 5.5 meters. It goes down through the following strata: 1) The surface residual soil stratum of Quaternary Period(Q). This stratum mainly consists of clay, and has a thickness of 21.2 meters. 2) Devonian Period Stratum. This stratum contains the following layers: (1) Tianziling upper subgroup limestone layer(D_3t^c). Rock in this layer is isotropic, close-grained and hard with grey, blocky and granopthyric textures. It has a thickness of 115.5 meters with a good stability. (2) Tianziling middle subgroup limestone layer (D_3t^b). The layer is composed of knotty, banded and blocky rock. Knotty rock contains soil highy and carbonilited locally, and has a poor stability when meeting with water; blocky rock has a fine texture and a good stability; the stability of banded rock is between that of knotty rock and blocky rock. The thickness of this layer reaches to 207 meters, (3) Tianziling lower subgroup limestone layer(D_3t^a). It is composed of oolitic, blocky and knotty limestone. Oolitic limestone is prone to fracturing when meeting with water. Its thickness is 144.76 meters. (4) Donggangling upper subgroup limestone layer(D_2d^b). It is grey dolositite with banded, fine grained and blocky characters. It has a good stability and a thickness of 118.24 meters.

New auxiliary shaft wall was supported by laying concrete over ± 0m, and by shotcrete under ± 0m. After it was put into use in 1984, splitting and peeling off locally on shotcrete lining occurred at times. In order to investigate its long − term stability, we cooperated with Fankou Lead-zinc Mine to undertake the research task, and established AE monitoring system in 1995 to monitor the stability of the shotcrete lining shaft wall between − 40m and − 400m. The system can monitor AE and measure the longitudinal sound velocity between every two adjacent levels.

3 Brief Introduction of Monitoring System

The system contains AE monitoring and sound velocity measurement. The monitoring range is from ± 0m to − 400m which is supported with shotcrete lining. It can be devided into two parts-emission part and receiving part. Emission part consists of an electric sparkling acoustic emission instrument with powerful power(DHH − 1) and two emitting geophones. It is used to produce an artifical acoustic emitting source to measure sound velocity propagating through rockmass. Receiving (or recording) part contains a 8-channel acoustic emission recording instrument (SDL-1), 8 geophones and a terminal computer. It is used to record artifical AE messages produced by DHH-1 and natural AE signals which occur in country rockmass of shaftwall. All the instruments are developed and produced by our own unit.

8 receiving geophones are installed on each of the eight levels on the shaft wall from −80m to −360m. Holes 55mm in diameters are perpendicular to the shaft wall with a depth of 1.8m in which receiving geophones are placed. Butter is used as a coupling agent between geophones and holewall, foam rubber plugs as a noise suppressant are mounted at the mouth of the holes. Two emitting geophones are mounted in two holes with 2 meters in depth under the receiving ones at −120 level and −280 level respectively. This two holes with a diameter of 80mm are heterotropic to the shaft wall with a 12° dipping angle to horizantal. Water is used as a coupling agent between emitting geophones and holewall. Geophones are connected with the instruments with coaxal shielded cables. In order that the provisional channels could be used wihle the normal channels are out of work, a channel exchange box are used between receiving geophones and SDL − 1. 8 levels can then be devided into two groups, each serves four levels such as −80, −120, −160, −200 and −240, −280, −320, −360 to monitor AE respectively. It can provide 3 spare channels and the other channel is used as the power line of the box. The monitoring station is located on the surface of the ground. Figure 1 shows the block diagram of the system.

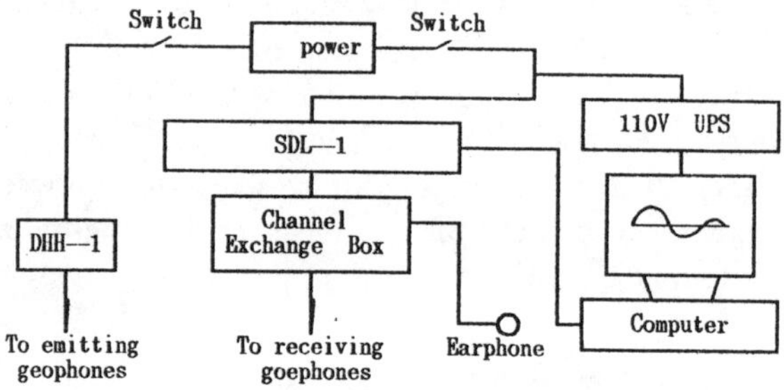

Figure 1: Block diagram of the monitoring system

4 Monitoring Results

The monitoring system are constructed and adjusted successfully in July, 1995, and was formally put into use the following month. We undertook the monitoring duty from July 1995 to June 1996, and the monitoring task was handed over to mine in August, 1996. The monitoring period is for 3 years till August 1998, it is now under the charge of mine . Here gives the data monitored from August 1995 to May 1996 by us.

4.1 AE monitoring results

In order to investigate the fracturing state of the shaft wall, we monitored the natural AE counts which occurred in country rockmass. Every level was monitored separately. It was found that natural AE counts occurred only at −80 level, −240 level and −280 level so far. Other levels are tranquil and have no AE messange recorded . Monthly average AE count rates monitored at this three levels from Augest 1995 to May 1996 are showed in Figure 2. AE counts occurred monthly at −80 levels and discontinuously at other two levels.

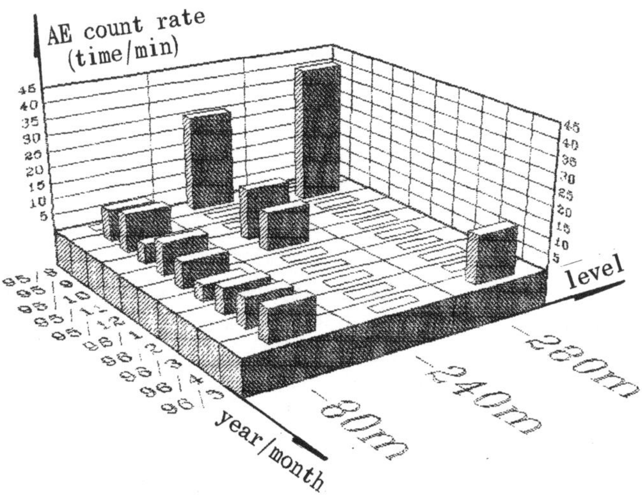

Figure 2: The relationship between AE count rate and time

4.2 Results of sound wave velocity measurement

Sound velocity measurement is carried out at the same time as AE monitoring is. By triggering DHH-1, emitting geophone will create an AE signal, and it then can be recorded at adjacent levels by receiving geophones. For example , we can pick up AE signal at −80 level and −160 level when emitting geophone at −120 level produces an AE. By analysing the wave shapes, we can obtain the time it goes through the distance between emission geophone and receiving one. Figure 3 presents this analysis and gives out the times it reaches − 80 level and − 160 level. The sound vlocity between two adjacent levels could be calculated by following formula:

$$V_{ij} = \triangle S_{ij} / \triangle T_{ij}$$

where i, j —— numbers of geophones between two adjacent levels;

V_{ij} —— sound wave velocity between i and j level;

$\triangle S_{ij}$ —— distance between i and j levels;

$\triangle T_{ij}$ —— times for sound transmitting from i to j level.

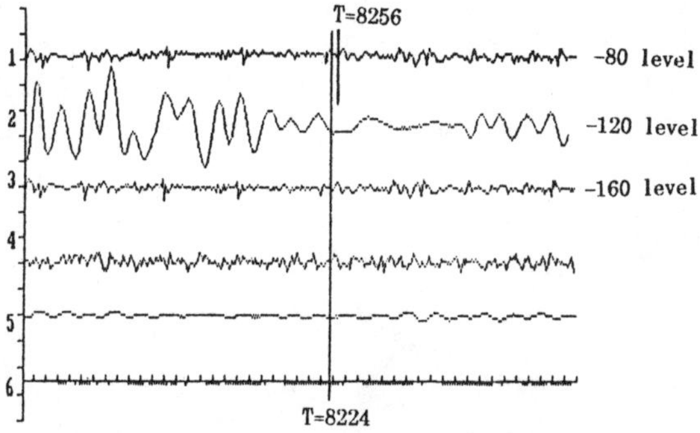

Figure 3: Diagram of sound wave analysis

The calculating results of sound wave velocity determined by (1) with respect to monitoring time are shown in Figure 4.

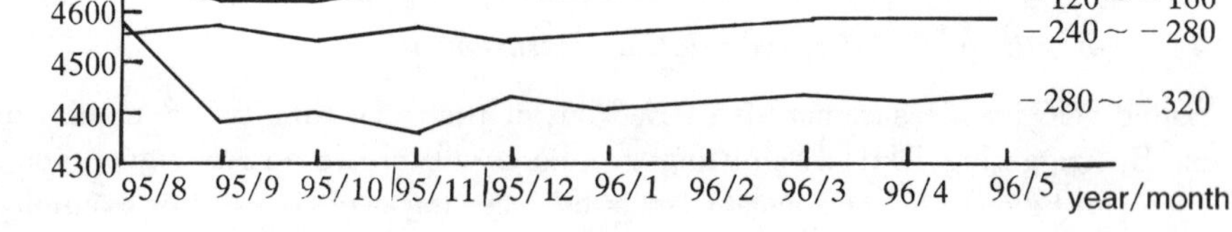

Figure 4: Sound velocities between every two adjacent levels

5 STABILITY ANALYSIS

5.1 Grey-cusp catastrophe coupling model

Based on the AE count rate, a grey-cusp catastrophe coupling model to analyse the country rockmass stability of shaft wall can be established. This method can make full use of adventages of dealing with data by grey theory. Random data can be changed closely to regular data with the shape of "s". Then abrupt disaster can be analysed by catastrophic theory. It overcomes the disavanages they own when they are used separately.

Considering $X^{(0)}$ as an original number series, $X^{(1)}$ as a producing series after AGO (accumulated generation operation), that is:

$$X^{(0)} = \{X^{(0)}_{(1)}, X^{(0)}_{(2)}, \cdots, X^{(0)}_{(n)}\}$$

$$X^{(1)} = \{X^{(1)}_{(1)}, X^{(1)}_{(2)}, \cdots, X^{(1)}_{(n)}\}$$

$X^{(1)}_{(k)}$ and $X^{(0)}_{(j)}$ have the following relationship:

$$X^{(1)}_{(k)} = \sum_{j=1}^{k} X^{(0)}_{(j)} \qquad (2)$$

$X^{(1)}$ can be expressed in terms of exponential function, namely, grey model of GM(1, 1). Then it can be expanded in power series according to Taylor's formula. Take the first five terms approximately , we can get the required ploynomial:

$$X^{(1)}_{(t)} = A_0 + A_1 t + A_2 t^2 + A_3 t^3 + A_4 t^4 + A_5 t^5 \qquad (3)$$

The derivative of $X^{(1)}_{(t)}$ with respect to the time is:

$$Y = \frac{dx^{(1)}_{(t)}}{dt} = A_1 + 2A_2 t + 3A_3 t^2 + 4A_4 t^3 + 5A_5 t^4 \qquad (4)$$

Substituting Yi monitored and monitoring time t_i into (4), we now can establish the liner system of equations as follows:

$$Y_i = A_i t_i \qquad (5)$$

On the basis of formula (5), we can solve the coefficiency Ai by Gass – Jardan's method.

With regard to the cusp catastrophe model of elementary catastrophic theory, no matter what kind forms of potential function U(t) are, expand U(t) can be expanded into a power series according to Taylor's formula. The expansion in series of U(t) is restricted to the first four terms, we get

$$U(t) = a_0 + a_1 t + a_2 t^2 + a_3 t^3 + a_4 t^4 \qquad (6)$$

Let $t = z_t - q$, and substitute into (6), we get

$$U(z_t) = b_0 + b_1 z_t + b_2 z_t^2 + b_3 z_t^3 + b_4 z_t^4 \qquad (7)$$

where $b_0 = a_4 q^4 - a_3 q^3 + a_2 q^2 - a_1 q + a_0$

$$b_1 = -4a_4q^3 + 3a_3q^2 - 2a_2q + a_1$$
$$b_2 = 6a_4q^2 - 3a_3q + a_2$$
$$b_3 = a_3 - 4a_4q$$
$$b_4 = a_4$$

where $q = a_3/4a_4$. Then Eq. (7) can be expressed as follows:

$$U(z_t) = b_4z_t^4 + b_2z_t^2 + b_1z_t + b_0 \tag{8}$$

In accordance with catastrophic theory, rewrite (8) to standard format of the cusp catastrophe model, that is

$$U(z) = \frac{1}{4}z^4 + \frac{1}{2}az^2 + b \tag{9}$$

where $a = b_2/\sqrt{b_4}$, $b = b_1/\sqrt[4]{4b_4}$

Then, the following equilibrum curved surface function $U'(z) = 0$ can be easily gotten:

$$z^3 + az + b = 0 \tag{10}$$

Bifurcation set function with which controlling variables should meet is:

$$4a^3 + 27b^2 = 0 \tag{11}$$

Now let :

$$D = 4a^3 + 27b^2$$

We get the following unstable failure criterion:

$$D \leqslant 0 \tag{12}$$

Substituting $a_0 = A_1$, $a_1 = 2A_2$, $a_2 = 3A_3$, $a_3 = 4A_4$, $a_4 = 5A_5$, into (6), then we can analyse the stability of shaft wall in the light of the procedure metioned above.

Based on the data obtained from −80 level, the calculating relults and stability evaluation are showed in table 1. The value of June in the table is the predicting result. Here −240 level and −280 level are neglected because of their less data obtained.

Table 1: D values and stability

Year/Month	96/1	96/2	96/3	96/4	96/5	96/6
D	385009	1684.3	1293.4	1273	5485	2220.3
Stability	Stable	Stable	Stable	Stable	Stable	Stable

5.2 *Sound velocity method*

It is recognized that sound velocity reflects the failure degree of rockmass. Sound velocity decreases with the increase of the number of microcracks. The coefficient completeness of rockmass (K value) can be made use of to assess rockmass condition. K value can be obtained by following formula:

$$K = (V_{mp}/V_{cp})^2 \tag{13}$$

here V_{mp}—— longitudinal velocity of rockmass between two adjacent levels;

V_{cp}——longitudinal velocity of intact rock, 5641.0m/s.

Bringing the values of V_{mp} and V_{vp} into(13), we can calculate the range of K. In accordance with rockmass classification method, and also we can evaluate the stability by K values. The calculating results and stability evaluation are showed in table 2. The results in table 2 indicate that the rockmass around the shaft lies in a good condition.

Table 2: K values and stability

Range of levels	Range of velocities m/s	K	Rockmass quality	Stability
-80～-120	4880～4900	0.75～0.76	Good	Stable
-120～-160	4620～4690	0.67～0.69	Fair	Stable
-160～-200	4880～4910	0.75～0.76	Good	Stable
-200～-240	4820～4850	0.73～0.74	Fair	Stable
-240～-280	4560～4600	0.65～0.66	Fair	Stable
-280～-320	4380～4585	0.60～0.65	Fair	Stable
-320～-360	4710～4690	0.69～0.70	Fair	Stable

6 Conclusion

(1) The establishment of the monitoring system is successful. It has been proved that this system is a new effective method and has a satisfactory performance. It overcomes disadvantages of traditional method.

(2) The monitoring results show that there are AE activities at -80, -240 and -280 levels. It also indicates that there are microcracks producing in rockmass around the shaft wall.

(3) Theoretical analysis by grey-cusp catastrophic coupling model indicates that the unstable failure of the shaft wall can't occur at present. At the same time, the coefficient completeness of rockmass shows that the country rockmass is in a good condition. Since AE counts monitored come from a large range of the country rockmass, stability analysis is also concerned with large range rockmass around the shaft. The existence of AE activity predicts that there will be possibility of splitting or peeling of the supporting material on the surface of shaft wall. Hence, the supplementary observation should be done periodically by special workers.

References

1.Shulin Li, Proc. Of the 3rd Conf. On Soil and Rock Mech. In Eastern China Press of Huazhong Uni. Of Sci.and Techn, 345-350 (1995).
2.Chuan Tang &Xiaohe Xu, Chinese J. Rock Mech. And Engn., Vol 9,No 2: 29-39 (1990).
3.Bernard J. Stebley, B.T. Brady & Tim J. Swendseid, Rockbursts and Seismicity in Mines.A.A.Balkema,Rotterdam,259-262 (1990).

Intelligent Rock Mechanics

PREDICTING ROCK FRACTURING BEHAVIORS UNDER CREEP CONDITIONS IN TERMS OF INTELLIGENT ROCK MECHANICS

XIA-TING FENG
College of Resources and Civil Engineering, Northeastern University, Shenyang 110006, P R China
E-mail: lnwang@mail.neu.edu.cn

M. SETO
Safety Engineering Department, National Institute for Resources and Environment, 16-3 Onogawa, Tsukuba, 305, Japan
E-mail: Seto@nire.go.jp

The "bottle-neck" problems including in the limited data range and somewhat or not well understanding of the mechanism exist in prediction of rock micro-fracturing behaviors. It is extremely hard or impossible to obtain accuracy prediction using traditional mathematical models. This paper proposed a new methodology in terms of intelligent rock mechanics. We focus on (1) Rock micro-fracturing behaviors are recognized by acoustic emission (AE) measurement at laboratory and in the field; (2) Nonlinear dynamic models describing the rock micro-fracturing behaviors are established by learning of neural networks using the measured data; (3) Rock micro-fracturing behaviors of novel cases can be predicted by using the obtained models. Based on the methodology, creep behavior of rock micro-fractures of Oshima granites have been conducted by acoustic emission (AE) measurement at laboratory. It was found from fundamental investigation the time distribution of rock micro-fractruing behaviors under creep condition have fractal structures described by $C(t) \propto t^D$, where D is the correlation exponent, t is time, $C(t)$ is correlation integral. The neural model **NN**(11,17,2) was found to describe reasonably well time behaviors of micro-fracturing in rock. Good agreement is seen to exist between AE behaviors (the cumulative AE events and the cumulative AE counts) predicted by the neural models and those measured in experiments for novel cases.

1 Introduction

Accuracy understanding and description of micro-fracturing behaviors in rock under creep condition are vital in rock and earthquake engineering. Some researchers have given their efforts on these points by using acoustic emission (AE) monitoring. Good relations have been reported between inelastic strain and AE rate (Mogi[1], Scholz[2], Lockner *et al.*[3], Mizutani *et al.*[4], Lockner *et al.*[5]). The correlation can be used to relate earthquake aftershock sequences to transient creep behavior in rock. Aftershock sequences have been found to decay with time according to an empirical relation $N(t) = K/(t+c)^p$ known as Omori's law. Here K, q and p are constants, t is time and $N(t)$ is the acoustic emission rate. In an unconfined creep laboratory test on basalt sample,

Hirata[6] treated individual burst of AE events as a mainshock-aftershock sequence. He found that the aftershock sequences underwent a transition from obeying an exponential decay law $N(t) = K\exp(-pt)$ to obeying Omori's law with the *p*-value in Omori's law showing a small decrease as the sample approached failure. Other researchers analyzed the overall AE rate (rather than individual burst sequences) during transient or primary creep in constant stress experiment. Locker *et al.*[3] conducted creep tests on granite and sandstone samples at 100 MPa confining pressure and stress levels from 45 to 95% of failure strength. They found a decreasing decay with increasing stress level. They showed that during transient creep, AE rate obeyed $N(t) = 1/(K_b^{-1} + qt)$, where K_b and q constants.

If a time interval is considered during which inelastic strain rate remains proportional to AE rate, then Omori's law is identical in form to the transient creep rate dependence for rock reported by Lomnitz[7] and Lockner[8].

Rock or rock masses, being geological bodies, posses complicated mechanical characteristics. As a consequence of numerous factors relating to fracturing processes, the micro-fracturing behaviors of them is often high nonlinear and dynamic. Such complex behavior has proved to be very difficult to characterize via conventional mathematical modeling approaches. Now a question raised is that how to make accuracy prediction of this time behavior.

Intelligent rock mechanics (Feng *et al.*[9]) appears to offer an attractive alternative that is worth exploring for such rock systems. In intelligent rock mechanics, learning and nonlinear dynamic processing capability of neural networks are emphasized and various governing factors are considered as a whole, without limiting their quantity if sufficient sets of example are available. The present study uses this methodology to model micro-fracturing behaviors of Oshima granites under creep condition in terms of intelligent rock mechanics.

2 Methodology

In terms of intelligent rock mechanics, an integrated intelligent methodology is proposed for modeling and predicting micro-fracturing behaviors of rock under creep condition (Fig.1). In this methodology, the time distributions of acoustic emission behaviors in rock micro-fracturing process under creep condition were firstly measured at laboratory or in the field. Using fractal theory, time distribution of micro-fractures in rock is investigated to find whether it is self-similar or not. Starting from this self-similarity, nonlinear dynamic models can be recognized to describe AE behaviors of rock micro-fracturing process by using the measured acoustic emission data to train neural network. This procedure is auto-progress and checked by another set of AE data to obtain an optimal model. Rock micro-fracturing behaviors of novel cases under similar condition are then predicted by using the obtained models. If predictions for novel cases is reasonable, the novel cases will be added to learning sample pattern set to re-train neural network. Thus ability of the obtained models can be auto-improved.

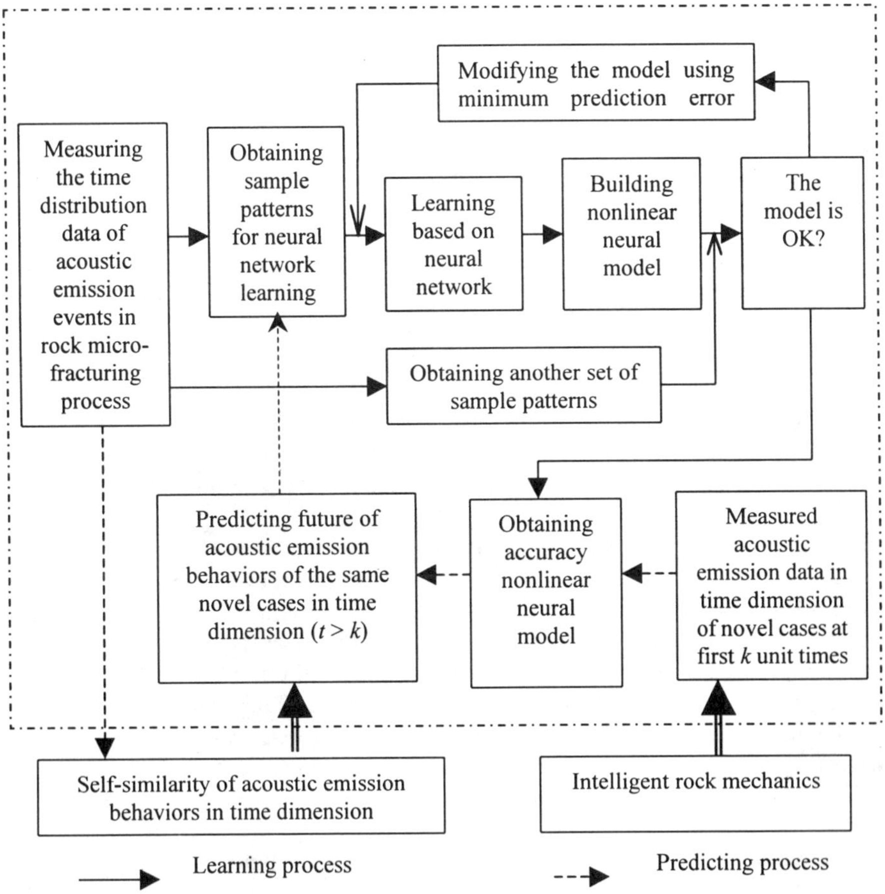

Fig.1 A methodology on modeling rock micro-fracturing behaviors under creep condition in terms of intelligent rock mechanics.

3 Data acquisition process

Double torsion tests under creep condition were carried out on Oshima granites at the laboratory of National Institute for Resources and Environment in Japan. The specimen was a plate with a notch that was cut along with central line that is parallel to the long axial of the specimen (Fig.2). Up to the distribution of weak planes, three types of the plate specimens have been investigated (Table 1). The specimens underwent a constant weight. The maximum load was set 70-90% strength of failure. Displacement was measured by one displacement transducer touched the upper loading plate.

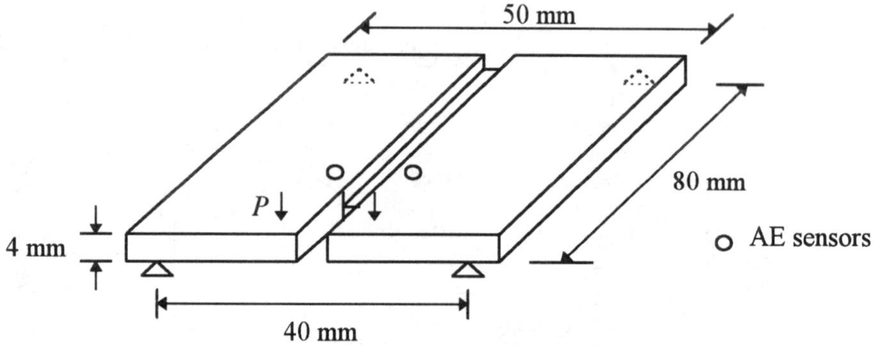

Fig.2 Specimen of double torsion test, the distance between two loading points *P* is 8mm. There is a notch with 1mm width ad 1mm height on central line of the plate

In order to measure AE signals, two PAC nano-30 sensors were mounted on the surface of the specimen (Fig.2). The frequency range of AE event recorded was set at range of 20 -1200 kHz. Signals from the AE sensors were amplified by a pre-amplifier with 40 dB gain, passed through a frequency filter (50-1200 kHz) and a post-amplifier of 40 dB gain inside the system. The amplitude threshold was set to be 45 dB. A MISTRAS-2001 system was used to record AE parameters including rise time, amplitude, event, count, energy, and duration. Each parameter was recorded as a time series (e.g. Fig.3).

4 Self-similarity of rock micro-fracturing behaviors in time

Self-similarity of the micro-fracturing behaviors of rock can be investigated utilizing fractal theory. After conducting a creep test on Oshima granite at 40 MPa confining pressure and about 85% fracture strength, Hirata *et al.* [9] found that the spatial distribution of hypocenters of acoustic emission is a fractal and a systematic decrease in D from 2.75 in transient creep to 2.25 in tertiary creep. Plots of the AE event locations show that this gradual decrease in dimension corresponds to the development of two diffuse clusters in the central region of the sample.

In the present study, a fundamental investigation is conducted to find whether the time distribution of micro-fracture in rocks also has a fractal structure or not. It starts at calculating correlation integrals $C(t)$ for the AE event distributions $(p_1, p_2, \cdots, p_N)$ in time dimension during rock micro-fracturing processes. The formula can be written as

$$C(t) = \frac{2}{N(N-1)} N(t)(T < t) \tag{1}$$

where $N(t)$ $(T< t)$ is the number of pairs (p_i, p_j) with a time interval smaller than t, and N is a total number of AE events during a selected time range to be analyzed.

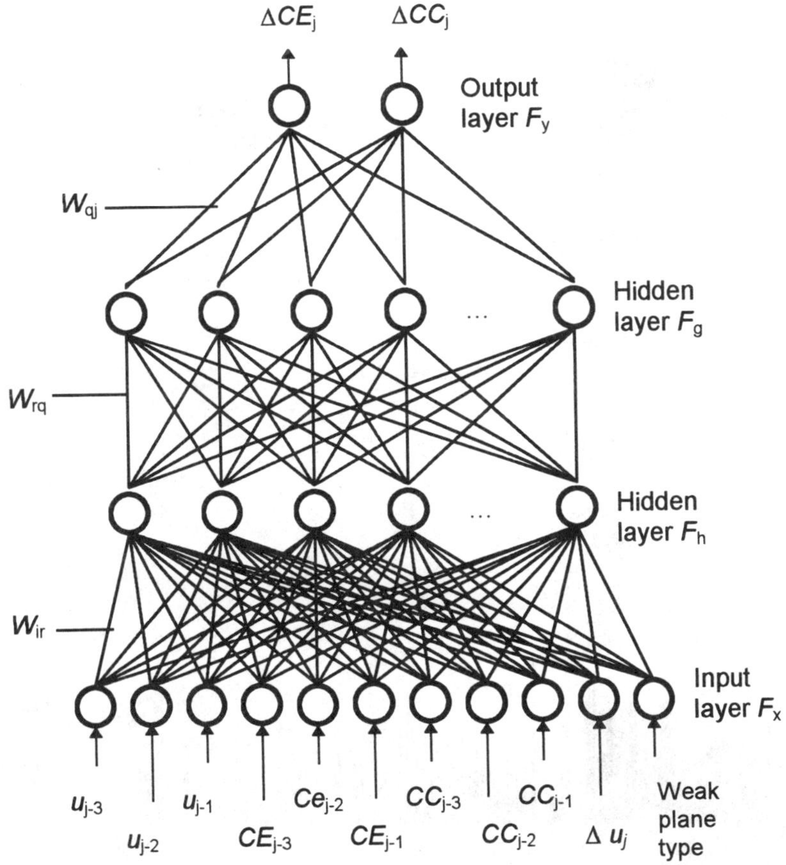

Fig.3 A feed-forward neural network representing AE behavior during rock micro-fracturing process

The fractal structure can be expressed by

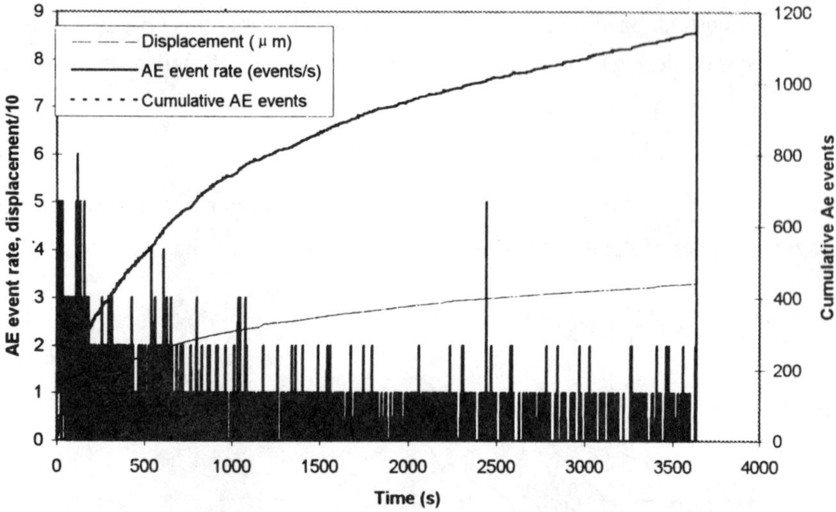

(a) AE event rate and cumulative events

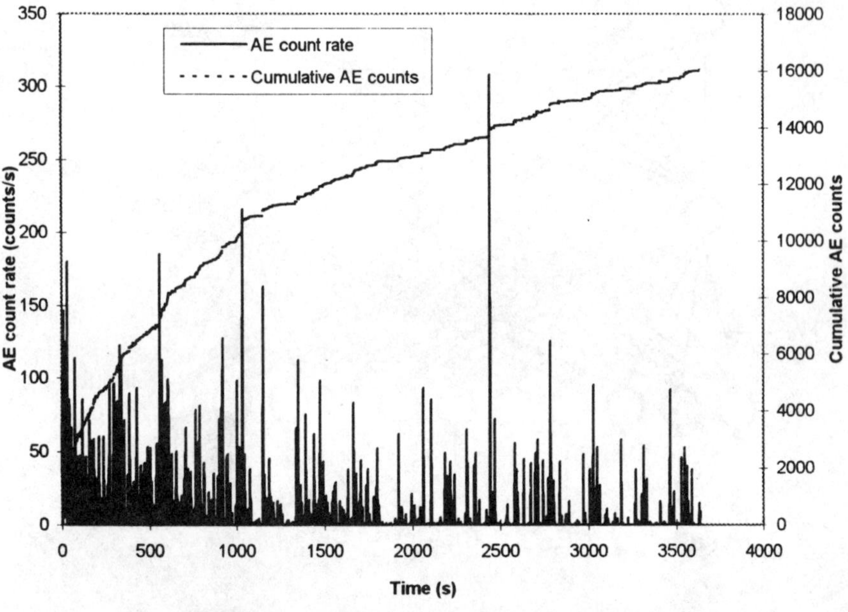

(b) AE count rate and cumulative counts

Fig.4 AE behaviors of rock under creep condition for case 5

$$C(t) \propto t^D \tag{2}$$

where D is a kind of fractal dimension called the correlation exponent.

The fractal dimensions D can be estimated from the slope of the plot. The results are listed in Table 1.

Table 1 Sample data and fractal dimensions

No.	Weak plane type***	Channel no.	Load (kg)	Total of AE events	Fractal dimension	Coefficient of determination R^2
1	B	1	12.2096	246	0.94	0.9973
2	B	2	12.2096	112	1.0	0.9969
3	B	1	12.2096	784	1.0	0.99984
4	B	2	12.2096	340	0.91	0.9979
5	B	1	13.7996	1153	0.99	0.99994
6	B	2	13.7996	597	0.93	0.99991
7**	C	1	12.2096	150	0.96	0.99384
8**	C	2	12.2096	26	0.74	0.9481
9*	A	1	13.1506	105	0.93	0.9985
10*	A	2	13.1506	34	0.81	0.9862

* Experimental data from these tests were treated as cases to check "over-training" of networks

** Experimental data from these tests were treated as novel cases and were used to test the trained neural networks

***Distribution of weak planes. A: Rift plane is parallel to the notch and hardware plane is vertical to the notch; B: Grain plane is parallel to the notch and rift plane is vertical to the notch; C: Hardware plane is parallel to the notch and grain plane is vertical to the notch;

5 Establishment of nonlinear neural model

Modeling on rock micro-fracturing behaviors performed on neural networks is to find a neural network $\mathbf{NN}(n, h_1, h_2, 2)$ (Fig.4) to represent the relationship between AE event increment $\Delta CC_j, \Delta CE_j$ with their previous values $(CC_{j-1}, CC_{j-2}, \cdots, CC_{j-k}, CE_{j-1}, CE_{j-2}, \cdots, CE_{j-k})$ as follows

$$\mathbf{NN}(n, h_1, h_2, 2): R^n \to R^m$$
$$(\Delta C\hat{C}_j, \Delta C\hat{E}_j) = \mathbf{NN}(n, h_1, h_2, 2)(CC_{j-1}, CC_{j-2}, \cdots, CC_{j-k}, CE_{j-1}, CE_{j-2}, \cdots, CE_{j-k}, u_{j-1}, u_{j-2}, \cdots, u_{j-k}, WP, \Delta u_j) \quad (j = k+1, k+2, \cdots,) \tag{3}$$

where $\Delta C\hat{C}_j, \Delta C\hat{E}_j$ are computed values that approximate the measured AE count increment ΔCC_j and AE event increment ΔCE_j at time j respectively. CC_{j-k}, CE_{j-k} are the measured AE counts and AE events at time $j-k$. $u_{j-k}, \Delta u_j$ are the measured displacement at time $j-k$ and displacement increment at time j. *WP* is type of weak plane. $\mathbf{NN}(n, h_1, h_2, 2)$ is the model to be determined, which approximates the actual relationship. n, h_1, and h_2 are the number of input nodes, number of nodes at the first hidden layer and the second hidden layer, which are to be learned by using experimental data to train neural network shown in Fig.3, $n = 3k + 2$.

An improved BP algorithm (Feng et al.[9]) was used to train the networks to obtain the models. This improved BP algorithm uses the minimum-error principle for the extrapolated predictions of novel cases--instead of the minimum-error criterion for learning sample predictions--to determine whether the neural network learning should finish or not.

The learned $\mathbf{NN}(n, h_1, h_2, 2)$ can be used to obtain "true" prediction of the testing cases. It builds the whole predicted curve of cumulative AE counts and cumulative AE events from $t = 0$ condition. The measured AE events and counts at first k unit times are used as input series to predict their increments $\Delta C\hat{C}_j, \Delta C\hat{E}_j$. These calculated increments are feed back to calculate $C\hat{C}_j, C\hat{E}_j$ (e.g. $C\hat{C}_j = CC_{j-1} + \Delta C\hat{C}_j$, $C\hat{E}_j = CE_{j-1} + \Delta C\hat{E}_j$ (if $j = k + 1$) or $C\hat{C}_j = C\hat{C}_{j-1} + \Delta C\hat{C}_j$, $C\hat{E}_j = C\hat{E}_{j-1} + \Delta C\hat{E}_j$ (if $j > k+1$). The input series for each next step prediction is then renewed by adding the predicted values $C\hat{C}_j, C\hat{E}_j$ and kept at the same length of k unit time.

6 Applications

The results of a series of AE tests listed in Table 1 were used to obtain and test the model. Cases were divided randomly into three data sets. The first data set of tests was used in training the neural networks to obtain the models $\mathbf{NN}(n, h_1, h_2, 2)$, while the second data set, identified with asterisks, was used for "over-training" check to obtain optimal model. And the third data set, identified with double asterisks, was used to test the performance of the obtained neural models in generalizing to testing cases. The data of cumulative AE events and counts as well as their increments were taken logarithm to the base 10, and then all data including load data in column were respectively transferred into the interval [0, 1] before being modeled or used for prediction.

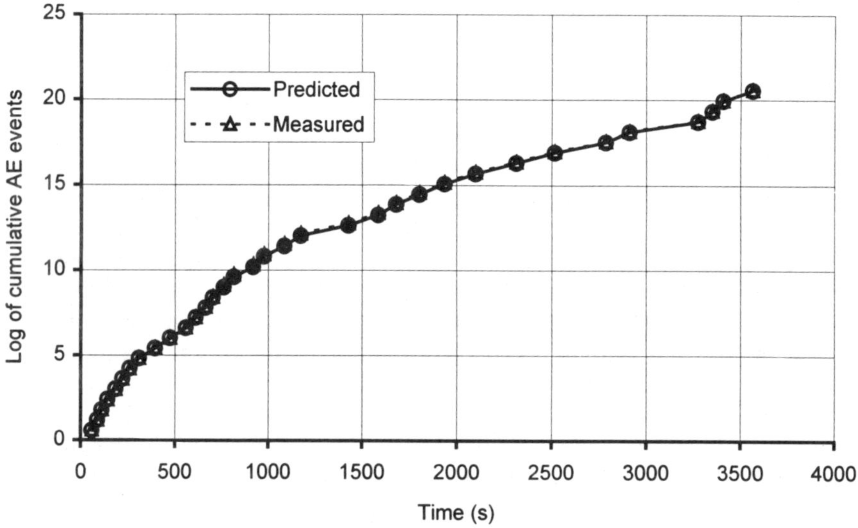

(a) Cumulative AE events

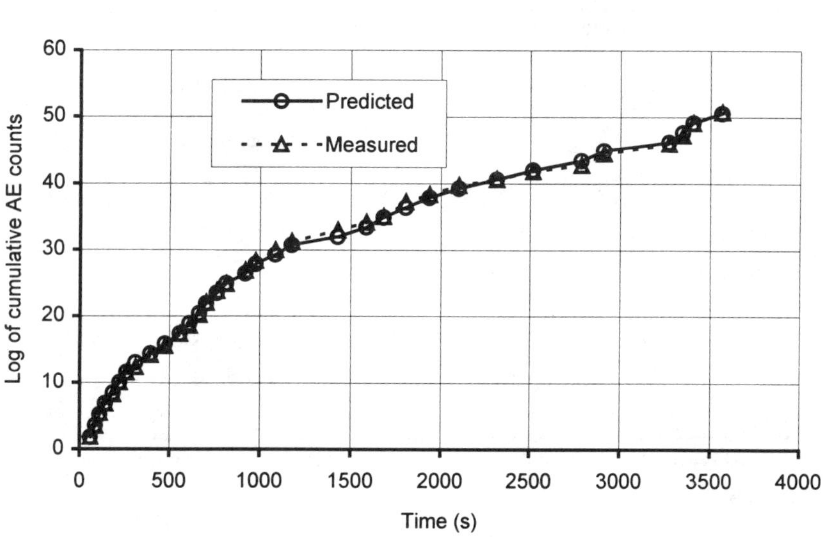

(b) Cumulative AE counts

Fig.5 Comparison of measured AE behaviors and predicted by the model NN(11,17,2)

A number of network architectures (h_1 and h_2), iterations i, and history points k were evaluated in developing a neural model by using the improved learning algorithm described above. With the aim to minimize the errors on the extrapolated predictions for testing cases, the best models were recognized. Table 2 shows the obtained neural model and the accuracy of the extrapolated predictions for "true prediction" cases. It was found that AE behaviors in the rock micro-fracturing process can be represented reasonably well by neural networks if three history points (i.e., k=3) are included. To validate these results, two of the more common statistical parameters, generally used to determine model accuracy and performance, are used. These parameters are mean absolute deviate (MAD) and mean absolute percent error (MAPE). MAD is the average of the absolute differences between the forecasted values and the experimental values, and MAPE is the absolute differences between the forecasted and experimental values expressed as a percentage of the experimental values. Statistical analysis of the results shown in Table 2 indicate that R^2 values for the testing set 7 for cumulated AE events and for cumulated AE counts (Fig.5) were both about 1. The MAPE was 0.58 and 1.4668, respectively. The MAD was 0.0529 and 0.317, respectively. These results demonstrate that the neural model **NN**(11,17,2) developed in this work can predict the AE behaviors for novel cases with a high accuracy of approximately 98%.

Table 2 Predicting performance of the obtained neural models for novel cases

Testing sample 7	The neural model used to predict	MAD	MAPE	Linear regression between extrapolated predictions and experimental results	Coefficient of determination R^2
For cumulative AE events	NN(11,17,2)	0.0529	0.5807	y = 1.00331*x+ 0.0129276	0.999944
For cumulative AE counts	NN(11,17,2)	0.3165	1.4668	y = 1.00235*x- 0.0080148	0.999227

y is experimental measurements, x is predicting results of the neural model

7 Conclusions

This paper investigated applicability of intelligent rock mechanical method to modeling of rock micro-fracturing behaviors under creep condition. Since rock micro-fracturing process under creep condition has self-similar structure, the model learned from measured data can be generalized to predict AE behaviors for novel cases. The modeling results indicate that if the network architecture, learning iteration and parameters, and history point *k* are determined reasonably, this method can give its predictions for novel cases with high accuracy. In the cases of the present study, the model **NN**(11,17,2) gave extrapolating predictions of AE behaviors in the time dimension for novel cases with accuracy of about 98%.

Further work will aim at modeling the creep behaviors under other soaking conditions (e.g. water, chemical solutions), and finding a practical method to describe, predict, and evaluate the AE behaviors of the fracturing process in rock masses. An integrated intelligent system including fracture mechanical modeling, damage mechanical modeling will be developed.

Acknowledgment

Japanese Governmental ITIT program for Xia-Ting Feng and National Natural Science Foundation of China under Grant No.59604001 financially supported this research. These supports are gratefully acknowledged. The authors would like to give their appreciate to Prof. Lin Yunmei for her kindly help.

References

1. Mogi, K., *Bull. Earthquake Res. Inst. Tokyo Univ.*, *40*,125-173 (1962).
2. Scholz, C. H. *J. Geophys. Res.* 73, 1417-1432 (1968).
3. Lockner, D. A. and Byerlee, J. D., *Bull. of the Seismological Society of Amer.* 67, 247-258 (1977).
4. Mizutani, H. Yamada I., and Masuda K. *Earthquake Prediction Research* 3. 595-605 (1985).
5. Lockner, D. A. and Byerlee, J. D. , Kuksenko V., Ponomarev A. and Sidorin A. *Nature* 350, 39-42 (1991).
6. Hirata T. *J. Geophys. Research* **92**, B7, 6215-6221 (1987).
7. Lominitz, C. *J. Geology* 64 473-479 (1985).
8. Lockner, D. A., *Int. J. Rock Mech. Min. Sci. & Geomech. Abstr.* **30**, 7, 883-899 (1993).
9. Hirata, T., Satoh T. and Ito K. *Geophys. J. R. astr. Soc.* 90, 367-374 (1987).
10. Feng X.-T, et al. *Intl. J. Rock Mech. Min. Sci.* **34**, 1, 135-141 (1997).
11. Grassberger, P., *Phys. Lett.*, *97*, 227-230 (1983)

Rock Stress

NEW APPROACH TO RECONSTRUCTION OF LARGE SCALE *IN SITU* STRESSES AND MODULI

A.N. GALYBIN, A.V. DYSKIN AND R.J. JEWELL

Department of Civil Engineering, The University of Western Australia, 6907, Australia

E-mail: galybin@cyllene.uwa.edu.au

A measuring scheme for near surface measurements of stress using the cylindrical jack technique is proposed that allows the simultaneous reconstruction of the stress components and of the elastic moduli of the rock mass at a scale considerably larger than for the conventional methods such as overcoring. The scheme allows interpretation of the measurements near the surface of excavation in both isotropic and orthotropic rocks, with a plane of symmetry parallel to the plane of measurements. It is based on two groups of extensometers symmetrically arranged around the borehole and not interfering with the drilling operations. An elastic analysis has shown that the proposed scheme is theoretically viable.

1. Introduction

The determination of *in-situ* stresses and deformation moduli at the scale of interest to Mining Engineering is one of the main challenges in Rock Mechanics practice. Small-scale measurements obtained by the overcoring technique have considerable scatter of the measured values (see for instance an example considered by Wiles and Kaiser[1]) and are subjected to the scale effect (Cuisiat and Haimson[2]). As a result, the particular stress measurements become non-representative, which requires a big number of small-scale measurements to be conducted at a variety of locations in order to extrapolate results to a larger scale.

Stress measurements at a larger scale can, in principle, be obtained by the flat-jack method in which the scale of the measurements is determined by the slot dimensions. However the method is laborious since each slot allows the determination of only one stress component.

The under-excavation technique described by Wiles and Kaiser[1] is capable of producing large scale stress measurements if the deformation moduli of the rock are known at the scale of excavation. Therefore, a technique for the determination of large-scale moduli is required. The knowledge of large-scale moduli is also needed to calculate stress concentrations as is used for the evaluation of stability of the excavation. However, lab testing allows only the small-scale determination of the moduli (the determination of moduli at many locations does not guarantee representativeness because of the presence of larger-scale discontinuities). The existing methods of large-scale *in situ* modulus determination are expensive.

The cylindrical jack method proposed by Dean and Beatty[3] allows a number of displacement measurements in different directions around a borehole after the drilling and subsequent pressurising phases and hence, potentially, recovery of the full 2-D stress tensor and elastic moduli. In the variant employed by Dean and Beatty[3] however only radial displacements were measured and the measuring pins were located relatively close to the borehole which restricted the scale of stress measurements and affected the accuracy (eg, Goodman[4]). Also, the measurement system interfered with the drilling operations and Poisson's ratio had to be obtained from separate tests.

A new measuring arrangement for simultaneous large-scale stress and modulus determination free from these shortcomings has been proposed by Galybin et al[5]. This paper proposes a further modification of the scheme which allows the usage of standard extensometers of equal sizes.

2. Measuring scheme for the cylindrical jack method

2.1. *Existing scheme*

Dean and Beatty [3] used the following scheme: (1) holes were drilled for pins at the vertices of an octagon; (2) pins were inserted and deformometers located to measure distances in four radial directions; (3) the deformometers were removed and a central hole drilled; (4) the cylindrical jack was inserted into the hole and the deformometers reinstalled; (5) contractions between the radially-situated pins were measured; (6) changes in radial distances were monitored during pressurising of the hole; (7) the values of stresses and the shear modulus were obtained from a least squares analysis.

Mathematically the method was based on the Kirsch solution (eg, Goodman[4]) for a circular hole in an elastic isotropic plane under plane-strain conditions. The changes in displacement due to drilling the hole of radius R and then pressurising it to a pressure p can be presented in the following combined form

$$u_r(r,\theta)+iu_\theta(r,\theta)=\frac{R^2}{2\mu r}\left\{\frac{\sigma_1+\sigma_2}{2}-Q+\frac{\sigma_1-\sigma_2}{2}\left[\kappa\, e^{-2i(\theta-\varphi)}+\left(1-\frac{R^2}{r^2}\right)e^{2i(\theta-\varphi)}\right]\right\} \tag{1}$$

where (r, θ) are the polar coordinates with the origin situated in the hole centre, $u_r + iu_\theta$ is the complex expression for the components of the vector of displacements in polar coordinates, μ is the shear modulus, $\kappa=3-4\nu$, ν is Poisson's ratio, σ_1 and σ_2 are the remote principal stresses, φ is the angle between the first principal direction and the x-axis of the reference coordinate frame, (x, y) as shown in Figure 1 and Q is the internal pressure created by a cylindrical jack.

The expression (1) contains five unknown parameters: moduli of deformability (μ, κ) and the characteristics of the *in situ* stress, $(\sigma_1, \sigma_2, \varphi)$. In order to recover these parameters it is sufficient, in principle, to perform four independent displacement measurements in different directions after drilling the hole (Q=0) and at least one measurement (one displacement component) after pressurising the hole. The difference in the displacements before and after the pressurising has the form

$$\left[u_r(r,\theta)+iu_\theta(r,\theta)\right]_{before}-\left[u_r(r,\theta)+iu_\theta(r,\theta)\right]_{after}=\frac{R^2Q}{2\mu r} \tag{2}$$

Therefore, by knowing the pressure magnitude, p, and the difference of the radial component of displacement, it is possible to determine the shear modulus directly (see also Dean and Beatty[3]), while the tangential component should remain unchanged which could be used for controlling the quality of measurements.

2.2. *Proposed modified measuring scheme*

As one can see from equation (1), if radial and circumferential displacements were measured it would be possible to determine all elastic moduli. This can be done by measuring the displacements (elongation/contraction) along specially chosen lines, eg, by utilising vibrating wire strain gauges or a strain gauged frame.

The minimum number of measurements depends upon rock characteristics. For isotropic rock at least four extensometers oriented in different directions are required. This will provide four independent measurements of the deformation change after the hole is drilled and another four after pressurising to provide a fourfold redundancy for the determination of the shear modulus. For the determination of the shear modulus, the best orientation of the extensometers would be in radial directions according to expression (2). However, as may be seen from expression (1), the radial displacement measurements (θ=const) alone are not sufficient to separate σ_1-σ_2 and κ from their product. Therefore some non radial measurements are required.

A similar situation can be observed for anisotropic rock. The only difference is that the number of measurements should correspond to the number of elastic moduli. For orthotropic rock there should be not less than six measurements performed in different directions after drilling the hole, plus extra measurement after pressurising.

For this reason, and because of natural errors of measurement, the number of measurements should be greater than the minimum required to allow redundancy. The measurement system should also not interfere with the drilling and pressurising operations. To satisfy these constraints, the symmetrical arrangement shown in Figure 1 has been suggested. In this arrangement N measurements are obtained in radial directions (lines A_1-C_1, A_2-C_2,...,A_N-C_N- in Fig.1(a)) and N measurements in the circumferential directions (lines B_1-B_2, B_2-B_3,..., B_N-B_1 in the figure). The number of measuring pairs, N, is to be chosen to satisfy the redundancy requirement and to provide the desirable accuracy of the stress and moduli determination. Choice of the extensometer base, d, is a compromise between the desire to increase the scale of measurements and the extensometer sensitivity.

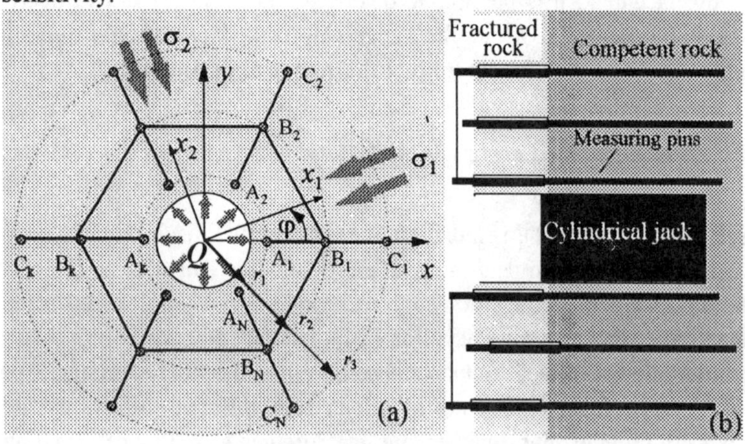

Figure 1. Proposed measuring scheme.

The proposed measuring procedure mainly follows the Dean and Beatty scheme but with the essential difference that extensometers do not have to be removed during the drilling and pressurising the hole. This obviously increases the accuracy.

To avoid the influence of the fractured zone around the excavation, pins should be inserted deep enough to be located in competent rock, Fig. 1(b). Otherwise, the results of measurements may reflect properties of the fractured zone.

Let $P_m(x_m, y_m)$ be the coordinates of pins inserted into the rock before drilling the hole, and $\boldsymbol{l}_{mn}= P_m-P_n$ be a vector connecting any two pins. After drilling the hole, let the displacement vector $\mathbf{V}(P)=\mathbf{V}(x,y)=\{u(x,y),v(x,y)\}$ be known at any point $P(x,y)$. Then small length variations, δ_{mn}, of the length $\boldsymbol{l}_{mn}$ has the form

$$\delta_{mn} = \Delta\mathbf{V}_{mn}\boldsymbol{l}_{mn} \tag{3}$$

Here

$$\boldsymbol{l}_{mn} = \{\cos\alpha_{mn}, \sin\alpha_{mn}\}, \qquad \Delta\mathbf{V}_{mn} = \mathbf{V}(P_n) - \mathbf{V}(P_m) \tag{4}$$

where α_{mn} is the angle between the vector $\boldsymbol{l}_{mn}$ and the positive direction of x-axis.

For the arrangement shown in Fig. 1(a), the relative elongation between pins can be expressed through a single index. Complex coordinates of the pins are

$$A_k = r_1\gamma_k, \quad B_k = r_2\gamma_k, \quad C_k = r_3\gamma_k, \quad \gamma_k = \exp\left(2i\pi kN^{-1}\right), \quad k = 1\ldots N \tag{5}$$

Hence, for circumferential extensometers (4) takes the form

$$\boldsymbol{l}_k = i\exp\left(i\pi(2k+1)N^{-1}\right), \qquad \Delta\mathbf{V}_k = \mathbf{V}(B_{k+1}) - \mathbf{V}(B_k) \tag{6}$$

Similarly for radial extensometers one has

$$\boldsymbol{l}_{k+N} = \exp\left(i\pi 2kN^{-1}\right), \qquad \Delta\mathbf{V}_{k+N} = \mathbf{V}(C_k) - \mathbf{V}(A_k) \tag{7}$$

The elongations/contractions between the pins, vector δ, can now be expressed by a single formula containing conjugated displacement vectors

$$\delta_k = \mathrm{Re}\left(\boldsymbol{l}_k\overline{\mathbf{V}}_k\right), \quad k = 1\ldots 2N \tag{8}$$

The next step is to express the displacement vectors $\mathbf{V}_k$ through stress components and moduli (it is assumed that rocks are linear elastic and homogeneous).

3. Procedure for stress and moduli reconstruction

3.1. Isotropic rock

For isotropic rock the stress-displacement dependence is given by (1). When the new parameters given below are introduced

$$P = \frac{\sigma_2 + \sigma_1}{2}, \quad \sigma = \frac{\sigma_2 - \sigma_1}{2}\cos 2\varphi, \quad \tau = \frac{\sigma_2 - \sigma_1}{2}\sin 2\varphi \tag{9}$$

the displacement vector in Cartesian coordinates can be expressed as follows

$$\mathbf{V}(x,y) = \frac{R^2}{2\mu r}e^{i\theta}\left[P - Q - \kappa e^{-2i\theta}(\sigma + i\tau) - \left(1 - R^2 r^{-2}\right)e^{2i\theta}(\sigma - i\tau)\right] \tag{10}$$

Reconstruction of the shear modulus. First, the shear modulus can be recovered on the basis of equation (2). If the number of readings taken during pressurising the hole is m for each radial extensometer, then one has a $(Nm\times 1)$ linear system of the parameter μ^{-1}. Solution of the system can simply be obtained by using the method of linear regression. Let **Q** be the vector of pressures applied by the jack, and **L** be the vector of corresponding extensometer readings for which the components are

$$L_k = \left[u_r(r_0 + d) - u_r(r_0)\right]_{before} - \left[u_r(r_0 + d) - u_r(r_0)\right]_{after} \tag{11}$$

where k=1...N. Then, the solution for the shear modulus has the form

$$\mu = 0.5 d r_0^{-1}(r_0 + d)^{-1}\mathbf{Q}^T\mathbf{Q}\left(\mathbf{Q}^T\mathbf{L}\right)^{-1} \tag{12}$$

It is evident, that due to a large volume of data involved the shear modulus can be obtained with the highest accuracy.

Reconstruction of the *in situ* stresses and Poisson's ratio. Assuming that the shear modulus has already been found, the jack pressure can be assumed to be zero, Q=0. The problem then contains four unknown parameters: P,σ,τ,κ. The number of readings available is equal to $2N$. If the number of independent readings is greater than 4, then there is sufficient redundancy, and the problem of simultaneous determination of the moduli and stresses can be reduced to the problem of minimization of a proper functional in the space of four variables.

Let δ be a vector of readings and $\mathbf{W}(P,\sigma,\tau,\kappa)$ be 2N-vector obtained by formulae (8) in which vectors $\mathbf{V}_k$ are given by the right-hand side of formula (10). Then solving the problem

$$F(P,\sigma,\tau,\kappa) = \left\|\delta - \mathbf{W}(P,\sigma,\tau,\kappa)\right\| \rightarrow \min \tag{13}$$

gives the stress and modulus reconstruction. Here ||...|| stands for a norm.

3.2. *Orthotropic rocks*

Orthotropic rocks (foliated rocks or rocks with orthogonal sets of fractures) will now be considered. In this 2-D consideration it is assumed that the surface of measurement is parallel to a plane of symmetry(ie. to a fracture or foliation plane).

Displacements in an orthotropic plane with a circular hole can be expressed through complex potentials Φ_1, Φ_2 as follows (Lekhnitskii[6])

$$u(x,y) = 2\,\mathrm{Re}\left[p_1\Phi_1(x,y) + p_2\Phi_2(x,y)\right],\quad v(x,y) = 2\,\mathrm{Re}\left[q_1\Phi_1(x,y) + q_2\Phi_2(x,y)\right]$$
$$p_k = \left(\mu_k^2 - \nu_1\right)E_1^{-1},\quad q_k = -\mu_k\nu_1 E_1^{-1} + \mu_k^{-2}E_2^{-1} \tag{14}$$
$$\mu_k^2 = \nu_1 - 0.5E_1G^{-1} - (-1)^k\left[\left(\nu_1 - 0.5E_1G^{-1}\right)^2 - E_1E_2^{-1}\right]^{1/2}$$

Here E_1, E_2 are Young's moduli along the principal directions of elasticity, ν_1 is the Poisson's ratio in x-direction, G is the shear modulus which characterizes changes of angles between principal directions.

If the plane is subjected to biaxial compression at infinity (σ_1, σ_2) and internal pressure Q inside the hole, the potentials have the following form

$$\Phi_k(x,y) = A_k\zeta_k(x,y)^{-1},\quad k = 1,2 \tag{15}$$

Here A_1, A_2 are constants as follows

$$A_1 = -\frac{R}{2}\frac{(P+Q)(i-\mu_2) - (\sigma + i\tau)(i+\mu_2)}{\mu_1 - \mu_2},\quad A_2 = \frac{R}{2}\frac{(P+Q)(i-\mu_1) - (\sigma + i\tau)(i+\mu_1)}{\mu_1 - \mu_2} \tag{16}$$

Parameters P, σ, τ are determined in (9). Functions ζ_k have the following form

$$\zeta_k(x,y) = \frac{x + \mu_k y + \sqrt{(x+\mu_k y)^2 - R^2(1+\mu_k)^2}}{R(1 - i\mu_k)} \tag{17}$$

Parameters P,σ,τ and μ_1, μ_2, $a_{11}=E_1^{-1}$, $a_{12}= -\nu_1E_1^{-1}$ are determined by minimizing the following functional

$$F(\mu_1,\mu_2,a_{11},a_{12},P,\sigma,\tau) = \left\|\delta_j - 2\,\mathrm{Re}\sum_{k=1}^{2}\left(p_k\cos\theta + q_k\sin\theta\right)A_k\eta_{k,j}\right\| \tag{18}$$
$$\eta_{k,j} = \zeta_k\left(x_{j+1},y_{j+1}\right)^{-1} - \zeta_k\left(x_j,y_j\right)^{-1}$$

4. Examples of the stress and modulus reconstruction

In order to illustrate the proposed technique the following examples have been considered for both isotropic and orthotropic types of rocks. In these examples the moduli and the stress states have been specified, the ideal readings of the extensometers computed and then distorted by introducing random errors to check how well the original stress state and the Poisson's ratio can be reconstructed.

The readings were modeled as the ideal ones computed for the above stress states using formulae (6)-(9) and then distorted by adding to each of them an independently generated random error uniformly distributed within the range of (-10%, 10%). For ideal readings δ_j, the readings used for the reconstruction had the components $\delta_j(1+0.1\xi_j)$, where ξ_j, j=1,..,2N were independent random variables uniformly distributed over the interval (-1, 1).

The functionals in formulae (13), (18) written for the Euclidean norm were minimized using a procedure built into MATHCAD 6 PLUS with the initial value obtained from the exact values of the parameters randomly disturbed within ±20%.

For each number of the measuring pairs, 50 sets of readings were independently generated and fitting parameters were reconstructed for every set of readings. The averages over 50 sets of readings and the variation coefficients (the standard deviation over the average) were then computed.

In both cases the following 2-D stress tensor was used: the principal compressive stresses at infinity, σ_1=-32MPa, σ_2=-18MPa, inclined to the x-axis at an angle φ=20°. Also R=0.04 m, d=0.2 m. The length r_2 depends on the number of extensometers and their length according to the equation $d=2\ r_0 \sin(\pi/N)$. Other distances shown in Fig. 1 are $r_1=r_2-d/2$, $r_3=r_2+d/2$.

The following numbers of measuring pairs have been tried: N=5,6,7,8.

4.1. *Isotropic rocks*

The chosen moduli were: E=81.9GPa, μ=33.5Gpa which yielded κ=2.11 (ν=0.222). This data corresponds to a granite (Lama and Vutukuri[7]).

The principal stresses and the principal direction computed from the average values of the reconstructed P, σ, τ as compare to the original ones, as well as the variation coefficients representing the relative errors are shown in Table 1 (the accuracy of the determination of the shear modulus will obviously be better since it used all 2N measurements for determination of this single parameter).

It can be seen that average characteristics are reconstructed with high accuracy in all cases. An examination of the relative errors shows that a reasonable compromise between the accuracy of reconstruction and the minimization of the number of measurements could be achieved with N=6.

Table 1. Average reconstructed stresses and modulus κ against the original ones (last column in the table). Variation coefficients (%) are given in brackets.

N	5	6	7	8	original
P/MPa	-25.0 (2.0)	-25.0 (2.4)	-25.0 (1.6)	-25.0 (1.5)	-25.0
σ/MPa	5.28 (8.9)	5.39 (8.0)	5.34 (5.1)	5.41 (4.5)	5.36
τ/MPa	4.39 (11.5)	4.54 (6.7)	4.57 (6.9)	4.54 (4.5)	4.50
κ	2.15 (6.9)	2.09 (5.7)	2.09 (5.0)	2.07 (4.7)	2.11

4.2. Orthotropic rock

The assumed moduli, E_1=63.4GPa, E_2=20GPa, G=7.9Gpa, ν_1=0.067 correspond to a schist (Lama and Vutukuri[7]). Two values of the cylindrical jack pressure were used: Q=0 and 10 MPa.

Table 2 compares the reconstructed principal stresses and the principal direction with the original ones (the variation coefficients are shown in brackets). It can be seen that the errors do not vary much when N changes from 5 to 8, which suggests that the minimum number of 2*2*5=20 measurements already gives sufficient redundancy. It can be seen that reconstruction of the parameter a_{12} characterizing the Poisson's ratio is subjected to large errors, despite average values which are close the exact one.

Table 2. Average reconstructed stresses and moduli for orthotropic rock against the original ones. Variation coefficients (%) are given in brackets.

N	5	6	7	8	original
$-i\mu_1$	2.78 (2.2)	2.75 (1.3)	2.70 (1.8)	2.74 (1.9)	2.73
$-10i\mu_2$	0.65 (2.9)	0.64 (1.9)	0.65 (2.9)	0.65 (2.6)	0.65
$E_1^{-1}10^4$MPa	1.52 (3.1)	1.60 (3.9)	1.63 (3.1)	1.58 (3.0)	1.58
$\nu_1 E_1^{-1}10^5$MPa	1.4 (122)	0.85 (186)	0.87 (143)	0.93 (133)	1.08
P/MPa	-25.0 (3.1)	-24.8 (2.3)	-2.4.8 (2.4)	-2.51 (2.0)	-25.0
σ/MPa	5.5 (7.5)	5.13 (5.3)	5.24 (7.2)	5.36 (5.4)	5.36
τ/MPa	4.7 (8.9)	4.47 (6.5)	4.45 (5.3)	4.47 (5.7)	4.50

5. Conclusions

The proposed measuring scheme allows the simultaneous reconstruction of the stress components and the elastic moduli of the rock mass at a scale considerably larger than the conventional methods such as overcoring. Its advantage is that the extensometers do not interfere with the drilling operations and therefore do not have to be dismantled, which will increase the accuracy of the measurements. The measuring system could be left in place after the test and monitored to estimate the effect of mining operations on the rock structure and the local stress distribution.

The technique proposed for measurements of surface stress and moduli values can form the basis for complete, large scale *in situ* stress reconstruction. This can be achieved by combining the results of the stress measurements at different locations on the excavation walls by means of back analysis of the solution for the excavation. In order to do this for an elastic rock mass its deformation moduli are required and they are obtained simultaneously with the stress measurements by the method proposed.

Acknowledgment. The authors acknowledge the support of the Large Australian Research Council Grant A89701124.

6. References

1. Wiles T.D. and Kaiser P.K. *In situ* stress determination using the under-excavation technique. Parts I, II. *Int. J. Rock Mech. Min. Sci. & Geomech. Abstr.*, **31**, 439-446, 447-456 (1994).
2. Cuisiat F.D. and Haimson B.C. Scale effects in rock mass stress measurements. *Int. J. Rock Mech. Min. Sci. & Geomech. Abstr.*, **29**, 99-117 (1992).
3. Dean A.H. and Beatty R.A. Rock stress measurements using cylindrical jacks and flat jacks at North Broken Hill Limited. *Broken Hill Mines Monograph* No 3, Australian Inst. Min. Metal. Melbourne Australia, 1-8 (1968).
4. Goodman R.E. *Introduction to Rock Mechanics*, John Wiley & Sons (1989).
5. Galybin, A.N., A.V. Dyskin and R. J. Jewell. A measuring scheme for determining *In situ* stresses and moduli at large scale. *Int. J. Rock Mech. Min. Sci. & Geomech. Abstr.* **34**: 157-162. (1997).
6. Lekhnitskii, S.G.. *Anisotropic plates.* NY: Gordon and Breach, (1968).
7. Lama, R.D and V.S. Vutukuri, *Handbook of Mechanical Properties of Rocks.* V. II. Trans Tech Publ., Switzerland (1978).

THEORETICAL AND PRACTICAL ANALYSIS OF *IN-SITU* ROCK STRESS ON THE UNDERGROUND WORKINGS IN JINCHUAN MINES

CHANGSHOU TAI
Beijing Central Engineering and Research Institute for Non-ferrous Metallurgical Industries
12 Fuxing Avenue,Beijing 100038,P.R.China

The underground workings of two mines of Jinchuan Non-ferrous Corp. deform and destruct seriously during development and production afterwards. Such phenomenon appeared due to the acting of *in-situ* rock stress. The pressure happened at the initial stage of the *in-situ* rock stress was most destructive. It occurred after exploding and lasted two to three months. The permanent support should be designed to satisfy the requirements in this serious period and avoid the failure of openings. Following the initial stage, there was a long attenuating process of the underground pressure. At this stage, the pressure usually appeared at the weak parts of the working and the main factors to be taken into account in deciding the support form are the arch pressure of rock falling, the pressure of rock dilation ,the effect of underground water etc. The happened pressure in this stage is easier to deal with than that caused by the releasing of the *in-situ* rock stress. The underground pressure caused by the *In-situ* rock stress does not seem to increase in accordance with the enlargement of the cross section of openings.

1 Introduction

The Jinchuan ore field has been called one of the national treasures in China. The total ore reserves are approximately 0.5 billion tons which contain more than 5 million tons of nickel and some other precious and rare metals as well, such as gold, silver, platinum, ruthenium, lawrencium etc. The shallower deposit of No.1 mine had been excavated by open pit and closed in 1990,while the deeper is being mined some 300m below surface by underground excavation until present. The production capacity of underground excavation is 3000t/d. The ore body of No.2 mine is familiar due to its high grade, big size, deep location and intensive fractures of both ore body and surrounding rock mass. At present, the second phase project has been completed in No.2 Mine. Some underground workings are located at 800m below the ground surface and the planned capacity is 8000t/d. Various problems of ground control and corresponding support of excavations are encountered during construction and production in these two mines.

In order to deepen the knowledge of the ground pressure, Jinchuan Company has carried out many studies on geo-technique and therefore obtained a lot of valuable scientific data, The mechanical competence of the ore body and its surrounding rock is generally poor. Except granite(V) and thicker marble(IV_2)formation strata, the values of RQD of the rest rocks are below 10,most of them even vary from 0.2 to 4.00. The Portodyakonov's indexes of hardness are below 6 and most of them only vary between 0.2~3.00. More than 70% of the underground workings are excavated in these poor rock formations.

The prevailing direction of the major *in-situ* rock stress is nearly perpendicular to the strike of strata, generally N 40°E. The horizontal stress is bigger than the vertical. The magnitude of major stress is approximately 3 MPa at the level 30m below the ground surface and increases to 30 MPa at 300-500m below surface with a peak value of 50 MPa. The magnitude of difference between horizontal major stress and vertical minor stress increases with the depth. The horizontal major stress increases about 4.5 MPa every 100m deeper,. The distribution of major stress magnitude and direction presented by the finite element analysis method in Jinchuan mine is shown in Fig.1. Fig.2 shows the relation between the depth and the ratio of the average horizontal stress to vertical stress.

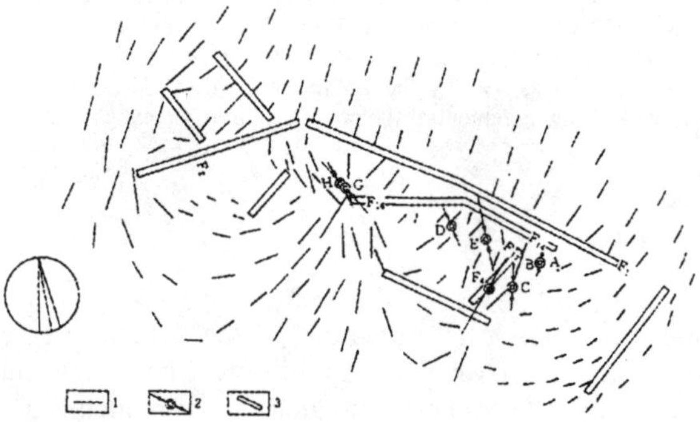

1---major compressive stress

2---direction of compressive stress measured at spots of field

3---main faults

Fig.1 The distribution of major stress obtained from finite element analyses in Jinchuan mine (The direction of additional extra force is N 15°E)

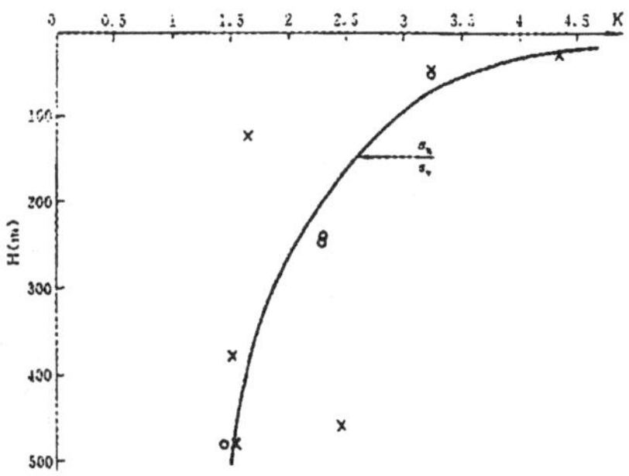

Fig.2 Depth versus ratio of average horizontal stress to vertical stress

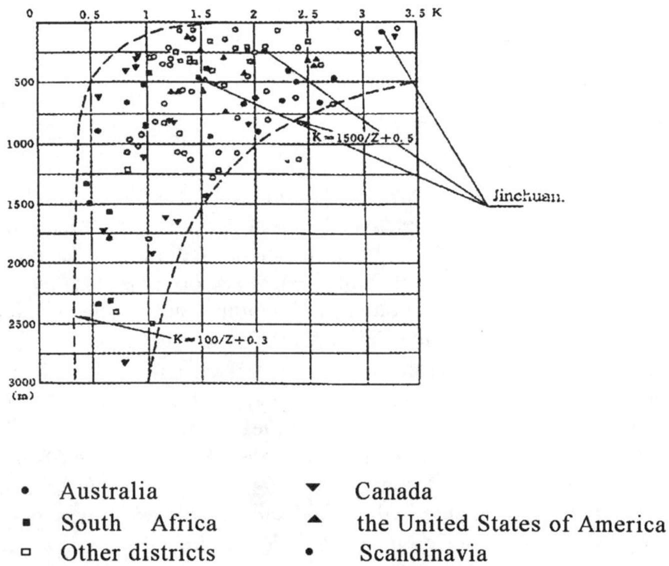

Fig. 3 Ratio of the average measured horizontal stress to the vertical stress

Fig. 3 is the comparison of the ratio of horizontal stress to vertical stress of Jinchuan mine with other countries

Fig. 3 shows that the ground stress in Jinchuan area has similar regularity as other countries in the world within the depth less than 500m , the ratio of horizontal stress to vertical stress ranks in the middle. Therefore the reason which cause the intensive ground pressure in Jinchuan mine is complex. Not only because of high stress, but also because of the characteristics of rock mass. As stated above, both ore body and surrounding rock are intensively fractured, the average strength of rock can not suffer the stress happened in the rock mass after drilling and blasting.

2 Action of ground pressure on underground workings

A variety of underground pressure phenomena, arising in Jinchuan mine workings, are contributed by many factors, mainly the ground stress. After excavation of openings ,the strain energy released to the space of excavation and the ground pressure redistributed in the surrounding rock of openings. The process could be illustrated as follows:

2.1 Initial ground pressure

According enormous experiences obtained from the practice of Jinchuan mine's development, the engineers have a common sense and knowledge that the initial ground pressure should be resisted not immediately after the excavation, otherwise all of the efforts will be in vain. Here was a obvious case, which could give us very interesting lesson. While one of the famous team of construction was invited to take part in the construction of Jinchuan's second phase project, because they are experienced and successfully passed Da-Yao-Shan tunnel of Jing-Guang railway(a very complex railway engineering) ,as well as through a lot of engineering with various accidents and different complex geological circumstances, they supposed that the Jinchuan engineering be pale before Da-Yan-Shan tunnel and they will success in these underground openings. A big chamber of over $50m^2$ cross section was excavated by them, they felt such engineering was easy for them and they finished it quickly by using double reinforced concrete support simultaneously following the working face. This team was highly awarded. However ,before long after the excavation ,the walls in both side of chamber dilated, the roof cracked and sank, the bottom of chamber distended as well. The chamber narrowed from 9m to 7 m. and its height reduced from 5.5m to less than 4 m. People paid great efforts to control the swelling of floor and breakage of roof.

Two years later, this chamber was reinforced again ,The floor and the engineering under this chamber were reinforced. The roof and side walls were re-excavated and supported again with original supporting system. From then on, neither distortion nor cracks appeared and the chamber are very stable. The equipment installed are still there without measurable movements. If this chamber had been excavated by smooth blasting and by using temporal metal mesh, rock-bolting and shotcrete lining ,and if the permanent support such like the double reinforced concrete framed support had been carried out three months later, this permanent support would have not been damaged. Later, the construction team has learnt a lesson from this experience, they successfully excavated an underground garage with the cross section from 20m to 120m by using the permanent support after the relaxation of the ground stress and no damages ever happened since then.

The regularities of initial ground stress and the measures to prevent initial stress could be summarized as follows:

1. The initial ground pressure, accompanying the excavations ,usually lasts for 2-3 months;
2. Any kind of rigid support should be used only after the relaxation of stress;
3. The principle of support must be "flexible or yielding support at first and rigid support at the end of relaxation of ground stress";
4. During this period, the underground working deforms quickly, at a velocity of 4-6mm/d for the first few days. In thin-medium thick marble strata(III_3), the anchor plate on the rock surface bears a force of 30 KN after 16 hours and this value increased to 70 KN after 21 hours. In stripe mixing rocks(IV), it is increased to 60 KN after 21 hours. In order to compensate the convergence of tunnel, the openings should be designed 300mm larger than the required.

2.2 The attenuation of ground pressure

After the initial ground pressure, there will be a stage of the attenuation of ground pressure, that means: the rock stress will be continuously relaxed for a long time with slowly increase of ground pressure. At this stage ,no matter the rate of deformation slowed down, but it still continued, even 2-3 years after excavation.

According the design ,concrete was recommended to use for the floor of workings, it was never expected that the whole engineering became stable after the concrete floor was finished. Such phenomena were obviously in both 1238 and 1138m sublevels. In the No.2 belt slope, along the 300m long slope, the pressure always occurs in a weak places of the working. Sometimes , the roof and the walls were in good condition, but the bottom dilated; in other situation, the bottom very stable, but the roof and the side walls were destroyed in some extent. In practice, taking the advantages of these phenomena that is, let the ground pressure released from the less important point on purpose while keep the rest stable. For example, in the lower 30m of No.2 belt slope, the floor at the pavement side dilated and the belt side is still stable, then it was determined to let the ground pressure continue being released at the pavement side, which was dealt with one year later. During this period, the floor of this slope at the pavement side continue to dilate while the roof, walls and the belt roadway side of the slope have been entirely preserved (keep stable). If the floor at the pavement side was reinforced immediately, it was likely that all ground pressure would be moved to the belt roadway side, and destroyed the base of belt, so the pavement side was act as a vulnerable part of the slope to protect the other parts. This measure is very useful in mines having similar phenomena.

2.3 Ground pressure effects on workings with different size

The destruction of the openings under the ground pressure at the first phase mining process was deeply impressed. At that time ,the stability of double-track roadway was better than that of single-track roadway. It was argued that the horizontal stress was bigger than vertical stress and the longitudinal axis of section of double-track roadway was horizontal, so the capacity of bearing pressure of double-track roadway was greater. Owing to that, it was required that the width-to-high radio of the roadway section designed should be equal to or approximate to the ratio of horizontal-to vertical stress. In other hand, Many people worried about the success of the second phase project because a lot of big trackless equipment were applied which made the accesses 2~4 times larger than that of the first phase. However, during the second phase design of roadways, proper width-to-vertical ratio of the section of the roadway was not to be considered as a critical principle because of the development cost. Now the second phase project has been finished. Unexpected, the situation of roadway is even better than that of the first phase, and the encountered difficulties are not so many as those in the first phase. Therefore people obtain a new conclusion, that is the underground pressure caused by the *in-situ* rock stress does not seem to increase in accordance with the enlargement of the cross section of the openings. The explanation of such wonderful phenomena could be concentrated to two points:

1. The ratio of average horizontal stress to vertical stress is different while the depths are different. The ratio decreased along with the increase of depth.(Fig.2), so the effects of width-to-height of cross section were obviously minimized at the second phase project (at 800m from ground surface)
2. The ground stress at different levels or different places of the same level is

different, but it is a constant at a specific point. When the roadway is drove, there will be appeared free space in the excavation. The ground stress ,which is the energy contained inside the strata, will release to this free space. If the free space is small, it is obviously that the releasing energy is more concentrated and the pressure is bigger while a big free space, shown relatively little pressure, The big section is favorable to release energy. This may be the reason which can be used to explain the pressure phenomena in the workings with larger section at the second phase project to compare with the pressure with small section at the first and second phase project in Jinchuan mine.

2.4 After virgin stress being released

The virgin stress will not recover after releasing, at least keep such condition for a long period. This fact can be proved during rehabilitation of the abandoned engineering. For example, during rehabilitation the big chamber mentioned above, which was destroyed by the intensive initial stress no needy to use additional strong support. It was because the virgin stress has released and recovering was not happened during production. Based on this regularity, it is not necessary to consider the virgin stress for the rehabilitation of workings. However, there exist other kind of pressure, which could be caused by the roof failure, rock swelling, underground water, weathering of surrounding rock and et al. The engineer should have a comprehensive knowledge about the case in practice, otherwise the design of support might be not appropriate with the stress condition. Actually, the workings deformation is a process of relaxation of the stress, which can be easily deal with ,if someone manager its regularity, just as the example illustrated above. That is: use the bulging behavior of opening floor to protect the roof and wall of opening, because the floor bulging phenomenon could be dealt with more easily than the rehabilitation of roof or both side walls, moreover there is no safe problem. The principle for protecting openings from rock failure can be shortly described as follows: reinforce some important places as protective object and set aside some unimportant places of workings not to be reinforces as the measures of protection. The ground pressure can not destroy the strengthened support points, so it will be released from the unimportant weak points as expected. After the release of energy, the unimportant point will be repaired .This is a successful and economical way proved by Jinchuan mine's experiences.

3 Support principles in Jinchuan mine

The ground pressure acts by stages, therefore the corresponding support should be designed to match with the condition of rock mass also by stages.

To deal with the initial ground pressure after excavation, rock bolting and shot concrete with wire mesh support is mainly used. In some huge chambers(section of over 100m^2),long anchor ropes are added to make good use of self-bearing ability of rock stratum and resist the high stress. At the same time, proper deformations of surrounding rocks should be permitted for the purpose of relaxation of rock stress.

After the initial stress being released, if working is in good condition or only a local part of it is destroyed, rock-bolting and shot concrete with wire mesh support can be still used for the permanent support. If working is seriously damaged, even converged too much, then enlargement will be needed. At this time, cast in-place concrete or reinforced concrete support are consider appropriate. Such principle of support has been successfully used for a long time in Jinchuan mine and it is very valuable due to basing on the plenty experiences and theoretical analysis.

Meanwhile, ore passes are lined with steel plates and the main ore pass is lined with rails or Mn steel plates. Huge chambers are supported with double-layer reinforced concrete after the relaxation of initial ground stress.

4 Conclusion

From above analysis, it can be concluded that the energy is released from rock strata on the underground openings by stages. When a large amount of energy is released in the initial stage, resistance is helpless, but this stage is relatively short. Then the velocity of increase of rock pressure is largely decreased. The increase of pressure in second stage is slowly and gradually ,but such increase does not stop completely for a long time. The pressure in this stage is usually appeared in the weak part of workings.

The underground pressure caused by the releasing of the *in-situ* rock stress does not seem to increase along with the cross section of workings, this conclusion is especially important for Jinchuan mine to build big section openings and use the large mechanized equipment in production.

Once the earth stress has released, it is impossible to recover in a relatively long period. This phenomenon is very significant in the mine production process.

Further theoretical analysis are still required for understanding of the mechanism of stress relaxation

Numerical Analysis

A METHOD TO CALCULATE STRESSES AND DEFORMATIONS IN 3D LAYERED STRATA

A.M. LINKOV
Institute for Problems of Mechanical Engineering (IPME),
Russian Academy of Science,
Bolshoy Prospect 61, V.O., St.-Petersburg, 199178, Russia
E-mail: link@mech.ipmc.ru

N.A. FILIPPOV
St-Petersburg Academy for Engineering Economics
Marata 27, St.-Petersburg, 121002, Russia

L.A. MILOVA
St-Petersburg Mining University
21 Linia 2, St.-Petersburg, 199026, Russia

V.V. ZOUBKOV
State Research Institute of Rock Mechanics and Surveying (*VNIMI*)
Sredny Prospekt 82, St.-Petersburg, 199026, Russia
E-mail: vzoubkov@pop3.rcom.ru

The importance of accounting for the layered structure of rock masses around openings is emphasized. Layers with different elastic properties and various types of contact interaction are discussed. A simplified mathematical formulation is given for 3D extraction openings in seam or reef deposits within a layered strata composed of layers with plane parallel boundaries. This formulation makes use of the geometrical feature of the problem. An efficient method is suggested for numerical solution of the resulting mathematical problem. It combines 2D Fourier transform with reduction to matrix difference equations of the second order. As shown, the sweep method provides an effective and stable solution of the problems under consideration. All the key formulae are given in an explicit form. A computer program is developed and results of numerical tests are presented. They confirm high efficiency of the method discussed. Its potential in facilitating the analysis of mining problems is illustrated by an example from practice.

1 Introduction

Rock masses often have a layered structure. Layers having different properties and various contact interaction strongly influence both virgin stresses and stresses induced by mining. There are many publications on this subject (e.g. works by Salamon[1], Erganov et al[2], Kolchin and Favermann[3], Nikishin and Shapiro[4], Shevliakov[5], Filippov[6],

Obolashvili[7], Maier and Novati[8], Linkov and Filippov[9], Salamon[10]). A comprehensive review of methods used and results obtained is given by Linkov and Filippov[9]. There is no need in reproducing here this analysis of the present state of the art. Note only that the difference equations approach suggested by Linkov and Filippov[9] seems the most efficient. Consequently, the paper aims to develop this approach as applied to rock mechanics problems. Specifically, the method can solve 3D problems for extraction openings in seam or reef deposits within a layered strata, which consists of layers with plane parallel boundaries.

2 Problem formulation

Consider a layered strata with plane parallel boundaries of layers (Fig. 1). The layers are supposed to be either horizontal with the upper layer at the earth's surface or inclined but located at a depth sufficient to neglect the influence of this surface. There may be extraction openings in some layers (seams or reefs). It is assumed that the sizes of the openings in the plane of a layer are greater then their height. Thereby, one may consider such openings as mathematical slits Hacket[11], Linkov[12], Crouch and Starfield[13], Petukhov et al[14], Napier and Stephansen[15]. As a result of this simplification, it is sufficient to solve problems for assemblies of layers between openings and match boundary conditions at two surfaces of an assemble, which corresponds to contacts with adjacent assembles and openings.

The primary task is to solve a problem for an assemble under prescribed tractions (displacements) assumed known on its external boundaries. This results in dependence between tractions and displacement on the external boundaries of the assemble. Using the mentioned matching conditions, one arrives at a system with respect to unknown values of displacements (tractions). Its solution provides complete data on tractions and displacements on boundaries of assemblies. Having them, one again considers an assemble and finds numerical values of displacements, strains and stresses in its layers. The layers are assumed to be elastic.

Mention should be also made on contact conditions for surfaces of layers within an assemble. In sufficiently general form they may be written as

$$\Delta \boldsymbol{u} = \boldsymbol{L}\boldsymbol{p} + \Delta \boldsymbol{u}_0, \tag{1}$$

where $\Delta \boldsymbol{u} = \boldsymbol{u}^- - \boldsymbol{u}^+$ is a vector of displacement discontinuities; $\boldsymbol{p}$ is a traction vector; $\boldsymbol{L}$ is a prescribed matrix of contact interaction; $\ddot{A}\boldsymbol{u}_0$ is a prescribed vector of contact distortions. The sign plus (minus) corresponds to that layer with respect to which the normal is outward (inward). Condition (1) incorporates as particular cases ideal contact ($\boldsymbol{L}=0$, $\ddot{A}\boldsymbol{u}=0$), smooth contact, and delamination. Referring to increments of values, it can describe irreversible contact deformations. It also can be used in successive steps when accounting for nonlinear contact interaction. In this case the "distortion" term $\ddot{A}\boldsymbol{u}_0$ is changed in iterations.

It is clear that the main problem is to find a solution for a 3D assemble of elastic layers under prescribed tractions (or displacements) on the external boundaries and given contact conditions (1) on the interacting surfaces of layers.

3 Method of solution

Consider 3D problems for layers with plane parallel boundaries (Fig. 1). Introduce the global coordinates x_1, x_2, x_3 and local coordinates x_1^j, x_2^j, x_3^j for each layer j ($j = 1, 2, ..., n$), as shown in Fig. 1.

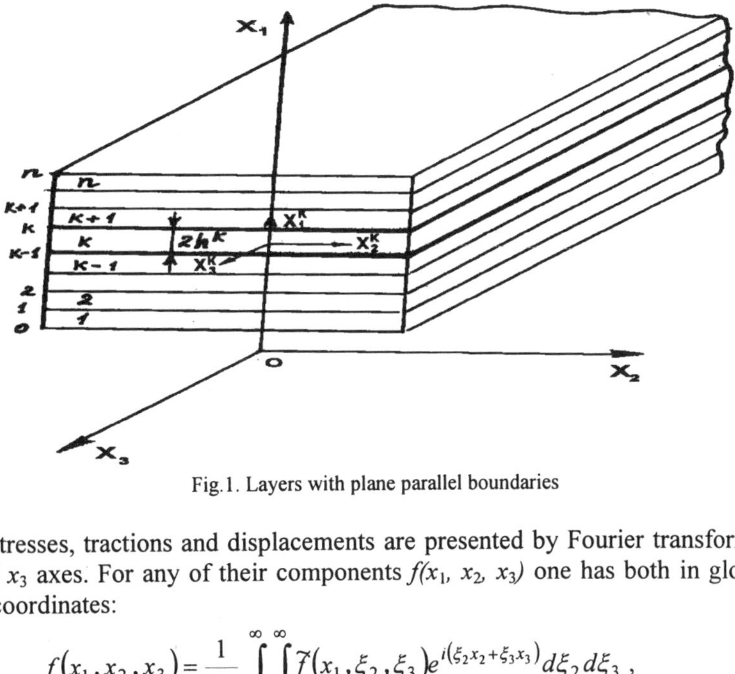

Fig.1. Layers with plane parallel boundaries

Stresses, tractions and displacements are presented by Fourier transforms over x_2 and x_3 axes. For any of their components $f(x_1, x_2, x_3)$ one has both in global and local coordinates:

$$f(x_1, x_2, x_3) = \frac{1}{2\pi} \int_{-\infty}^{\infty}\int_{-\infty}^{\infty} \bar{f}(x_1, \xi_2, \xi_3) e^{i(\xi_2 x_2 + \xi_3 x_3)} d\xi_2 d\xi_3 , \tag{2}$$

where $\bar{f}$ is the image of the function f at the plane ξ_2, ξ_3. It is found by means of the formula inverse to (2):

$$\bar{f}(x_1, \xi_2, \xi_3) = \frac{1}{2\pi} \int_{-\infty}^{\infty}\int_{-\infty}^{\infty} f(x_1, x_2, x_3) e^{-i(\xi_2 x_2 + \xi_3 x_3)} dx_2 dx_3 \tag{3}$$

For brevity, the images of stresses $\bar{\sigma}_{ij}$, tractions $\bar{p}$, displacements $\bar{u}$ also will be called stresses (tractions, displacements) when there may be no confusion in regard to their meaning. The images satisfy a system of ordinary differential equations, which appear after the double Fourier transform (2) is applied to the initial partial differential equations.

A solution may be written as a sum of symmetric and anti-symmetric parts:

$$\tilde{\sigma}_{ij} = \tilde{\sigma}_{ij}^{s} + \tilde{\sigma}_{ij}^{a}, \quad \tilde{u}_i = \tilde{u}_i^{s} + \tilde{u}_i^{a} \tag{4}$$

It is easy to see that for arbitrary tractions $\tilde{\sigma}_{11}$, $\tilde{\sigma}_{12}$, $\tilde{\sigma}_{13}$ and displacements $\tilde{u}_1$, $\tilde{u}_2$, $\tilde{u}_3$ their symmetric and anti-symmetric parts are expressed as

$$\tilde{\boldsymbol{\sigma}}^{s}(x_1) = \frac{1}{2}\begin{pmatrix} \tilde{\sigma}_{11}(x_1) + \tilde{\sigma}_{11}(-x_1) \\ \tilde{\sigma}_{12}(x_1) - \tilde{\sigma}_{12}(-x_1) \\ \tilde{\sigma}_{13}(x_1) - \tilde{\sigma}_{13}(-x_1) \end{pmatrix}; \quad \tilde{\boldsymbol{\sigma}}^{a}(x_1) = \frac{1}{2}\begin{pmatrix} \tilde{\sigma}_{11}(x_1) - \tilde{\sigma}_{11}(-x_1) \\ \tilde{\sigma}_{12}(x_1) + \tilde{\sigma}_{12}(-x_1) \\ \tilde{\sigma}_{13}(x_1) + \tilde{\sigma}_{13}(-x_1) \end{pmatrix};$$

$$\tilde{\boldsymbol{u}}^{s}(x_1) = \frac{1}{2}\begin{pmatrix} \tilde{u}_1(x_1) - \tilde{u}_1(-x_1) \\ \tilde{u}_2(x_1) + \tilde{u}_2(-x_1) \\ \tilde{u}_3(x_1) + \tilde{u}_3(-x_1) \end{pmatrix}; \quad \tilde{\boldsymbol{u}}^{a}(x_1) = \frac{1}{2}\begin{pmatrix} \tilde{u}_1(x_1) + \tilde{u}_1(-x_1) \\ \tilde{u}_2(x_1) - \tilde{u}_2(-x_1) \\ \tilde{u}_3(x_1) - \tilde{u}_3(-x_1) \end{pmatrix}.$$

For an isotropic medium the general symmetric and anti-symmetric solutions are obtained from formulae given by Obolashvili[7]. For traction vectors on the planes parallel to contacts one has

$$\tilde{\boldsymbol{\sigma}}^{s} = \boldsymbol{D}^{s}\boldsymbol{A}^{s}, \quad \tilde{\boldsymbol{\sigma}}^{a} = \boldsymbol{D}^{a}\boldsymbol{A}^{a} \tag{5}$$

where

$$\boldsymbol{D}^{s}(x_1,\xi_2,\xi_3) = \begin{pmatrix} ch z_1 - z_1 sh z_1 & i\xi_2 x_1 sh z_1 & i\xi_3 x_1 sh z_1 \\ -i\xi_2 x_1 ch z_1 & -sh z_1 - \left(\xi_2^2/\xi\right) x_1 ch z_1 & -\left(\xi_2\xi_3/\xi\right) x_1 ch z_1 \\ -i\xi_3 x_1 ch z_1 & -\left(\xi_2\xi_3/\xi\right) x_1 ch z_1 & -sh z_1 - \left(\xi_3^2/\xi\right) x_1 ch z_1 \end{pmatrix};$$

$$\boldsymbol{D}^{a}(x_1,\xi_2,\xi_3) = \begin{pmatrix} sh z_1 - z_1 ch z_1 & -i\xi_2 x_1 ch z_1 & -i\xi_3 x_1 ch z_1 \\ -i\xi_2 x_1 sh z_1 & ch z_1 + \left(\xi_2^2/\xi\right) x_1 ch z_1 & \left(\xi_2\xi_3/\xi\right) x_1 sh z_1 \\ -i\xi_3 x_1 sh z_1 & \left(\xi_2\xi_3/\xi\right) x_1 sh z_1 & ch z_1 + \left(\xi_3^2/\xi\right) x_1 sh z_1 \end{pmatrix};$$

$$\xi = \sqrt{\xi_2^2 + \xi_3^2}; \quad z_1 = \xi \cdot x_1; \quad \boldsymbol{A}^{s} = \begin{pmatrix} A_1^s \\ A_2^s \\ A_3^s \end{pmatrix}; \quad \boldsymbol{A}^{a} = \begin{pmatrix} A_1^a \\ A_2^a \\ A_3^a \end{pmatrix};$$

A_1^s, A_2^s, A_3^s, A_1^a, A_2^a, A_3^a are arbitrary constants within a layer; they serve to satisfy prescribed conditions on the boundaries of a layer.

Three other components $\tilde{\sigma}_{22}$, $\tilde{\sigma}_{33}$, and $\tilde{\sigma}_{23}$ are defined by similar formulae. They are of need only at the final step of calculations, when we want to know all the stresses within layers. As these formulae are rather lengthy and not of immediate use, they are not presented here. They are easily obtained from the formulae given.

For displacements, one has

$$\tilde{\boldsymbol{u}}^{s} = \boldsymbol{Q}^{s}\boldsymbol{A}^{s}, \quad \tilde{\boldsymbol{u}}^{a} = \boldsymbol{Q}^{a}\boldsymbol{A}^{a} \tag{6}$$

where

$$\boldsymbol{Q}^{s}(x_1,\xi_2,\xi_3) = \frac{1+\nu}{E}\frac{1}{\xi}\begin{pmatrix} 2(1-\nu)shz_1 - z_1chz_1 & -i(1-2\nu)(\xi_2/\xi)shz_1 + i\xi_2x_1chz_1 & -i(1-2\nu)(\xi_3/\xi)shz_1 + i\xi_3x_1chz_1 \\ -i(1-2\nu)(\xi_2/\xi)chz_1 - i\xi_2x_1shz_1 & 2\nu(\xi_2^2/\xi^2)chz_1 - (\xi_2/\xi)\xi_2x_1shz_1 & 2\nu(\xi_2\xi_3/\xi^2)chz_1 - (\xi_2/\xi)\xi_3x_1shz_1 \\ -i(1-2\nu)(\xi_3/\xi)chz_1 - i\xi_3x_1shz_1 & 2\nu(\xi_2\xi_3/\xi^2)chz_1 - (\xi_3/\xi)\xi_2x_1shz_1 & 2\nu(\xi_2\xi_3/\xi^2)chz_1 - (\xi_3/\xi)\xi_2x_1shz_1 \end{pmatrix};$$

$$\boldsymbol{Q}^{a}(x_1,\xi_2,\xi_3) = \frac{1+\nu}{E}\cdot\frac{1}{\xi}\begin{pmatrix} 2(1-\nu)chz_1 - z_1shz_1 & i(1-2\nu)(\xi_2/\xi)chz_1 - i\xi_2x_1shz_1 & i(1-2\nu)(\xi_3/\xi)ch\xi x_1 - i\xi_3x_1shz_1 \\ -i(1-2\nu)(\xi_2/\xi)shz_1 - i\xi_2x_1chz_1 & -2\nu(\xi_2^2/\xi^2)shz_1 + (\xi_2/\xi)\xi_2x_1chz_1 & -2\nu(\xi_2\xi_3/\xi^2)sh\xi x_1 + (\xi_2/\xi)\xi_3x_1chz_1 \\ -i(1-2\nu)(\xi_3/\xi)shz_1 - i\xi_3x_1chz_1 & 2\nu(\xi_2\xi_3/\xi^2)shz_1 + (\xi_3/\xi)\xi_2x_1chz_1 & -2\nu(\xi_3^2/\xi^2)sh\xi x_1 + (\xi_3/\xi)\xi_3x_1chz_1 \end{pmatrix};$$

$\boldsymbol{A}^{s}$, $\boldsymbol{A}^{a}$ are the same as in (5).

The first (second) of the formulae (5), (6) allow us to express the symmetrical (anti-symmetrical) part of displacements through the symmetrical (anti-symmetrical) part of tractions:

$$\tilde{\boldsymbol{u}}^{s} = \boldsymbol{R}^{s}\tilde{\boldsymbol{\sigma}}^{s}, \quad \tilde{\boldsymbol{u}}^{a} = \boldsymbol{R}^{a}\tilde{\boldsymbol{\sigma}}^{a}, \tag{7}$$

where $\boldsymbol{R}^{s}(x_1) = \boldsymbol{Q}^{s}(\boldsymbol{D}^{s})^{-1}; \quad \boldsymbol{R}^{a}(x_1) = \boldsymbol{Q}^{a}(\boldsymbol{D}^{a})^{-1}.$ (8)

Introduce tractions $\tilde{\boldsymbol{p}}_t^j$, $\tilde{\boldsymbol{p}}_b^j$ and displacements $\tilde{\boldsymbol{u}}_t^j$, $\tilde{\boldsymbol{u}}_b^j$ on the top and bottom boundaries of the j-th layer:

$$\tilde{\boldsymbol{p}}_t^j = \tilde{\boldsymbol{\sigma}}^j(h^j,\xi_2,\xi_3), \quad \tilde{\boldsymbol{p}}_b^j = \tilde{\boldsymbol{\sigma}}^j(-h^j,\xi_2,\xi_3),$$
$$\tilde{\boldsymbol{u}}_t^j = \tilde{\boldsymbol{u}}^j(h^j,\xi_2,\xi_3), \quad \tilde{\boldsymbol{u}}_b^j = \tilde{\boldsymbol{u}}^j(-h^j,\xi_2,\xi_3).$$

Then, from (7), (8), applied to the j-th layer, after some algebra, one obtains equations for images:

$$\tilde{\boldsymbol{u}}_t^j = \boldsymbol{R}_{tt}^j\tilde{\boldsymbol{\sigma}}_t^j + \boldsymbol{R}_{tb}^j\tilde{\boldsymbol{\sigma}}_b^j, \quad \tilde{\boldsymbol{u}}_b^j = \boldsymbol{R}_{bt}^j\tilde{\boldsymbol{\sigma}}_t^j + \boldsymbol{R}_{bb}^j\tilde{\boldsymbol{\sigma}}_b^j, \quad j = 1, \ldots, n \tag{9}$$

where in this case $\boldsymbol{R}_{tt}^j$, $\boldsymbol{R}_{tb}^j$, $\boldsymbol{R}_{bt}^j$, $\boldsymbol{R}_{bb}^j$ are square matrices of the 3rd order. They are defined through $\boldsymbol{R}^{js}(h)$ and $\boldsymbol{R}^{ja}(h)$ as

$$\begin{aligned} \boldsymbol{R}_{tt}^j &= \left(\boldsymbol{R}^{js}+\boldsymbol{R}^{ja}\right)/2; \quad \boldsymbol{R}_{tb}^j = -\left(\boldsymbol{R}^{js}-\boldsymbol{R}^{ja}\right)_1/2; \\ \boldsymbol{R}_{bt}^j &= \left(\boldsymbol{R}^{js}-\boldsymbol{R}^{ja}\right)'/2; \quad \boldsymbol{R}_{bb}^j = -\left(\boldsymbol{R}^{js}+\boldsymbol{R}^{ja}\right)_1'/2. \end{aligned} \tag{10}$$

where the subscript '1' means that the first column of the matrix in brackets is taken with the sign 'minus'; the subscript 'prime' means that the first row is taken with the sign 'minus'.

The decomposition on symmetric and anti-symmetric parts significantly simplifies the derivation of the matrices in (9). It allows us to use matrices of the third order instead of the sixth order when obtaining the third order matrices $\boldsymbol{R}_{tt}^j$, $\boldsymbol{R}_{tb}^j$, $\boldsymbol{R}_{bt}^j$, $\boldsymbol{R}_{bb}^j$. For isotropic layers, after some lengthy algebra, it follows

$$\boldsymbol{R}_{tt}^j = \frac{1+\nu^j}{E^j\xi}\begin{pmatrix} a_{11} & a_{12} & a_{13} \\ -a_{12} & a_{22} & a_{23} \\ -a_{13} & a_{23} & a_{33} \end{pmatrix}, \quad \boldsymbol{R}_{tb}^j = \frac{1+\nu^j}{E^j\xi}\begin{pmatrix} b_{11} & b_{12} & b_{13} \\ b_{12} & b_{22} & b_{23} \\ b_{13} & b_{23} & b_{33} \end{pmatrix},$$

$$\boldsymbol{R}_{bt}^j = -\left(\boldsymbol{R}_{tb}^j\right)_1', \quad \boldsymbol{R}_{bb}^j = -\left(\boldsymbol{R}_{tt}^j\right)_1',$$

where the explicit expressions for the coefficients of the matrices are

$$a_{11} = \left(thz + th^3z + z/ch^4z\right)\left(1-\nu^j\right)/\Delta, \quad a_{12} = i\xi_2 d_2, \quad a_{13} = i\xi_3 d_2,$$

$$a_{22} = \left(1+th^2z\right)/thz - \xi_2^2 d_3, \quad a_{23} = -\xi_2\xi_3 d_3, \quad a_{33} = \left(1+th^2z\right)/thz - \xi_3^2 d_3,$$

$$b_{11} = \left[thz/ch^2z + z\left(1+th^2z\right)/ch^2z\right]\left(1-\nu^j\right)/\Delta, \quad b_{12} = i\xi_2 d_1, \quad b_{13} = i\xi_3 d_1,$$

$$b_{22} = \left(-1+\xi_2^2 d_2\right)/\left(thz\cdot ch^2z\right), \quad b_{23} = \xi_2\xi_3 d_4/\left(thz\cdot ch^2z\right), \quad \xi = \sqrt{\xi_2^2+\xi_3^2},$$

$$b_{33} = \left(-1+\xi_3^2 d_4\right)/\left(thz\cdot ch^2z\right), \quad d_1 = \left(1-\nu^j\right)\left(2h^j/\Delta\right)\left(thz/ch^2z\right),$$

$$d_2 = \left[1-2\left(\nu^j th^2z - z/ch^4z\right)/\Delta\right]/\xi, \quad \Delta = th^2z - z^2/ch^4z, \quad z = \xi\cdot h^j,$$

$$d_3 = \left[\left(\nu^j th^2z - z^2/ch^4z\right)\left(1+thz\right) + \left(1-\nu^j\right)\cdot z\cdot thz/ch^4z\right]/\left(\Delta\cdot\xi^2 thz\right),$$

$$d_4 = \left[\nu^j th^2z - z^2/ch^4z + \left(1-\nu^j\right)\cdot z\cdot thz\cdot\left(1+th^2z\right)\right]/\left(\Delta\cdot\xi^2\right).$$

The equations (7) are supplied with boundary conditions. These are equilibrium conditions for surfaces in contact

$$\tilde{\boldsymbol{p}}_b^j = \tilde{\boldsymbol{p}}_t^j = \tilde{\boldsymbol{p}}^j \qquad (j = 1, \ldots, n) \tag{11}$$

and the law of interaction (1), which is written as

$$\tilde{\boldsymbol{u}}^{j+1} - \tilde{\boldsymbol{u}}^j = \boldsymbol{L}^j \tilde{\boldsymbol{p}}^j + \Delta \tilde{\boldsymbol{u}}_0 \qquad (j = 1, \ldots, n) \tag{12}$$

The system (1), (11), (12) may be efficiently solved by reduction to the three-point difference equations as suggested by Linkov and Filippov[9]:

$$\boldsymbol{A}^j \boldsymbol{p}^{j-1} - \boldsymbol{C}^j \boldsymbol{p}^j + \boldsymbol{B}^j \boldsymbol{p}^{j+1} + \boldsymbol{F}_p^j = 0 \quad j = 1, \ldots, n-1 \tag{13}$$

where $$\boldsymbol{A}^j = -\boldsymbol{R}_{tb}^j, \quad \boldsymbol{C}^j = \boldsymbol{L}^j + \boldsymbol{R}_{tt}^j - \boldsymbol{R}_{bb}^{j+1}, \quad \boldsymbol{B}^j = \boldsymbol{R}_{tt}^{j+1} \tag{14}$$

$$\boldsymbol{F}^j = \left(\boldsymbol{u}_{b0}^{j+1} - \boldsymbol{u}_{t0}^j\right) - \Delta \boldsymbol{u}_0^j \tag{15}$$

The difference equations represented by (13) are solved under the usual conditions on the boundaries $j = 0$ and $j = n$:

$$-\boldsymbol{C}^0 \boldsymbol{p}^0 + \boldsymbol{B}^0 \boldsymbol{p}^1 + \boldsymbol{F}^0 = 0 \tag{16}$$

$$\boldsymbol{A}^n \boldsymbol{p}^{n-1} - \boldsymbol{C}^n \boldsymbol{p}^n + \boldsymbol{F}^n = 0 \tag{17}$$

where coefficients are defined by (13), (14) where now $\boldsymbol{R}_{tb}^0 = 0$, $\boldsymbol{R}_{tt}^0 = 0$, $\boldsymbol{R}_{bb}^{n+1} = 0$, $\boldsymbol{R}_{bt}^{n+1} = 0$, $\boldsymbol{u}_{t0}^0 = 0$, $\boldsymbol{u}_{b0}^{n+1} = 0$, so that $\boldsymbol{C}^0 = \boldsymbol{L}^0 - \boldsymbol{R}_{bb}^1$, $\boldsymbol{B}^0 = \boldsymbol{R}_{bt}^1$, $\boldsymbol{F}^0 = \boldsymbol{u}_{t0}^1 - \Delta \boldsymbol{u}_0^0$, $\boldsymbol{A}^n = -\boldsymbol{R}_{tb}^n$, $\boldsymbol{C}^n = \boldsymbol{L}^n + \boldsymbol{R}_{tt}^n$, $\boldsymbol{F}^n = -\boldsymbol{u}_{t0}^n - \Delta \boldsymbol{u}_0^n$.

If tractions $\boldsymbol{p}^n = \boldsymbol{f}^n$ are prescribed on the boundary $j = 0$, we have $\boldsymbol{C}^n = \boldsymbol{I}$, $\boldsymbol{B}^n = 0$, $\boldsymbol{F}^n = \boldsymbol{f}^n$. If tractions are given on the boundary $j = n$, then $\boldsymbol{A}^n = 0$, $\boldsymbol{C}^n = \boldsymbol{I}$, $\boldsymbol{F}^n = \boldsymbol{f}^n$.

The most efficient method to solve system (16), (17) is the sweep method. It combines two virtues (Godunov and Ryabenkii[16], Samarski and Gulin[17]):

(i) a small number of arithmetic operations, and

(ii) it is insensitive to rounding errors.

(The latter property is known as stability Godunov and Ryabenkii[16], Samarski and Gulin[17].) These virtues have been clearly demonstrated by Linkov and Filippov[9], where Fourier series were employed in a 2D problem. It was stated that calculations with double precision provide the accuracy of 6 digits up to 1000 layers with relative thickness of 0.001.

For the right-hand sweep method, one successfully finds the sweeping matrices $\boldsymbol{\alpha}^j$ and vectors $\boldsymbol{\beta}^j$ in direct run (Samarski and Nikolaev[18]):

$$\begin{aligned} &\boldsymbol{\alpha}^1 = \left(\boldsymbol{C}^0\right)^{-1} \boldsymbol{B}^0, \quad \boldsymbol{\beta}^1 = \left(\boldsymbol{C}^0\right)^{-1} \boldsymbol{F}^0 \\ &\boldsymbol{\alpha}^{j+1} = \left(\boldsymbol{C}^j - \boldsymbol{A}^j \boldsymbol{\alpha}^j\right)^{-1} \boldsymbol{B}^j \qquad (j = 1, \ldots, n-1) \\ &\boldsymbol{\beta}^{j+1} = \left(\boldsymbol{C}^j - \boldsymbol{A}^j \boldsymbol{\alpha}^j\right)^{-1} \left(\boldsymbol{F}^j + \boldsymbol{A}^j \boldsymbol{\beta}^j\right) \quad (j = 1, \ldots, n) \end{aligned} \tag{18}$$

Traction vectors are then found in reverse run by

$$p^n = \beta^{n+1}, \quad p^{j-1} = \alpha^j p^j + \beta^j \quad (j = n, \ldots, 1) \tag{19}$$

The physical meaning of the sweeping matrices α^j ensures that they involve well-posed mathematical problems. This meaning is quite simple: coefficients of the matrix α^j present the tractions which arise from unit tractions at the j-th contact under homogeneous conditions $F^k = 0$ of contacts $k = 0, 1, \ldots, j - 1$.

The matrices R^j_{tt}, R^j_{tb}, R^j_{bt}, R^j_{bb} are calculated for each fixed pair (ξ_2, ξ_3). After this, one apples the sweep method to solve the system with respect to the images $\tilde{p}^j$ of the tractions. As mentioned below, it provides high accuracy. Having $\tilde{p}^j$ for a pair (ξ_2, ξ_3), one may use (4) to find its symmetric and anti-symmetric parts $\tilde{p}^{js}$ and $\tilde{p}^{ja}$. After this, the inverted equations (5), taken for $x^j_1 = h^j$, give constants A^{js} and A^{ja}. Consequently, (4), (5), (6) provide images of tractions and displacements within layers for the fixed pair (ξ_2, ξ_3). Having found these images for a set of pairs (ξ_2, ξ_3) used in the inverse transform, one finds tractions and displacements themselves by means of the inverse formula (2). Thus, the problem for 3D layered media becomes solved.

4 Numerical tests and an illustrative example

A computer program was written and tested on the basis of the method described by Zoubkov et al[19]. To evaluate the accuracy of calculations the authors compared sweep matrices for assemblies of isotropic layers with the same equal thickness and elastic properties under ideal contact between layers but with a different number of layers. For instance, the results for four layers were compared with those for eight layers: the sweep matrices for the middle of a package have to coincide in both cases.

Calculations with ordinary precision showed that for layers with a thickness greater 0.001 of the length of the Fourier expansion area, the coefficients of the sweep matrices coincide within 5 significant digits. When using calculation with double precision, eight correct digits were obtained. Such accuracy is quite sufficient for practical applications.

The developed computer program allows us to find stresses, strains and displacements for any number of layers with arbitrary Young's moduli and Poisson's ratios. As an illustration, Fig. 2 presents a practical scheme and results for a layered strata with a seam, which is mined out by two steps, one ahead of the other. The plan of the extracted area is shown in Fig. 2a. In Fig. 2b, the cross-section AA (BB) is presented. Stresses are normalized by the virgin normal component.

The normalized stresses σ_1 are plotted in Fig. 2c for the cross-section AA and in Fig. 2d for the cross-section BB. The scale in the x_2 direction is enlarged to make clear the changes in stresses near the step; the area of the major concern is the dashed line in Fig. 2b. The graphs in Fig. 2c, 2d present σ_{11} for three values of the ratio E_1/E_2: 1, 5, 10. The case $E_1/E_2 = 1$ corresponds to homogeneous rock.

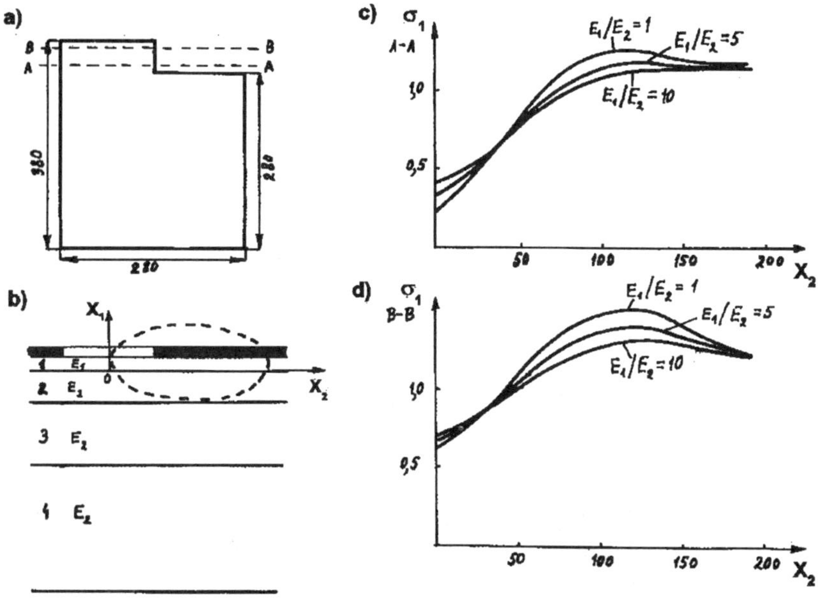

Fig. 2. Scheme and results of calculations
a) plan of extracted area; b) the cross-section AA; c) normalized stresses σ_1 for the cross-section AA; d) normalized stresses σ_1 for the cross-section BB.

It may be seen that the maximum vertical stress σ_{11} is greater in the cross-section BB then that in AA. It is due to decrease in unloading generated by the mined out area. The growth in the stiffness of the first layer decreases the unloading as well: a hard layer serves as a shield for the compliant layers beneath it. The same "shield" effect appears also in smoother curve of abutment pressure.

The case of a number of mined out seams is included. Note also that an application of moduli changing in time (Amusin and Linkov[20]) provides possibility to account for visco-elastic properties of layers.

Acknowledgement. Two of the authors (A.M.Linkov and V.V.Zoubkov) appreciate the financial support of the Russian Fund of Fundamental Investigations provided when finalizing this work (Grant № 97-05-64191).

References

1. M.D.G. Salamon, *The Influence of Strata Movement and Control on Mining Development and Design*. (University of Durham, Ph. D. thesis, 1962).
2. J.S. Erganov et all, *Stability of Horizontal Openings in Inclined Layered Rocks*. (Alma-Ata: Nauka, 1971). (In Russian).
3. G.B. Kolchin and A.A. Favermann, *The Theory of Elasticity for Inhomogeneous Bodies*. (Kishinev Sticenci, 1972). (In Russian).
4. V.S. Nikishin and G.S. Shapiro, *Problems of Elasticity Theory for Multi-layered Media*. (Moscow: Nauka, 1973). (In Russian).
5. Y.A. Shevliakov, *Matrix Algorithms in the Theory of Elasticity for Inhomogeneous Media*. (Kiev, "Vischa shkola", 1977). (In Russian).
6. N.A. Filippov, *Development of a Method for Evaluation of Stresses and Protected Zones in a Layered Rock Massif*. (Ph. D. thesis, All-Union Research Institute for Mine Geomechanics and Surveying (VNIMI), Leningrad, 1977). (in Russian).
7. E.I. Obolashvili, *Fourier Transform and its Application in the Theory of Elasticity*. (Tbilici, Metcierba, 1979). (In Russian).
8. G. Maier and G. Novati, *Int. J. the Numerical and Analytical Methods in Geomechanics*, **11**, 435 (1987).
9. A. Linkov and N. Filippov, *Meccanica*, **26**, 195 (1991).
10. M.D.G. Salamon, *J. South African Inst. Min. Metall.*, **91**, 9 (1991).
11. P. Hackett, *Trans. Inst. Min. Engrs.*, **118**, 421 (1959).
12. A.M. Linkov, *Some Problems of Fracture Mechanics* (with Application to Rock Mechanics). (The 2-nd Degree Doctor thesis. St-Petersburg State University, St-Petersburg, 1977). (In Russian).
13. S.L. Crouch and A.M. Starfield, *Boundary Element Method in Solid Mechanics*. (George Allen & Unwin, London, 1983).
14. I.M. Petukhov et al, *A Theory of Protective Seams*. (Nedra, Leningrad, 1983). (in Russian).
15. J.A.L. Napier and S.J. Stephansen, Proc. Int. Symposium on the Application of Computational Mechanics in the Mineral Industries. Johannesburg, S. A. Inst. Mining and Metall., **1**, 3 (1988).
16. S.K. Godunov and V.S. Ryabenkii, *Difference Schemes*. (North-Holland, 1987).
17. A.A. Samarskii and A.V. Gulin, *Numerical Methods*. (Nauka, Moscow, 1989). (in Russian).
18. A.A. Samarskii and E.S. Nikolaev, *Numerical Methods for Grid Equations*. (Nauka, Moscow, 1978).
19. V.V. Zoubkov et al, On Evaluation of Stress-strain State of 3D Layered Massif. (VINITI, № **1737-B90**, 1988). (in Russian).
20. B.Z. Amusin and A.M. Linkov, *Izv. AN SSSR, Mekhanika Tverdogo Tela*, **9**, 139 (1974).

MANIFOLD METHOD AND ITS APPLICATIONS TO ROCK ENGINEERING

Weiyuan ZHOU, Qiang YANG and Ruoqiong YANG

Dept. of Hydraulic Eng., Tsinghua University,
100084 Beijing, PR China
E-mail: dhhlqb@mail.tsinghua.edu.cn

The manifold method is an approximate method for solving differential equations defined in a domain with arbitrary shapes. It is applicable to any boundary value problems with moving least square interpolants. This method requires only nodal data. No element connectivity is needed. Lagrangian multipliers are used to ensure the essential boundary conditions in smooth mathematics states. Weighted orthogonal basis functions are constructed so that the requirement for solving equations at each quadrature point is eliminated. In this paper, the manifold method is described. The newly developed "manifold method", based on finite cover systems, has the potential to meet more engineering requirements. Dr. Gen-hua Shi [1] proposed this method. In his paper "numerical manifold Method", the term "manifold" is defined as a generalization of the differential manifold which is the main subject of differential geometry, topology, differential topology and modern algebra of mathematics. The difference of the manifold and the traditional differential manifold is the following: the global functions of the differential manifold are highly differentiable and entirely defined irrelevant with the covers; the defined global functions of the manifold are based on covers and only piecewise differentiable, mostly discontinuous on the contact interfaces. Manifolds connect many individual folded domains together to cover the entire material domains. The basic theory of the manifold method is expressed and explained.

1 Introduction

In geotechnique engineering, it has become clear that in numerical analysis, mesh generation is a far more time-consuming and expensive task than assembly and solution of the finite element equations. It requires a lot of time for preparation of data. Three dimensional meshing is still an extremely burdensome task and the conversion of solid models to finite element data is time-consuming and often introduces numerous ambiguities.

Nayrole *et al.* [2] have recently introduced a very interesting and promising method which they call the diffuse element method. In this method, only a mesh of nodes and a boundary description is needed to developed the Galerkin Equations. The interpolating functions are polynomials which are fit to the nodal values by least-square approximation. The finite element mesh is totally unnecessary in this method.

In the domain of composite materials, the corresponding integral representation of the solution of the boundary-value problem is proximate with the account of interpolants of the derivatives of some orders which is defined within a cover.

Numerical examples makes it clear that by manifold method stress concentrations and stress intensity factors at discontinuities could be computed very conveniently. For crack propagation and coalescence of fissures, it is very advantageous to compute their behaviours without remeshment around crack tips.

The manifold method was developed with element-free method to cope with coupled problems involving discontinua and continua. It may be widely used in solving geotechnique problems which generally imply discontinuous joints in rock materials such as, slope slides, foundation stability and underground engineering.

2 Manifold structure and covers on polygon domains

Set M forms a differential manifold structure if defining M a set of limited and countable charts. A point in set M corresponds a chart in the domain. In Euclidean space, $q = q(q_1, q_2, \cdots, q_n)$, there exists a one-to-one mapping $f : W \to U$, subset $W \subset U$. Set M in U domain, so the map function is a open set as shown in Fig. 1.

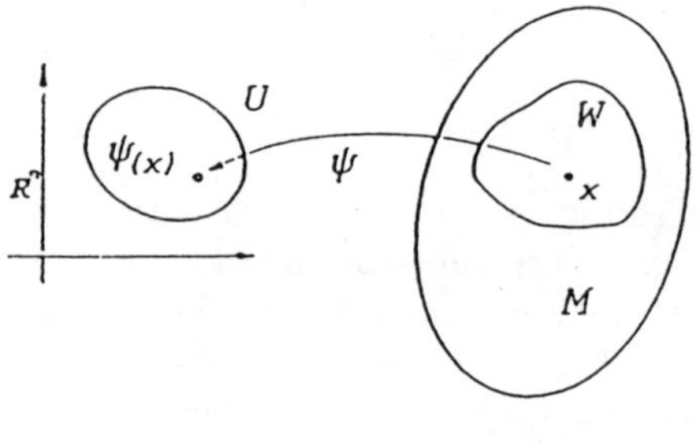

Figure 1: Map of manifold

The manifold method is adopted to find the approximate solution of a boundary value problem. At first, overlapped covers are set up in the polygon domain K. Interpolating functions are to be established, which satisfy the boundary values. For each cover, an approximate solution which may be certain elementary functions is established. As shown in Fig. 2, P_j, $j = 1, 2, \cdots, n_b$ are the vertices of the polygon K; P_q, $q = n_b + 1, \cdots, l$ are middle points of its sides in the polygon; P_m, $m = l+1, \cdots, M$ are inner points of K where $n_b < l < M$.

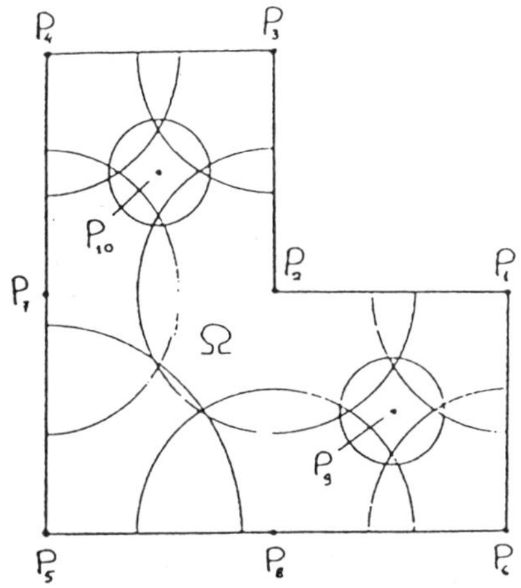

Figure 2: A multifold covering of the polygon K

An approximate solution is found by the manifold method for a polygon domain. The cover for K polygon is referred to countable covers which exhibit sectors, circular disks or semicircular disks.

Shape functions in covered domains could be used to express the mathematical solutions in the domain. They may be some basic functions fitted to special boundaries or derivatives along the interfaces between neighbor domains. As shown in Fig. 2, these expressions form the base of boundary solutions when using the manifold method.

3 Discrete coordinate functions defined in covered domains

It is different with F.E.M for implementation of manifold method to establish shape functions which are found by the way given below:

On assuming that u denotes the function within a cover, their values on fitted pointed are u_i, $i = 1, 2, \cdots n$. The function $u_k(x)$ is the approximate function of u,

$$u_k(x) = \sum_{i=1}^{n} \lambda(x) u_i \tag{1}$$

Covers are established over polygon K, which require a shape function. The shape function is found by the weighted residual method. The above mentioned shape function is expressed by piecewise continuous functions, which will lead to the structure stiffness matrix after the shape functions are set up.

In a cover, any number of points may be assumed to be fitted points for the solution in the domain. The value for the fitted points are given below:

$$u_i = u(x_i) \qquad i = 1, 2, \cdots n. \tag{2}$$

and their approximate solutions are

$$u_k = \sum_{i=1}^{n} \lambda_i(x) u_i \qquad i = 1, 2, \cdots n. \tag{3}$$

which may be expressed in terms of multi-variable polynomial, triangular polynomials, Legendre or Chilisheve polynomials.

In order to find u_k, a basic function $\Phi = \{\Phi_1, \Phi_2, \cdots, \Phi_m\}^T$ is defined in cover C, and then

$$u_k = \sum_{j=1}^{m} \Phi_j(x) a_j = \Phi^T(x)\mathbf{a} \tag{4}$$

where $\mathbf{a} = \{a_1, a_2, \cdots, a_m\}^T$ is one set of unknown parameters. At each nodal points in the domain,

$$u_k(x_i) = \sum_{j=1}^{m} \Phi_j(x_i) a_j \tag{5}$$

By moving least square method, $u_k(x_i)$ could be found for weighted function $W(x_i)$

$$\begin{aligned} &\min \textstyle\int_c W(x_i)[u_k(x_i) - u_i]^2 dc = \\ &\min \textstyle\int_c W(x_i)[\Phi^T(x_i)\mathbf{a} - u_i]^2 dc \end{aligned} \tag{6}$$

which derivative with respect to $\mathbf{a}$ leads to

$$\begin{aligned} &2W(x_i)[\Phi^T(x_i)\mathbf{a} - u(x_i)]\phi(x_i) = 0 \\ &W\Phi^T\mathbf{a}\Phi = Wu\Phi \\ &W\Phi^T\Phi\mathbf{a} = W\Phi u \end{aligned} \tag{7}$$

If set

$$S = W\Phi^T\Phi, \qquad T = W\Phi \tag{8}$$

then

$$Sa = Tu \Longrightarrow \mathbf{a} = S^{-1}Tu \tag{9}$$

and

$$u_k = \Phi^T S^{-1} Tu = \lambda u \tag{10}$$

The shape function λ are expressed as

$$\lambda = \Phi^T S^{-1} T \tag{11}$$

and

$$\begin{aligned} \lambda_{,i}^T &= \Phi_{,i}^T S^{-1} + \Phi^T (S^{-1} T_{,i}) \\ &= \Phi_{,i}^T S^{-1} T + \Phi^T (S_{,i}^{-1} T) + \Phi^T (S^{-1} T_{,i}) \end{aligned} \tag{12}$$

By means of shape functions, the unknowns u can be obtained consequently.

4 Galerkin method in solution of boundary problems

In domain K, its boundary problem based on the manifold method concept which is described above are easily solved by various numerical methods. Here Galerkin method is applied. Based on variational principles, the equation to be solved is:

$$Lu = f \qquad n \in k \tag{13}$$

According to the Galerkin method,

$$(Lu_0 - f, \Phi_i) = 0, \quad i = 1, 2, \cdots, n \tag{14}$$

The solution of eqn. (14) is equivalent to that of eqn. (13). Within the cover C, linear functions are adopted as

$$u_h = \sum_{l=1}^{k} \Phi_l u_l \tag{15}$$

It is clear u_h denote those unknowns to be solved linear equation. Eqn. (14) become

$$\sum_{l=1}^{k} (L\Phi_l, \Phi_k) u_l = (f, \Phi_k) \qquad k = 1, 2, \cdots, n \tag{16}$$

This formula gives weak form solutions, and they are obtained from each cover domains. It is shown as Fig. 3.

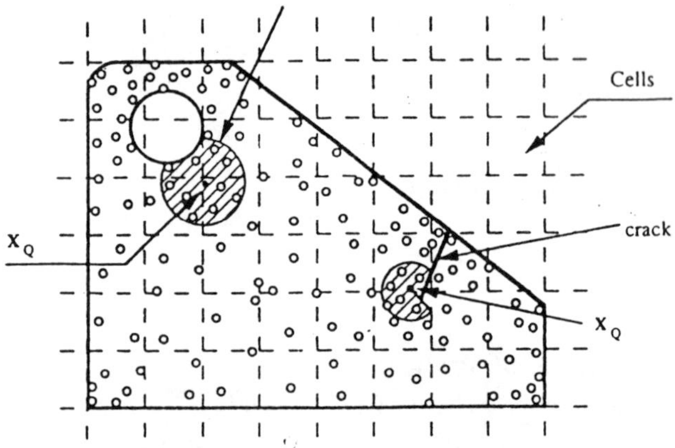

Figure 3: Covers for quadrature of structure

5 Geotechnique engineering applications

In a domain Ω, a boundary value problem can be defined as

$$\begin{aligned} \nabla\boldsymbol{\sigma} + \mathbf{b} &= \mathbf{0} \\ \boldsymbol{\sigma}\cdot\mathbf{n} &= \overline{\mathbf{t}} \text{ on stress boundary } \quad \Gamma_t \\ \mathbf{u} &= \overline{\mathbf{u}} \text{ on displacement boundary } \quad \Gamma_u \end{aligned} \tag{17}$$

In the domain Ω, the quadrature variational equation is

$$\begin{aligned} &\int_\Omega \delta(\nabla V^T):\boldsymbol{\sigma} d\Omega = \int_\Omega \delta V^T\cdot\mathbf{b}d\Omega + \int_{\Gamma_t}\delta V^T\cdot\overline{\mathbf{t}}d\Gamma \\ &+\int_{\Gamma_u}\delta\lambda^T\cdot(\mathbf{u}-\overline{\mathbf{u}})d\Gamma + \int_{\Gamma_u}\delta V^T\cdot\lambda d\Gamma \end{aligned} \tag{18}$$

where λ is introduced as Lagrangian multipliers. By using eqn. (15), the stiffness matrix is obtained as

$$\begin{aligned} &K_{ij} = \int_\Omega B_i^T D B_j d\Omega - \int_{\Gamma_u}\Phi_i SNDB_j d\Gamma \\ &-\int \Gamma_u B_i^T D^T N^T S\Phi_j d\Gamma \\ &f_i = \int_{\Gamma_t}\Phi_i F d\Gamma + \int_\Omega \Phi_i b d\Omega + \int_{\Gamma_u} B_i^t D^T N^T S\overline{\mathbf{u}}d\Gamma \end{aligned} \tag{19}$$

where

$$B_i = \begin{bmatrix} \Phi_{i,x} & 0 \\ 0 & \Phi_{i,y} \\ \Phi_{i,y} & \Phi_{i,x} \end{bmatrix} \tag{20}$$

$$N = \begin{bmatrix} n_x & 0 & n_y \\ 0 & n_y & n_x \end{bmatrix} \tag{21}$$

$$S = \begin{bmatrix} S_x & 0 \\ 0 & S_y \end{bmatrix} \tag{22}$$

$$S_x = \begin{cases} 1, u_x & \text{prescribed on} \quad \Gamma_u \\ 0, u_x & \text{unknown on} \quad \Gamma_u \end{cases} \tag{23}$$

$$S_y = \begin{cases} 1, u_y & \text{prescribed on} \quad \Gamma_u \\ 0, u_y & \text{unknown on} \quad \Gamma_u \end{cases} \tag{24}$$

A test example is a cantilever beam. From the elasticity theory, the exact solution is

$$\begin{aligned} &u_x = \tfrac{P}{6EJ}(y - \tfrac{d}{2})[(6l - 3x)x + (2 + \nu)(y^2 - 2Dy)] \\ &u_y = \tfrac{P}{6EJ}[3\nu(y^2 - 2Dy + \tfrac{D^2}{2})(l - x) \\ &+\tfrac{1}{4}(4 + 5\nu)D^2x + 3(l - \tfrac{x}{3})x^2] \end{aligned} \tag{25}$$

where

$$J = \frac{D^3}{12} \tag{26}$$

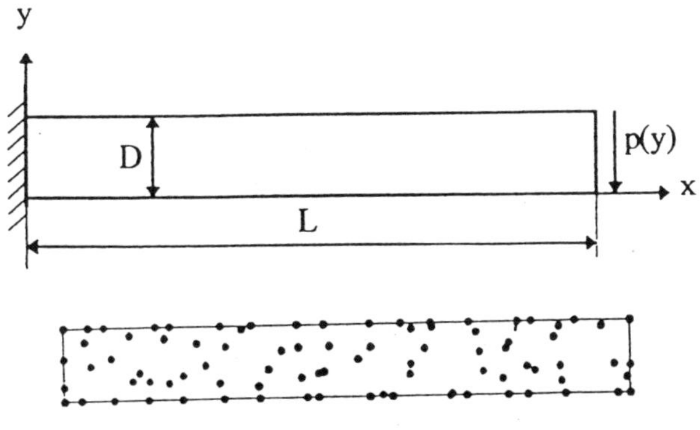

Figure 4: Description of the beam problem and nodal points

The configuration of the manifold method calculation is shown in Fig. 4 where $E = 3 \times 10^7$, $D = 1$, $L = 8$. The comparison of the manifold method, the finite element method and the elasticity theory is shown in Table 1. The result has illustrated that the manifold method has higher accuracy than the

Table 1: Ratio of deflection $W^{\text{num}}/W^{\text{exact}}$

method	$\nu = 0.25$	$\nu = 0.4999$
manifold method	0.999	0.999
F.E.M.	0.607	0.029

finite element method and can even reach high accuracy when the material is near incompressible state.

Here shown the stability analysis of the ship lock of Three Gorges Project in China on the Yangtse River. It is a five-stage lock with steep banks in two sides. A water head of 160m is applied on the dam after reservoir water impounding. The ship lock is going to be excavated in three years. As shown in Fig. 5, a layout of the ship lock and profile are presented. In the lock site, granite rocks with joints and faults have been found. A lot of research had been done about the stability of the lock. In the paper, manifold method is used to analyze the slope behaviour and its stability. It is clear that no elements data are required. Only nodal point data are provided. The slope stability contour and failure mode obtained by the manifold method are presented in Fig. 6 and Fig. 7, respectively.

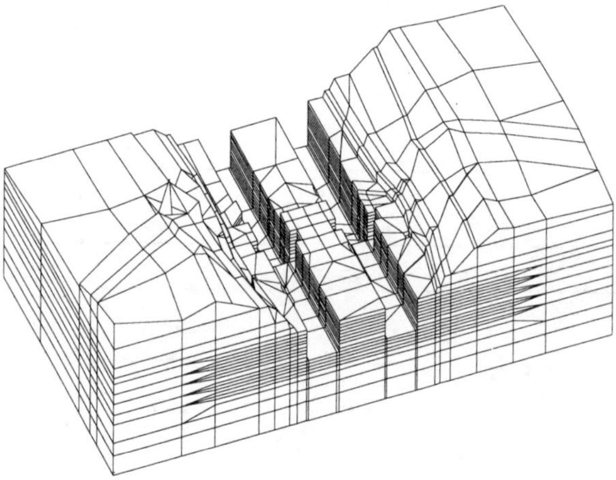

Figure 5: Layout of the Three Gorges ship lock

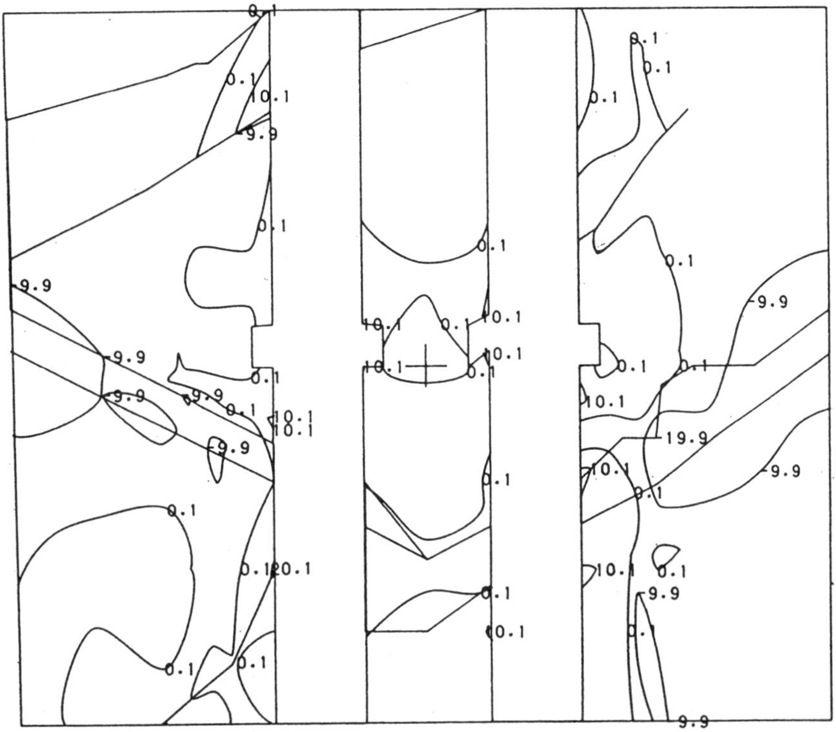

Figure 6: Point safety factor distribution of Three Gorges ship lock on the horizontal cross section at $\bigtriangledown$ 150.00

6 Conclusion

The starting point of the manifold method and element-free method is to simplify the analysis procedure and use coarse integration. The test example shows that such ways don't decrease the accuracy of the numerical result.

Since the interpolants in the method is for all domain, the smoothing step in F.E.M postprocessing procedure can be omitted. The manifold method and element-free method has great potential for geotechnique engineering problem because the mesh generation becomes totally unnecessary in the method.

Acknowledgments

The research work reported here is supported by National Key Projects on Basic Research and Applied Research. The work of Yan Gongri, Kou Xiaodong, Xiao Hongtian and Wang Zhiren is appreciated greatly.

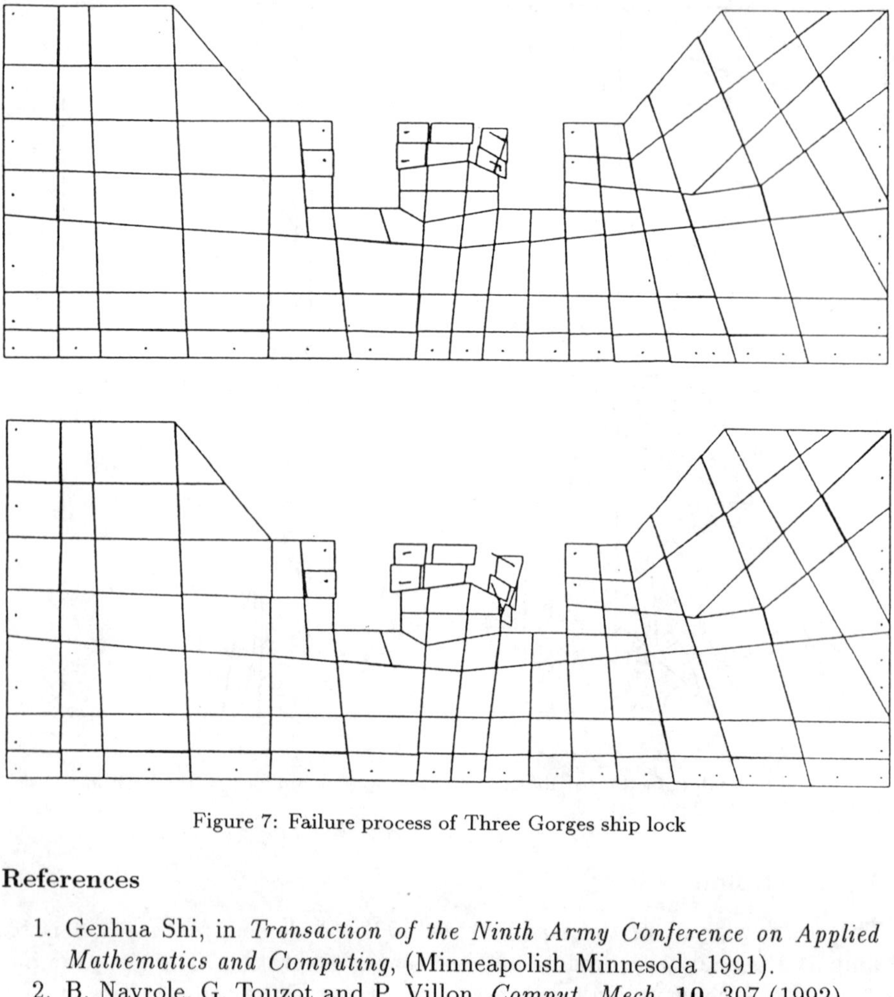

Figure 7: Failure process of Three Gorges ship lock

References

1. Genhua Shi, in *Transaction of the Ninth Army Conference on Applied Mathematics and Computing*, (Minneapolish Minnesoda 1991).
2. B. Nayrole, G. Touzot and P. Villon, *Comput. Mech.* **10**, 307 (1992).
3. T. Belyschko, Y.Y. Lu and L. Gu, *Int. J. for Num. Methods in Eng.* **37**, 329 (1994).
4. Lin, Dezhang and Mo, Haihong, Manifold method of material analysis. R.M.I., The Univ. of Oklahoma. (1994)
5. Y.Y. Lu, T. Belyschko and L. Gu, *Compu. Methods Appl. Mech. Enggr.* **113**, 397 (1994).
6. Genhua Shi, Manifold Method Fractured and Jointed Rock Masses. (Lake Tahoe, California, 1992)

AN ELASTO-PLASTIC FINITE ELEMENT ANALYSIS FOR THE STABILITY OF ROCKMASS ABUTMENT AND THE STRESS OF HIGH ARCH DAMS

QICHAO MA   GUANGYUAN LI   LAN QI
Department of Water Resources and Habour Engineering,
Tianjin University, Tianjin, 300072, P. R . China
E-mail: hegu @ public.tjuc.com.cn

In this paper, an analysis has been made for the stability of rockmass abutment and the stress of high arch dam by using the elasto-plastic FEM in combination with Xiaowan engineering. The evaluation of the stability of rockmass abutment and the stress of high arch dams are obtained by studying the all process (including the initial stress field, secondary stress field and the additional stress field affected by the function of the working loads) of the stress of dam foundation under the interaction between the dam and its foundation. Further, the influence and the function of the tectonic geostress on them are explored.

1 Introduction

At present, the rigid body limit equilibrium method is still used in engineering for the stability analysis of rockmass abutment, but it can only shows the critical state in sliding, and can not appear the evolution process of stress and deformation of structural surface, even more can not take account of the function of geostress factor. But the rockmass is a geological mass which is possessed of a certain structure and exists a certain geostress environment. The rockmass structure and the existing condition control the basic rule of rockmass deformation and failure. The geostress play the role in the failure process of rockmass from the preliminary state (the stress state coming to crack in the rockmass) progresses gradually to final state (limit equilibrium). In this paper an elasto-plastic FEM is adopted to analyze the stability of rockmass abutment and dam stress in combination with various unfavorable geological factors and loading condition. The evaluation of the stability of rockmass abutment and the dam stress are obtained by studying the all process (including the initial stress field, secondary stress field and the additional stress field affected by the function of the working loads) of the stress of dam foundation under the interaction between the dam and its foundation. Further, the influence and the function of the tectonic geostress on them are explored.

2 Mechanical Model

2 . 1 Elasto-plastic constitutive model

2．1．1 Criterion of yield

Rockmass is a frictional behavior material which has strain hardening (softening), and the tensile strength is lower. At present, the isotropic hardening-softening yield criterion of Drucker-Prager is usually applied to rock materials (Fig. 1). The expression is

$$f = \alpha I_1 + J_2^{1/2} - K = 0 \tag{1}$$

where $I_1=\sigma_{kk}$ is the first invariant of the stress tensor, $J_2=1/2S_{ij}S_{ij}$ is the second invariant of the stress deviatoric tensor, $\alpha=\mathrm{tg}\,\phi/(9+12\mathrm{tg}^2\phi)^{1/2}$, $K=3c/(9+12\mathrm{tg}^2\phi)^{1/2}$.

The parameters α and K are represented by cohesion (c) and friction angle (ϕ).

The yield surface defined by equation (1) is open pass at the axial direction of the hydrostatic pressure, it indicates that the rockmass can bear an infinite of hydrostatic pressure, but this case does not tally with the practical deformation state of the rockmass.

For this reason, a cap is added to the opening of the circular conic node

$$I_1 = I_1^a \tag{2}$$

where I_1^a is the position of cap, it indicates the limit of the sum of three principal compressive stress under the action of three dimensional compressive force of rockmass. The compound yield surface is composed of equation (1) and (2). The cap expands gradually with plastic bulk deformation, it is a model which integrates perfectly-plastic (shear deformation) with bulk strain-hardening, which is called D-P yield criterion possessed a cap. The yield situation of the geotechnical materials under high geostress condition may be reflected fundamentally.

2．1．2 Cracking criterion

The tensile strength of rockmass is much lower than the compressive strength. When the principal tensile stress σ_i (i=1, 2, 3)exceeds the tensile strength R_T permitted of the rockmass, the rockmass will be cracked.

After the rockmass cracking in tension, the principal tensile stress on the normal direction of fracture will be released.

2．2 Theoretical method of seepage calculation

The governing equation of three dimensional stable seepage may be expressed as

$$\frac{\partial}{\partial x}(k_x \frac{\partial H}{\partial x}) + \frac{\partial}{\partial y}(k_y \frac{\partial H}{\partial y}) + \frac{\partial}{\partial z}(k_z \frac{\partial H}{\partial z}) = 0 \tag{3}$$

According to variational principle, to solve the above definite solution problem is equivalent to find the limit value of the following functional,

$$\begin{cases} I(H) = \frac{1}{2}\iiint_a [k_x(\frac{\partial H}{\partial x})^2 + k_y(\frac{\partial H}{\partial y})^2 + k_z(\frac{\partial H}{\partial z})^2] dxdydz = \min \\ H(x,y,z) = H_a(x,y,z) \end{cases} \quad (4)$$

On the basis of fundamental principle of FEM, first, express the potential head of elements in terms of the potential head of nodes and then list the functional expression (4) of each element, finally, accumulate the functional of each element and solve the minimum value. By using the set of linear algebraic equations taking the nodal potential heads as the unknown quantity, and boundary condition. The potential head of each node within the seepage region (Ω) may be solved. When there is a free water surface (surface of saturation) within the seepage region, the surface of saturation can be determined through iteration.

2．3 Regression analysis for geostress field

The fundamental principle of regression analysis for geostress field is listed in reference (2). According to a small number of data actually measured in-situ and mathematical model of finite element, 3-dimensional geostress field may be found by using mathematics statistics principle in the regression analysis. According to the requirement take partial region within regression geostress field as sub-structure and exert displacements on the boundary of sub-region to model the geological tectonic movement and the influence of self-weight of rockmass beyond the sub-region.

3 Numerical Computation and Result Analysis

Xiaowan Hydropower Station is located at Yunnan province of the People's Republic of China. It is a dominant reservoir and large power station at the middle and lower reaches of Lancangjiang River. The dam with a maximum height of 292 m is a varied thickness parabolic double curvature arch dam, the total storage capacity is 152.32×10^8 m^3, the total installed capacity is 4200 MW.

3．1 Determination of the sliding boundary of rockmass abutment

According to the computational result of the rigid body limit equilibrium method, the severe structural surface, the upstream face opened and the downstream face hung in the air are chosed as the sliding boundary of the rockmass abutment.

The bottom slide plane: the occurrence of this plane is : strike N8° E, dip NW, inclination 32° on the left bank; N10° W, NE ∠ 32° on the right bank.

The lateral slide plane: The occurrence of this plane is: strike N10° E, dip NW, inclination 90° on the left bank; N10° W, NE ∠ 90° on the right bank. The combination of the bottom plane and the lateral plane constitute the benched slide boundary.

The upstream face opened: Owing to the effects of the upstream water pressure and other loads, the rockmass abutment near the upstream face of arch dam may be cracked along the joint in EW direction due to tensile stress. Therefore, we may chose the cracking plane as the upstream face opened. The downstream face hung in the air: The fault or crushed zone near the downstream of dam site may be regarded as the face hung in the air, accordingly, we may chose the fault F_5 as the downstream face hung in the air.

3. 2 Mesh dissection

Global model is divided into four sub-region: 1) The dam itself is dissected into eight layers along the elevation, two layers along the thick, and amounts to 176 elements. 2) The possible sliding body surrounded by weak structural planes is dissected into 83 elements. 3) The rockmass abutment beyond the sliding body is dissected into 1560 elements. 4) The sliding weak structural plane itself is dissected into 151 elements. The whole model is dissected into 1970 elements and 2503 nodes. The element meshes are shown in Fig. 2.

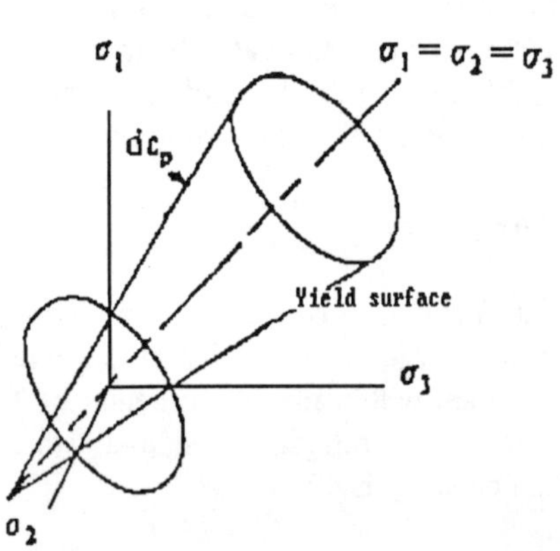

Fig. 1 Principal stress space

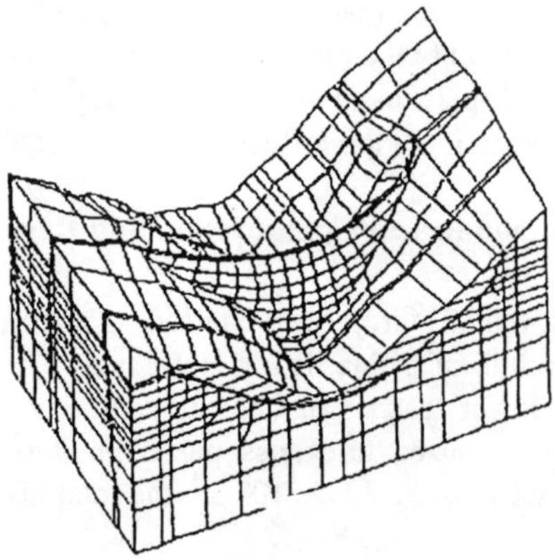

Fig. 2 Mesh dissection

3. 3 Mechanical parameters

The mechanical parameters adopted are listed in Table 1.

Assuming the weak structural plane is constituted from elasto-plastic material and obey the having cap D-P yield criterion and cracking criterion. Using the shear-friction factor $f'=1.0$; the cohesive force $C'=0.8$ MPa; the tensile strength $R_T=1.5$ MPa; $I_1^a=60.0$ Mpa, and assuming other sub-regions are constituted from elastic material.

Table 1 Mechanical parameters

Material classification	E(GPa)	μ	γ (kN/m^3)
Dam concrete	21	0.18	24.0
Class Ⅰ、Ⅱ rockmass	17~22	0.23	26.5
Class Ⅲ rockmass	12	0.27	26.0
Class Ⅳ $_a$ rockmass	5.5	0.28	25.5
Weak zone	3.0	0.29	25.0

3．4 Computational loads

The computational loads include upstream/downstream water pressure, seepage body force in the foundation and initial geostress.

Assuming the pour of the dam is completed once, the pouring process is not considered. The rising process of the water load is performed by the five stage impounding from reservoir empty to full. The failure computation has been performed with a material strength reduction factor.

In order to analyze the effect of tectonic geostress, there are two computational scheme: scheme 1 includes tectonic geostress in computation, and scheme 2 does not.

3．5 Computational results and analysis

3．5．1 Effect of tectonic geostress

As shown in Table. 2.

Table 2 Failure rate of structural surface

	K=1		K=2		K=3		K=4	
loading water level (m)	scheme 1	scheme 2	scheme 1	scheme 2	scheme 1	scheme 2	scheme 1	scheme 2
completion	12.23	6.64	46.3	20.20	66.23	38.36	76.81	54.01
1050	6.02	3.37	32.61	14.22	56.45	28.68	68.29	48.83
1130	3.08	2.59	28.43	11.42	52.22	28.76	64.38	41.41
1170	5.32	4.04	35.82	16.73	60.17	32.64	72.16	54.41
1210	10.84	6.94	38.35	21.17	63.24	46.82	79.22	59.78
1240	15.70	9.11	41.23	23.74	65.91	48.56	84.56	66.40

1) During the process of storage (from reservoir empty condition to full condition), the failure rate in the scheme 1 is about 18.9 % ~ 84.0 % greater than

that of the scheme 2.

2) While the dam is completed but does not store water yet, the failure rate of the scheme 1 is 12.23 %, among which 8.08% in the left bank, 4.15 % is in the right bank; the failure rate of scheme 2 is 6.64 %, among which 0.3 % is in the left bank, 6.34 % is in the right bank.

3) In the process of storing water to the normal water level, the variation of the failure rate (max./min. failure rate) of the scheme 1 n_1=5.1; and that of the scheme 2, n_2=3.5, the latter is about 70 % of the former.

4) Considering the potentiality of the stability for the rockmass abutment, according to the strength reserve computation, as the water level reaches the normal water level, the failure rate of the structural surface gradually increases with the strength of material decrease, the failure rate of the scheme 1 is 27.3 %~ 73.7 % greater than that of the scheme 2.

5) When the loads maintains constant, with the material strength of the structural surface decrease the function of the tectonic geostress gradually decreases. For example, while the material strength reserve coefficient K=2, the failure rate of the scheme 1 is 73.7 % greater than that of the scheme 2; while K=3, it is 35.2 %; while K=4, it is 27.3 %, it conforms to the rule of the release of geostress.

3.5.2 Failure mechanism of structural surface and working state (The scheme 1)

1) While the dam is completed but does not store water yet, the failure region of the scheme 1 is that the left bank is greater than the right bank, the bottom sliding plane is greater than the lateral sliding plane, the compression-shear failure is the major, the tensile failure only make up 20.5 %, the failure region occur mainly near the F_5 fault boundary in the left bank downstream.

2) In the storage-operating period, the failure region is that the left bank is greater than the right bank, the bottom sliding plane is greater than the lateral plane, the compression-shear failure is the major, the tensile failure is less than 1/10. The failure region expands gradually from downstream of the left bank at low water level to the heel near the bottom of valley at high water level, the heel occur the tensile failure.

3) The evolutional process of the failure for the structural surface is that the failure rate decreases gradually with the upstream water level stage by stage increase, when the water level exceeds elevation 1130m, the failure rate increases gradually, the failure rate of the reservoir full condition is greater than that of the reservoir empty condition.

4) In the reducing process of the material strength, the developmental tendency on failure of the structural surface is fundamentally consistent with the rule of the normal storage process.

5) When the loads maintains constant, with the material strength gradually decrease, the increment rate of yield failure of the structural surface is that as the

material strength reserve coefficient K ⩽ 3, the increasing speed is comparatively fast, as K > 3, it is comparatively slow.

3 . 5 . 3 Deformation of rockmass abutment

Table 3 is the displacements at midpoint of the arch abutment for each dominant elevation, it may be seen

Table 3 Displacements of arch abutment (cm) for the scheme 1

Elevation	left bank				right bank			
(m)	K=1	K=2	K=3	K=4	K=1	K=2	K=3	K=4
1245	5.24	5.25	5.34	5.36	4.48	4.52	4.61	4.82
1210	4.88	4.90	5.03	5.09	4.26	4.31	4.43	4.84
1170	4.40	4.44	4.59	4.82	4.16	4.21	4.37	5.14
1130	4.01	4.09	4.33	5.27	3.97	4.07	4.42	6.20
1090	3.60	3.75	4.25	6.17	372	3.91	4.79	7.87
1050	3.56	3.79	4.72	7.46	3.89	4.21	5.84	9.90
1010	3.90	4.35	6.04	9.61	4.23	4.74	7.10	11.84
975	4.29	5.04	7.44	11.59	4.38	5.14	7.95	12.89
953	4.33	5.38	8.35	12.97	4.32	5.40	8.50	13.38

1) The displacement of the abutment increases with the decrease of material strength, but the incremental speed at the base is greater than that at the crest. As the material strength reserve coefficient K increases from 1 to 4, the displacement at the crest of the left bank will increase from 5.24 cm to 5.36 cm, whereas at the base will increase from 4.33 cm to 12.97 cm; the rule of the displacement at the right bank is identical to that at the left bank.

2) From the distributive rule of the abutment deformation, it may be seen that the displacement at the crest is slightly greater than that at the base, the distribution is nearly uniform at normal water level, but with the material strength decrease, the displacement at the base is gradually greater than that at the crest, such as when K=4, the displacement at the base is about two point five times as great as the crest.

3) When the water load maintains constant, the displacement incremental speed gradually increases with the material strength decrease. After K > 2, the incremental speed is evidently accelerated, it can be considered that the rockmass abutment has been transformed from elastic state into plastic state.

4 Conclusions

1) Under the same loading, unloading and strength of materials, different initial geostress field may obtain different failure mechanism and stability of rockmass. Therefore, selecting rational initial geostress field is of great importance.

2) In this paper, an elasto-plastic FEM is adopted to model the process of the

initial geostress field, excavation of dam foundation, construction of dam, reservoir impoundment and strength of materials decrease stepwise, above mentioned cases, the developing process of the yield failure of the structural surface are reflected fundamentally.

3) Whether the tectonic geostress is favourable or unfavourable to the stability of dam and its foundation, mainly depends on magnitude, direction of the geological tectonic movement and occurrence, position of the structural surface.

4) The displacement of dam and its rockmass abutment gradually increases with the strength of the structural surface decrease. Due to the influence of failure in the structural surface, the increasing speed of displacement on the base is greater than the crest. It is equivalent to occur a rotation of the dam. This is the deformable behaviour of the foundation rockmass possessed continuous structural surface under the failure process.

5) From the yield failure state of the structural surface and the displacement of the dam and its rockmass abutment, it shows that the rockmass abutment of Xiaowan high arch dam is in elastic operating state, and cannot slide failure in normal operating condition. According to the strength reserve computation, as the strength reducing factor K=2, there is no unstable signs, the rockmass abutment is in stable; as K=4, both the displacement of the dam and the yield region of the structural surface increase remarkably. This phenomenon indicates that the dam appears remarkable unstable signs. if the function of tectonic geostress is not concerned, the stability of the rockmass abutment will increased.

References

1 Qichao Ma, et al., *Journal of Hydraulic Engineering*, No. 6, 74 (1992).

2 Huaizhi Guo, Qichao Ma, et al., *Journal of Geotechnical Engineering*, **5**, 64 (1983).

3 *Arch Dams* • *Accident and Safety of Large Dams*, by Naihua Ru, et al. (Water Conservancy and Hydroelectric Power, P. R. China, 1995).

APPLICATION OF THE FINITE THICK –PLATE STRIP METHOD IN THE UNDERGROUND BLAST–RESISTANT STRUCTURES

JUN WU QINGXIAN LI FUHOU LI

Luoyang Hydraulic Engineering Technology Research Institute, China, PC471023

The semi - analytical numerical method is the best supplement of the pure numerical method and analytical method as a kind of the combined calculation mode of numerical and analytical method. The principle and characteristics of the finite thick-plate strip method and its applications in the underground blast - resistant structures are emphatically described in this paper, based on the review of the development of the semi - analytical method. The application of examples shows that this method has less degree of freedom and higher precision. Meanwhile, it is especially suitable for the dynamic analysis of the blast-resistant underground composite sandwich structure.

1 Introduction

The semi - analytical numerical method is a combined method of analytical and numerical approach. Its main features are that it has advantages of the analytical and numerical methods and overcomes the shortages of these two methods to a great extent. Generally, the pure analytical method gives a theory solution with high precision and less calculation, but it is difficult to be matched with the large solution range and various problems under different conditions. The pure numerical method is on the contrary, its advantages are that it has large solution range and is suitable for the various complicated geometry, boundary, load and material property, and it is easy to be grasped as a generalized method, its shortage is that it gives merely an approximate solution, no matter what objects and what problems they are, They must all scattered in every direction and the polynomial interpolation is used to approximate the solution function of various kinds of problems which have not been distinctively treated to their properties, so some unnecessary work must be done in the numerical calculation. The above shortages of the pure numerical method will obviously improved by using the studied results of the analytical method to be partially instead of scattering and interpolation. From another point of view, the problems which can't be solved by the pure analytical method may also be solved now by using partial discrete interpolation and numerical solution method.

The semi - analytic numerical method has been proposed and developed only for the recent ten years, but it is being developed very fast. The finite thick-plate strip method discussed in this paper is the development of the generalized finite element and finite strip method. It uses number of steps in some directions and uses finite element in other direction. It is a semi-analytical numerical method with less degree of freedom and higher precision.

2 The Thick - plate Strip Method

All kinds of composite sandwich plates have the structure forms always used in the fields of aviation, shipping and machinery. They have been greatly applied in the underground blast - resistant structures in the recent years because of its light weight and high strength. However, the structure and material of parts of these plates are heterogeneous and anisotropy, therefore, the analytical solution can't be derived for most actual parts of engineering structures. In other hand, while the common used finite element numerical method is in using, often needs to cut along three directions of the structure, which occupies most internal storage and takes long time.

The finite strip method proposed in this paper uses an analytical function to express the length and thickness, as well as makes the division only along the width, therefore, the three - dimensional problem of generalized finite element and two - dimensional problem of generalized finite strip could be simplified to one dimensional problem for the structures such as composite sandwich plate. The internal storage and time for the structure analysis will be largely saved.

The theoretical foundation of the finite thick-plate strip method is the dynamic theory of thick plate and finite strip theory, its key question is to establish a suitable analytical function and a displacement function. It is already proved that the best convergence can be reached by using the characteristic function(e.g. thick - plate strip vibration mode function)of the homogeneous dynamic equation of single direction thick plate as a uniform analytical function to analyse the structure vibration problem such as shear deformation and rotation inertia. In such case the problem will be more simple.

2.1 The Vibration mode function of thick - plate strip

The vibration of composite sandwich structure mainly includes four mechanical characteristics: heterogeneous, anisotropy, shear deformation and rotation inertia effect. Heterogeneous is reflected in the thick – plate strip vibration mode function by the equivalent rigid coefficient and equivalent mass coefficient, anisotropy will be considered in establishing rigid matrix, the shear deformation and the rotation inertia effect will be introduced into the independent rotation angle vibration and related generalized mass, so the thick - plate strip vibration mode function could reflect the influence on the vibration mode of structure by different materials used in the various places of structure.

For the single direction thick plate, the fundamental dynamic equation of generalized rotation angle β and deflection w could be expressed as follows:

$$\begin{cases} \dfrac{\partial^4 W}{\partial y^4} - \left(\dfrac{6\rho}{5G} + \dfrac{\rho J}{D}\right)\dfrac{\partial^4 W}{\partial y^2 \partial t^2} + \dfrac{6\rho^2 J}{5GD}\cdot\dfrac{\partial^4 W}{\partial t^4} + \dfrac{\rho h \partial^2 W}{D \partial t^2} = \dfrac{q}{D} + \dfrac{6}{5Gh}\left(\dfrac{\rho J}{D}\cdot\dfrac{\partial^2 q}{\partial t^2} - \dfrac{\partial^2 q}{\partial y^2}\right) \\ \dfrac{5Gh}{6}\beta + \rho J\dfrac{\partial^2 \beta}{\partial t^2} + D\dfrac{\partial^3 W}{\partial y^3} + \dfrac{5Gh}{6}\cdot\dfrac{\partial W}{\partial y} - \dfrac{6\rho D}{5G}\cdot\dfrac{\partial^3 w}{\partial y \partial t^2} = -\dfrac{6D}{5Gh}\cdot\dfrac{\partial q}{\partial y} \end{cases} \tag{1}$$

where D, Gh, ρh are respectively the bending resistance, shear resistant rigidity and mass, rotation inertia density of the plate strip.

Equation (1) is used for the free vibration, its displacement vibration mode $Y(y)$and rotation angle vibration mode $\varphi(y)$ should satisfy the following equations

$$\begin{cases} Y'''' + \left(\dfrac{6\rho}{5g} + \dfrac{\rho J}{D}\right)W^2 Y'' + \left(\dfrac{6\rho^2 J}{5GD}W^4 - \dfrac{\rho h W^2}{D}\right)Y = 0 \\ Y''' + \left(\dfrac{6\rho}{5G}W^2 + \dfrac{5Gh}{6D}\right)Y'' + \left(\dfrac{5Gh}{6D} - \dfrac{\rho J}{D}W^2\right)\varphi = 0 \end{cases} \tag{2}$$

$$\begin{cases} Y(y) = A\sin\alpha_1 y + B\cos\alpha_1 y + Csh\alpha_2 y + Dch\alpha_2 y \\ \varphi(y) = \dfrac{\dfrac{d^3 Y(y)}{dy^3} + \left(\alpha_o^2\omega^2 + \dfrac{b_o^2}{a_o^2}\right)\dfrac{dY(y)}{dy}}{d_o^2\omega^2 - \dfrac{b_o^2}{a_o^2}} \end{cases} \tag{3}$$

where, $\dfrac{\alpha_1}{\alpha_2} = \dfrac{W}{\sqrt{2}}\sqrt{\sqrt{\left(a_o^2 - d_o^2\right) + \left(\dfrac{2b_o}{W}\right)^2} \pm \left(a_o^2 + d_o^2\right)}$, $a_o^2 = \dfrac{6\rho h}{5Gh}$, $b_o^2 = \dfrac{\rho h}{D}$, $d_o^2 = \dfrac{\rho J}{D}$,

four undetermined constants A, B, C, D could be determined according the boundary conditions in each end, then the four - order homogeneous linear algebraic equations can be established, the frequency equation can be deduced while following function is equal zero, that is:

$$F(\omega) = 0 \tag{4}$$

Solve frequency ω_m, two end α_{1m}, α_{2m} and relative each-order vibration mode $Y_m(y)$ and $\varphi_m(y)$.

For the vibration mode function of the common two - ends simple supported thick plate, they are:

$$\begin{cases} Y_m(y) = \sin \frac{m\pi}{l} y \\ \varphi_m(y) = \cos \frac{m\pi}{l} y \end{cases} \tag{5}$$

2.2 *Displacement Function*

At first, the plate is divided into several strips, which are connecting with each other by several discrete tie lines. The unknown parameters of tie lines should be studied in determination of displacement function. Because there are three generalized displacements inside the strips, the functions of three groups of unknown parameters in two side lines could be expressed in the following equations by using the vibration mode function of thick plate strip in the Y direction and algebraic polynomial in X direction(Fig.1).

$$\begin{cases} W = \sum_{m=1}^{r} \left[(1-\xi)W_{1m} + \xi W_{2m}\right] Y_m(y) \\ \beta_x = \sum_{m=1}^{r} \left[(1-\xi)\theta_{x1m} + \xi\theta_{x2m}\right] Y(y) \\ \beta_y = \sum_{m=1}^{r} \left[(1-\xi)\theta_{y1m} + \xi\theta_{y2m}\right] Y_m(y) \end{cases} \tag{6}$$

where: $\xi = \frac{x}{e}$, r are numbers of progression term, Y_m, φ_m are vibration mode functions of the thick plate strip determined by the end boundary condition, which can be written as:

$$\begin{cases} W = \sum_{m=1}^{r} Y_m(y) \sum_{R=1}^{s} N_k(\xi) W_{Rm} \\ \beta_x = \sum_{m=1}^{r} Y_m(y) \sum_{R=1}^{s} N_k(\xi) \theta_{xRm} \\ \beta_y = \sum_{m=1}^{r} \varphi_m(y) \sum_{R=1}^{s} N_k(\xi) \theta_{yRm} \end{cases} \tag{7}$$

$N_k(\xi)$ is polynomial mode function which is different according to the number of tie lines S in strip and can be expressed by the following matrix formula:

$$[f] = \sum_{m=1}^{r} [N]_m \{\delta\}_m = [N]\{\delta\} \tag{8}$$

where: $\{f\} = [W, \beta_x, \beta_y]^T$

$$\{\delta\} = [W_{1m}, \theta_{x1m}, \theta_{y1m} \cdots\cdots W_{sm}, \theta_{xsm}, \theta_{ysm}]^T$$

The derivation of other strain, internal force, rigid matrix and mass matrix is same as the finite element and finite strip method.

3 Application in the Blast – resistant Structures

The main problem for applying the finite thick plate strip method in the underground blast – resistant structure is how to consider the influence of soil alluvium above structure. The calculation method proposed in this paper is described as follows.

3.1 *Double Thick Strip*

The soil alluvium is identified along the main direction as a series of thick plate strips which are constituting double thick strips with the structure . This idea was first suggested by Cao Zhiyuan[2], in his method the displacement matrix array of the double thick strip element is given based on displacement coordinate condition on the joint boundary of structure and soil media, displacement function is constituted by the semi – analytical function and the dynamic balance equation which is established according to variation principle, and then a solution can be obtained. An improvement is made in this paper based on the above method, in which the new displacement polynomial of the double finite thick-plate strip is derived by uniform permutation of the degree of freedom of the double thick strip nodal line. The rigidity and mass matrix of the media and structure can be combined by layer, the balance equation of the system can be finally established based on the energy principle. Its displacement polynomial is building according following description.

The structure in soil is divided into a series of double thick strips, one of which is considered (Fig.2). According to the non – classical thick plate theory, assume the surface displacement between two layers as U_{so}, V_{so}, W_{so}, Y_{Ro}, V_{Ro}, W_{Ro} respectively and $W_{so} = W_{Ro} = W$, the generalized rotation angle in the layer is β_{xs}, β_{ys}, β_{xr}, β_{yr}. On the contact surface between two layers, there are :

$$\begin{cases} U_s = U_R \\ V_s = V_R \\ W_s = W_R = W \end{cases} \tag{9}$$

The displacement function of two layers can be written by combining formula (9) based on above description:

$$\begin{cases} U_{so}(\xi, y) = \sum_{m=1}^{r} y_m(y) \sum_{k=1}^{p} N_k(\xi) U_{km} \\ V_{so}(\xi, y) = \sum_{m=1}^{r} y^*_m(y) \sum_{k=1}^{p} N_k(\xi) Y_{km} \\ W_s(\xi, y) = \sum_{m=1}^{r} y_m(y) \sum_{k=1}^{p} N_k(\xi) W_{km} \\ \beta_{xs}(\xi, y) = \sum_{m=1}^{r} y_m(y) \sum_{k=1}^{p} N_k(\xi) \beta_{xskm} \\ \beta_{ys}(\xi, y) = \sum_{m=1}^{r} \varphi_m(y) \sum_{k=1}^{p} N_k(\xi) \beta_{yskm} \end{cases} \tag{10}$$

$$\begin{cases} \beta_{xr}(\xi, y) = \sum_{m=1}^{r} y_m(y) \sum_{k=1}^{p} N_k(\xi) \beta_{xrsk} \\ \beta_{yr}(\xi, y) = \sum_{m=1}^{r} \varphi_m(y) \sum_{k=1}^{p} N_k(\xi) \beta_{yrkm} \\ W_R(\xi, y) = W_s(\xi, y) \\ U_{Ro}(\xi, y) = U_{so} + \frac{H_R}{2}\beta_{xr} + \frac{H_s}{2}\beta_{xs} \\ V_{Ro}(\xi, y) = V_{so} + \frac{H_R}{2}\beta_{yr} + \frac{H_s}{2}\beta_{ys} \end{cases} \tag{11}$$

$y_m(y)$, $y^*_m(y)$, $\varphi_{m}(y)$ are the m step displacement of the thick plate strip and vibration mode function of rotation angle relative to the boundary condition at two ends of the structure. The derivation of the other formulae can be seen in reference, written by Zhang Luoqi[1].

3.2 *The Influence of lateral stress and strain*

In the underground structure, most media of the surrounding support structure belong to the loose backfill media with low elastic modulus, and under the action of conventional blast load, its forced deformation is very complicated and is not the strict three– dimensional, so, when the suggested double thick strip calculation model is used, the influence of lateral positive stress and strain should be corrected, otherwise, there will be some inconsistent. In this paper, when the material nonlinear calculation is carried out for this kind of structure, the calculation of lateral positive strain is made by using free – field pressure approximately instead of lateral positive stress and the relative formulae in the soil – media dynamic constitutive model (cap cover model), since the lateral pressure in the soil media layer is higher.

3.3 *The Treatment of Soil-media Boundary Condition*

The treatment of soil – media boundary condition should be consistent with the structure boundary condition. The length of underground structure is generally finite, but soil – media is infinite. Strictly speaking, soil strip boundary should be a various boundary, but this treatment will bring complexity to the analysis of problem. It can be seen from the given information of test that the blast load of conventional weapons attenuates very fast along the distance as well as its pressure and deformation are very small at boundary. This influence factor can be illuminated in the analysis calculation, so that it can be consistent with the structure boundary condition.

3.4 Calculation Example

There is a four – side simple supported composite sandwich plate, its side length is 1, thickness is h, the bearing load distributed uniformly along the length is q_o, the central deflection of its linear plastic solution is :

$$W_o = \alpha \frac{q_o l^4}{E h^3}, \qquad \alpha - \text{central deflection coefficient}$$

The plate is divided into two layers according their thickness which is H_s, H_r, the static linear elastic calculation is made, the calculation results of α values ar shown in table 1.

Table 1: The Calculation Result of α Values

$\frac{h}{l}$	analytic solution[3]	finite element solution[4]	thick plate strip solution
0.1	0.04632	0.04628	0.04636
0.2	0.05217	0.05202	0.05273
0.3	0.06192	0.06160	0.06144
0.4	0.07557	0.07500	0.07516

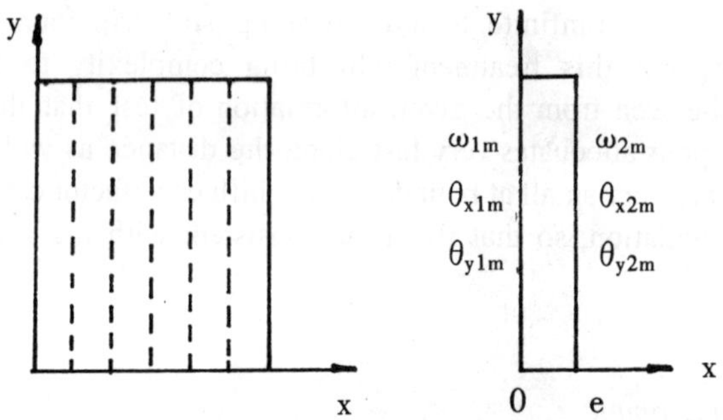

Figure 1　　Finite thick – plate strip model

4 Conclusion

The method used here in the analytical calculation of the underground blast-resistant structure was a creative way in semi-numerical analysis, which has the following advantages.

(1) Less degree of freedom, higher precision, the calculation matrix order using the finite thick - plate strip method is less n^2 times or n times than that using three-dimensional finite element method and the general finite strip method respectively. It is equivalent to n^4 and n^2 times of internal memory and time (n is the division element number in a direction).

(2) The displacement in the established element and independent rotational angle function are counted into the influence factors of shear deformation in the element rotation inertia and coupling effect etc, therefore, this method more truly reflects the stressed deformation of the composite sandwich structure.

(3) It can not only be convenient to consider the nonlinear influence of materials, but also calculate the laminated type protective structure by the laminated treatment of media and structures without increasing the degree of freedom.

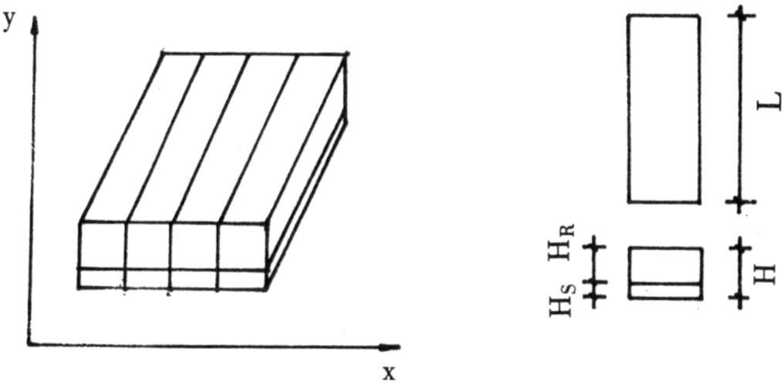

Figure 2 Double finite thick strip model

References

1. Zhang Luoqi, The finite strip method in structure analysis, Jaotong press, 1978.
2. Cao Zhiyuan, The double finite strip method for the underground structure, Calculation of structure dynamics and its applications, No.1,1984.
3. The Dynamics Research Institute of Academia science, Bending stability and vibration of the composite shell, Science press, 1980.
4. Xie Yiqian, etc. , The finite element method in the elastic and plastic dynamics, Machinery industry press, 1985.
5. Cao Zhiyuan, etc. ,Thick plate dynamic theory and its applications, science press, 1983.

Theoretical Analysis

FLOW AND DEFORMATION IN ANISOTROPIC FRACTURED POROUS MEDIA

M. BAI AND J.-C. ROEGIERS
Rock Mechanics Institute, The University of Oklahoma, Norman, OK 73019-1014, USA
E-mail: mao@uoknor.edu

The present study is mainly devoted to the analyses of fractured porous media with reference to fluid flow and associated rock deformations using the anisotropic dual-porosity poroelastic theory. After presenting new concepts in material heterogeneities and anisotropies using scale analogies, the analyses focus on mathematical formulations and discuss the related parameters. The former elicits the mechanical and hydraulic anisotropies, with the tensorial forms of elastic compliances and poroelastic coefficients, as well as compressibilities and permeabilities for the fractures and matrix blocks, respectively. The latter probes a few generalized stress-permeability relationships, suitable for modeling the fractured poroelastic media.

1 Introduction

Terzaghi[1] was the first person to propose an effective stress theory, which offers a sensible explanation for the time-dependent phenomenon of soil consolidation. During the past decades, the concept of effective stress law has proved to be quite useful in describing the mechanical behavior of various porous material, including rock. This law was adopted by Biot[2] in establishing the fundamental coupled equations for linear deformation-flow systems, formally known as poroelasticity.

Biot[3] further extended the theory of consolidation to relatively general anisotropic situations. The effects of anisotropy on mechanical and hydraulic properties have been known to be significant, particularly for getechnical materials such as shales, and sedimentary formations such as layered sandstones and limestones. Biot's anisotropic poroelasticity[3] was expressed into more general conceptual forms by Carroll,[4] who derived the constitutive relations for the effect of pore fluid pressure on the deformation of saturated rocks exhibiting mechanical anisotropy. Further generalization of the anisotropic poroelasticity was made by Thompson and Willis,[5] who provided the formulation for the homogeneous media using micromechanical concepts. From a practical point of view, Cheng[6] offered a simplified form for the concepts developed by the referenced individuals, suitable for certain engineering applications.

However, Biot's approach is unable to account for the mechanical and hydraulic behavior of heterogeneous rock masses, such as fractured porous media.

Barenblatt et al.[7] were recognized as the first to introduce the dual-porosity concept in characterizing flow in fractured porous media. Considering that this past model was theoretically rigorous but practically restrictive, Warren and Root[8] proposed some simplifications based on the physical implications. Since then, the dual-porosity concept has been applied extensively in many engineering and science fields for the analysis of fluid flow in fractured porous media.

It is understood that the elastic impact was not incorporated in the dual-porosity model originally proposed by Barenblatt et al.[7] The coupling between solid deformations and fluid flow in dual-porosity media was first investigated in the analysis of civil and petroleum engineering problems by Wilson and Aifantis,[9] using analytical means. Numerical endeavor was attempted by Khaled et al.[10] for consolidation cases. An alternative formulation of the so-called dual-porosity poroelasticity was given by Elsworth and Bai,[11] Applications of such alternative rationale in the engineering field was highlighted by Bai et al.[12]

Current conceptualizations appear to have covered the following four areas: (a) poroelasticity,[2] (b) anisotropic poroelasticity,[4, 5] (c) dual-porosity,[7] and (d) dual-porosity poroelasticity.[9, 11] However, the model of anisotropic dual-porosity poroelasticity remains to be developed. The mathematical formulation of such a model, along with the discussion on the general definitions of material heterogeneity, anisotropy and associated parameters, will be presented in this paper.

2 Material Heterogeneity and Anisotropy

By definition, homogeneity and heterogeneity are associated with spatial distributions and corresponding variations; while isotropy and anisotropy are related to directional distributions and corresponding changes, of a coupled system, respectively. In terms of these spatial and directional characteristics, the following four groups of porous media may be identified:

- Type 1: homogeneous and isotropic,
- Type 2: homogeneous and anisotropic,
- Type 3: heterogeneous and isotropic,
- Type 4: heterogeneous and anisotropic.

In the existing models, the first three types of porous media have been represented by the following counterpart models:

- Model 1: single-porosity poroelasticity,
- Model 2: anisotropic single-porosity poroelasticity,

- Model 3: dual-porosity poroelasticity.

To the knowledge of the authors, the theoretical model for the 4th type of porous media, i.e., anisotropic dual-porosity poroelasticity, is not available at present.

The heterogeneity and anisotropy mentioned so far is quite general. With reference to various scales of the examined domains, different heterogeneous and anisotropic characteristics of the fractured porous media may co-exist. Considering the following three types of domains and the related scales:

- Scale 1: macroscopic scale reservoir,
- Scale 2: fractured blocks on the mesoscopic scale, or,
- Scale 3: single fracture or porous block on the microscopic scale.

In the modeling, the various scales may correspond to the following respective spaces:

- Space 1: reservoir confined by the studied case boundaries,
- Space 2: fractured blocks defined as Representative Elementary Volume (REV), or,
- Space 3: a single fracture or a block delineated by the constitutive (conceptual) relations.

In the modeling, all scales and spaces should be considered; however, with reference to different aspects. The governing equations, along with boundary and initial conditions, are defined macroscopically. The modeling parameters (e.g., fracture spacing, matrix permeability, etc.) are characterized using the REV rationale. The constitutive laws (e.g., Darcy's law, effective stress law, etc.) are established using the microscopic scale without referring to any specific domains, since the single fracture or pores may be too tiny to be spatially significant.

The features of the heterogeneities and anisotropies in all three scales can be further divided between those for matrix blocks and those for fractures. As a result, it becomes a complex issue to distinguish different heterogeneities and anisotropies at various scales for different media.

In view of heterogeneities, for instance, micromechanical characteristics for the solid matrix may show anisotropic behaviors at grain-size levels, but at the same time may be envisioned as globally isotropic. By the same token, macroscopic heterogeneity may be exhibited in the fractured porous media where the fracture network may show random patterns or arbitrary densities in the simulated domain. Conversely, the individual fracture itself may be microscopically homogeneous due to uniform infilling or constant roughness, etc. Similarly, the porous matrix may be macroscopically or mesoscopically homogeneous in the reservoir domain or in the REV, but microscopically heterogeneous as a result of variations in the grain sizes or void structures.

With respect to the anisotropy, for example, natural fractures definitely introduce the macroscopic or mesoscopic anisotropies regardless whether the fractures uniformly distribute with equal apertures or not, because the fractures themselves impose the directional variations of material properties in the reservoir or in the REV domain. This is the primary reason why the fractured media is generally envisioned as equivalent to the heterogeneous media.[12] On the other hand, anisotropic characteristics within the single fracture may be microscopically different from those either on the macroscopic or on the mesoscopic scales, depending on the structures and material properties inside the fracture. Similar concepts may be applicable for the porous matrix. Even though the microscopic anisotropy at the grain scale may contribute to the general macroscopic or mesoscopic anisotropies at the reservoir or the REV scales, the natures of these anisotropies can be substantially different from each other. Significant differences may be demonstrated between macro- and meso-scales, with the latter cases frequently being considered as the averaged quantities from the former domain.

Despite these complexities and differences, it is certain that if all the fractures and matrix blocks are microscopically homogeneous and isotropic, then the domain of interest is homogeneous and isotropic. The reverse may not be true; i.e., the domain may look macroscopically or mesoscopically homogeneous and isotropic, but the fractures and the matrix blocks themselves can be microscopically heterogeneous and anisotropic.

In actuality, the terms of heterogeneity and isotropy are meant differently based on different mechanisms. They may apply to the characteristics of certain parameters such as porosity and permeability, or to mechanical aspects as a result of stress variations, or even to hydraulical aspects such as changes in regional as well as local flow conditions. For the nonisothermal situations, the terms can also be temperature-dependent.

For a relatively complete segregation of the material heterogeneities and anisotropies of the fractured porous media, 24 groups of different combinations may be formed from the four types of material properties, three different scales of domains , and two types of media. For a more comprehensive subdivision in view of the various mechanisms (i.e., physical, mechanical, hydraulic, and thermal), 96 groups of various combinations may be identified. Conceivably, the material heterogeneities and anisotropies can be exhibited in such a broad extent, especially for a coupled flow-deformation system, that there is no single model in the current technology which covers all aspects of heterogeneous and anisotropic properties. To narrow the subject of the study to an amenable scope, this paper merely examines the anisotropic characteristics of a heterogeneous (fractured) porous medium in the following two aspects:

- the general mathematical formulation at the macroscopic level; and,
- the parametric identification (e.g., the stress-dependent permeability) at the mesoscopic level.

3 Mathematical Formulation

For anisotropic single-porosity poroelastic media, the governing poroelastic equations have been developed by Biot [3] for a rather specific case and by Thompson and Willis [5] for a more general formulation.

The equation for the solid phase can be expressed as:

$$\nabla \cdot (\tilde{\Psi} : \tilde{\varepsilon} - \tilde{\alpha} p) = 0 \tag{1}$$

where $\tilde{\varepsilon}$ is the strain tensor, $\tilde{\Psi}$ is the fourth order elastic tensor, $\tilde{\Psi} = \tilde{C}^{-1}$, $\tilde{C}$ is the fourth order compliance tensor, $\tilde{\alpha}$ is the second order Biot's coefficient tensor, and p is the pore pressure.

The governing equation for the fluid phase can be written as:

$$\nabla \cdot (\frac{\tilde{k}}{\mu} \cdot \vec{\nabla} p) = \tilde{\delta} : \tilde{\alpha^*} \frac{\partial p}{\partial t} + \tilde{\alpha} : \frac{\partial \tilde{\varepsilon}}{\partial t} \tag{2}$$

where $\tilde{k}$ is the permeability tensor, μ is the fluid dynamic viscosity, $\tilde{\alpha^*}$ is the compressibility tensor, and $\tilde{\delta}$ is the unit tensor.

It is important to note that the parameters for anisotropic poroelastic media are tensors, instead of scalars, with the latter being the case for isotropic poroelasticity.

The compliance tensor $\tilde{C}$ is related to the elastic tensor $\tilde{\Psi}$ via the following relation:

$$\Psi_{ijkl} C_{klmn} = \frac{1}{2}(\delta_{im}\delta_{jn} + \delta_{in}\delta_{jm}) \tag{3}$$

where δ_{ij}[a] is the Kronecker symbol; $\delta_{ij} = 1$ for $i = j$ and $\delta_{ij} = 0$ for $i \neq j$.

The complete compliance tensor can be expressed as:

$$\tilde{C} = C_{ijkl} = \begin{bmatrix} C_{1111} & C_{1122} & C_{1133} & C_{1112} & C_{1113} & C_{1123} \\ C_{2211} & C_{2222} & C_{2233} & C_{2212} & C_{2213} & C_{2223} \\ C_{3311} & C_{3322} & C_{3333} & C_{3312} & C_{3313} & C_{3323} \\ C_{1211} & C_{1222} & C_{1233} & C_{1212} & C_{1213} & C_{1223} \\ C_{1311} & C_{1322} & C_{1333} & C_{1312} & C_{1313} & C_{1323} \\ C_{2311} & C_{2322} & C_{2333} & C_{2312} & C_{2313} & C_{2323} \end{bmatrix} \tag{4}$$

[a]Not to be confused with $\tilde{\delta}$, which is the unit second order tensor

This matrix is symmetric. For this reason, the independent parameters of the compliance tensor reduce to 21, rather than 36 for the completely anisotropic media. The number further reduces to 9 for orthotropic media; to 5 for transversely isotropic media; and, finally, to 2 for the completely isotropic media.

The Biot coefficient α, in tensor form, is given as follows:

$$\tilde{\alpha} = \alpha_{ij} = \begin{bmatrix} \alpha_{11} & \alpha_{12} & \alpha_{13} \\ \alpha_{12} & \alpha_{22} & \alpha_{23} \\ \alpha_{13} & \alpha_{23} & \alpha_{33} \end{bmatrix} \tag{5}$$

In view of symmetry, the total 9 parameters reduce to 6. This number further reduces to 3, 2 and 1 for orthotropic, transversely isotropic, and isotropic media, respectively.

The permeability tensor can be described as:

$$\tilde{k} = \begin{bmatrix} k_{11} & k_{12} & k_{13} \\ k_{21} & k_{22} & k_{23} \\ k_{31} & k_{32} & k_{33} \end{bmatrix} \tag{6}$$

Since the permeability tensor possesses identical number of terms as the Biot coefficient tensor, the numbers of the independent parameters are the same for different types of the porous media.

Though the compressibility tensor $\tilde{\alpha}^*$ maintains an identical number of parameters as the Biot coefficient tensor $\tilde{\alpha}$, it requires more specific definitions since it relates to the lumped compressibilities of both solid and fluid.

For modeling fluid flow and solid deformations of anisotropic fractured porous media, the previous formulation of the single-porosity poroelasticity needs to be extended to the dual-porosity one. The difference between the single- and dual-porosity poroelasticities rests not only on the fluid flow portion, but also on the solid deformation part. Flow within the matrix blocks and within the fractures are treated as two separate events. As a result, two flow equations must be given to characterize the two events. These two equations are linked through a common mass transfer term in each equation with opposite signs, indexing the interactive flow due to the pressure difference between the matrix and the fractures. The elastic displacements of the matrix blocks and the fractures are interwoven with merely the lumped effects being considered. In consequence, only one general equation (e.g., 1-D) is required to describe the solid equilibrium. In addition, a rigorous poroelastic formulation requires that the effect of pressure gradient is retained in the solid equilibrium, while that of the volumetric strain is kept in the flow continuity equation.

Based on the conceptual procedure proposed by Elsworth and Bai, [11] flow-dependent deformations within a dual-porosity poroelastic formulation for anisotropic media can be written as:

$$\nabla \cdot (\tilde{\Psi}^* : \tilde{\varepsilon} + \tilde{\alpha}'_{ma} p_{ma} + \tilde{\alpha}'_{fr} p_{fr}) = 0 \tag{7}$$

where the subscripts ma and fr denote the matrix and fractures, respectively; p_{ma} and p_{fr}, are the respective pressures of each continuum at each point in space; $\tilde{\Psi}^*$ is defined as:

$$\tilde{\Psi}^* = (\tilde{C}_{ma} + \tilde{C}_{fr})^{-1} \tag{8}$$

and $\tilde{\alpha}_{ma}$and $\tilde{\alpha}_{fr}$ are the second order pressure ratio factor tensors, which can be expressed as:

$$\tilde{\alpha}'_{ma} = \tilde{\Psi}^* : \tilde{C}_{ma} : \tilde{\alpha}_{ma} \tag{9}$$

$$\tilde{\alpha}'_{fr} = \tilde{\Psi}^* : \tilde{C}_{fr} : \tilde{\alpha}_{fr} \tag{10}$$

For the fluid flow in the anisotropic media, the general deformation-dependent fluid flow equations can be given as:

$$\nabla \cdot (\frac{\tilde{k}_{ma}}{\mu} \cdot \vec{\nabla} p_{ma}) = \tilde{\delta} : \tilde{\alpha^*}_{ma} \frac{\partial p_{ma}}{\partial t} + \Gamma(p_{ma} - p_{fr}) - \tilde{\alpha}_{ma} : \frac{\partial \tilde{\varepsilon}}{\partial t} \tag{11}$$

and,

$$\nabla \cdot (\frac{\tilde{k}_{fr}}{\mu} \cdot \vec{\nabla} p_{fr}) = \tilde{\delta} : \tilde{\alpha^*}_{fr} \frac{\partial p_{fr}}{\partial t} - \Gamma(p_{ma} - p_{fr}) - \tilde{\alpha}_{fr} : \frac{\partial \tilde{\varepsilon}}{\partial t} \tag{12}$$

where Γ is the interporosity flow coefficient.

For a dual-porosity poroelastic anisotropic medium, the number of required parameters is roughly doubled over that of its single-porosity counterpart since the matrix and the fractures possess individual groups of parameters. Specifically, with consideration of symmetry, the total independent parameters of the compliance tensors for both matrix and fractures are 42 for the completely anisotropic medium; 18 for orthotropic media; and, 10 for transversely isotropic media. For the pressure ratio factor tensors, permeability tensors, and compressibility tensors of the dual-porosity poroelastic media, these numbers reduce to 12, 6, and 4 for the completely anisotropic media, orthotropic media, and transversely isotropic media, respectively.

4 Parametric Study

Among many tensor-form parameters in the governing equations for the anisotropic dual-porosity poroelasticity, the permeability tensor can be considered as one of the most influential factors on fluid flow. The parametric analyses focus on probing the related mechanisms.

4.1 Permeability Tensor of Fractured Media

The permeability anisotropy can be attributed to the existence of heterogeneities such as natural fractures. Neglecting roughness and infilling state, fluid flow within a single fracture can be simplified as the flow through parallel plates.[13] The analytical expression of such flow conditions is characterized by certain critical parameters such as the gap between the plates (analogous to the fracture opening or aperture), along with a fracture-density index known as the fracture spacing. In reality, the fracture flow is primarily restricted to the two-dimensional fracture plane with only minimum fluid leak off perpendicularly to the fracture plane. As a result, the permeability tensor for the fractured media can be described as:

$$k_{ij} = \sum_{l=1}^{N} \frac{b_l^3}{12 s_l} (\delta_{ij} - n_i^l n_j^l) \tag{13}$$

where $\mathbf{n}$ is the unit vector parallel and perpendicular to the fracture plane, l is the index for l-th fracture with a spacing s.

In the three-dimensional domain of the fractured media, the permeability tensor can be described as:

$$k_{ij} = \sum_{l=1}^{N} \frac{b_l^3}{12 s_l} \begin{bmatrix} 1-(n_x^l)^2 & -n_x^l n_y^l & -n_x^l n_z^l \\ -n_y^l n_x^l & 1-(n_y^l)^2 & -n_y^l n_z^l \\ -n_z^l n_x^l & -n_z^l n_y^l & 1-(n_z^l)^2 \end{bmatrix} \tag{14}$$

For the two-dimensional scenario, the permeability tensor can be simplified as:

$$k_{ij} = \begin{bmatrix} k_{xx} & k_{xy} \\ k_{yx} & k_{yy} \end{bmatrix} \tag{15}$$

where,

$$k_{xx} = \sum_{l=1}^{N} \frac{b_l^3}{12 s_l} \sin^2 \psi_l \tag{16}$$

$$k_{xy} = k_{yx} = \sum_{l=1}^{N} \frac{b_l^3}{12 s_l} (-\sin \psi_l \cos \psi_l) \tag{17}$$

$$k_{yy} = \sum_{l=1}^{N} \frac{b_l^3}{12 s_l} \cos^2 \psi_l \tag{18}$$

and, ψ is the angle between the $x-$ direction and the normal direction to the fracture plane in a Cartesian coordinate system.

For the orthogonal flow in which the fracture plane coincides with the major axes, the terms k_{xy} and k_{yx} in equation (15) disappear.

4.2 Stress-dependent Permeability Tensor

Permeability is traditionally defined as a physical parameter which represents the effective flow cross-sectional area. However, field and experimental observations indicate that the permeability may not be a constant, but a function of the stress environment. Consequently, initial formation permeability anisotropies can be modified when subjected to in-situ stress changes. Even for the rare cases where the porous media originally exhibit an isotropic permeability, anisotropic conditions may be introduced when the porous media are subjected to external loading.

The permeability tensor and associated variations resulting from a stress perturbation can be represented by a change in fracture aperture (neglecting the sign for the fracture index), i.e.,

$$k_{ij} = b_0 + \Delta k_{ij} = b_0 + \frac{\Delta b_{i'j'}^3}{12s}(\delta_{ij} - n_i n_j) \tag{19}$$

where b_0 is the initial fracture aperture, Δk_{ij} is the permeability change, $\Delta b_{i'j'}$ is the tensorial fracture aperture perturbed by the change in elastic deformation, the subscript $i'j'$ is measured in the directions perpendicular to that of ij. For the orthogonal flow where the off-diagonal terms of the permeability tensor disappear, the permeability changes can be simply expressed as:

$$\Delta k_1 = \frac{(\Delta b_2 + \Delta b_3)^3}{12s} \tag{20}$$

$$\Delta k_2 = \frac{(\Delta b_1 + \Delta b_3)^3}{12s} \tag{21}$$

$$\Delta k_3 = \frac{(\Delta b_1 + \Delta b_2)^3}{12s} \tag{22}$$

Expressing the aperture change in any arbitrary direction for the orthogonal flow, the vectorial aperture Δb_i is equivalent to the change in solid displacement Δu_i; i.e.,

$$\Delta b_i = \Delta u_i \tag{23}$$

With the assumption that $s \gg b_0$, the aperture change in equation (23) may be expressed as:

$$\Delta b_i = \frac{s\ E\Delta\varepsilon_i}{sK_n + E} \tag{24}$$

where E is the elastic modulus, K_n is the normal fracture stiffness, and the normal strain vector $\Delta\varepsilon_i$ can be converted to the normal stress vector through the following expression:

$$\Delta\varepsilon_i = \frac{1}{E}[\Delta\sigma_i - \nu(\Delta\sigma_j + \Delta\sigma_k)] \tag{25}$$

where ν is Poisson's ratio.

If the influence of shear-induced dilatancy on fracture aperture is also considered, Δu_i in equation (23) should include both normal and shear displacements. As a result, equation (24) can be further modified as:

$$\Delta b_i = \frac{s\ E}{sK_n + E}\Delta\varepsilon_i + \frac{s\ G}{sK_{sh} + G}\Delta\gamma_{jk}\tan\phi_d \tag{26}$$

where K_{sh} is the shear stiffness of the fracture, G is the shear modulus, ϕ_d is the dilatational angle, and $\Delta\gamma_{jk}$ is the change of shear strains, which can be converted to the shear stresses as

$$\Delta\gamma_{jk} = \frac{1}{G}(\Delta\tau_{ij} + \Delta\tau_{ik}) \tag{27}$$

For the dual-porosity poroelastic formulation, the permeability of the matrix blocks also needs to be defined since separate flow equations are designed for the matrix and for the fractures. The stress-dependency of the matrix permeability may be associated with the changes in void spaces or grain sizes due to the stress changes, i.e.,

$$\Delta k_j = k_0\left[\sum_{i=1,i\neq j}^{3}\left\{1 \mp \frac{1}{2}\left[\frac{9(1-\nu^2)}{2}(\pi\Delta\varepsilon_i)^2\right]^{1/3}\right\}^2 - 1\right] \tag{28}$$

where k_0 is the original permeability.

For linear poroelastic formulation, the permeability is restricted to being independent of other primary variables. As a result, the stress-dependent permeability may result in a nonlinear poroelastic system, where the solution procedure should be completed in an incremental fashion with updated stresses at each spatial and temporal levels of evaluated pore pressure. For the steady state flow situations in which the solid displacements and pore pressures are decoupled, the sequential solution procedure may be adopted to calculating the stresses and the related stress-dependent permeability separately.

5 Concluding Remarks

In this paper, new definitions of material heterogeneities and anisotropies according to the three different levels of domain scales, i.e., macro-, meso-, and micro-scopic scales, are presented. The importance of such subdivisions is the ability to characterize the material heterogeneities and anisotropies in a more physically sensible manner, so as to minimize the ambiguity of defining these terms in general. For analyzing one of the heterogeneous and anisotropic scenarios at the macroscopic scale, the general formulation of the anisotropic dual-porosity poroelastic model is presented, which enables the flow and deformations in the anisotropic fractured poroelastic media to be simulated. A relevant example is given in the determination of the stress-dependent permeability, which provides an assessment on the parametric anisotropies on the mesoscopic scale.

Acknowledgments

Support of the National Science Foundation/State/Industry under the S/IUCRC program and under contract EEC-9209619 is gratefully acknowledged.

References

1. K.V. Terzaghi, *Acad. Wiss. Wien. Math. Nauturwiss.*, K1. Alot. 2A, **132**, 105 (1923).
2. M.A. Biot, *J. Appl. Phys.* **12**, 155-164 (1941).
3. M.A. Biot, *J. Appl. Phys.* **26**, 182 (1955).
4. M.M. Carroll, *J. Geophys. Res.* **84**, 7510 (1979).
5. M. Thompson and J.R. Willis, *J. Appl. Mech., ASME* **58**, 612 (1991).
6. A.H.D. Cheng, *Int. J. Rock Mech. Min. Sci.* **34**, 199 (1997).
7. G.I. Barenblatt *et al*, *J. Appl. Math. Mech.* **24**, 852 (1960).
8. J.E. Warren and P.J. Root, *Soc. Pet. Engng. J.*
9. R.K. Wilson and E.C. Aifantis, *Int. J. Engng. Sci.* **20**, 1009 (1982).
10. M.Y. Khaled *et al*, *Int. J. Num. Anal. Meth. Geom.* **8**, 101 (1984).
11. D. Elsworth and M. Bai, *J. Geotech. Engng., ASCE* **118**, 107 (1992).
12. M. Bai *et al*, *Int. J. Rock Mech. Min. Sci. & Geomech. Abstr.* **30**, 1185 (1993). **3**, 245 (1963).
13. J. Bear, *Dynamics of Fluids in Porous Media*, American Elsevier Pub., Boulder (1972).

SIMPLE CONSTITUTIVE EQUATIONS FOR BRITTLE ROCKS, BASED ON STRAIN-DEPENDENT STIFFNESS

Y. FUJII and Y. ISHIJIMA
Division of Structural and Geotechnical Engineering, Hokkaido University, N13W8, Sapporo 060-8628, JAPAN
E-mail: fujii@rock.eng.hokudai.ac.jp

It was found that the trace of the axial strain, the circumferential strain, and the axial stress of brittle rock under compression is located roughly on a plane. Constitutive equations representing axial stress and confining pressure by axial and circumferential strains were derived based on this observation assuming that one of the two stiffness coefficients is a function of strain and its value decreases with rock expansion. A numerical calculation supports this assumption. The constitutive equations do not contain any conventional strength constants such as tensile strength, compressive strength, the angle of internal friction, etc. Stress-strain curves obtained with the constitutive equations simulate Class II as well as Class I strain-softening behavior. Stress-strain curves for actual brittle rock were also simulated by adjusting the constants.

1 Introduction

The goal of the research reported in this paper is to construct constitutive equations which can represent reasonable axial stress-lateral strain relationships as well as various axial stress-axial strain relationships for brittle rock. First, axial strain-circumferential strain-axial stress relationships of brittle rock are represented in a three-dimensional coordinate system defined by these quantities. Next, constitutive equations representing axial stress and confining pressure by axial and circumferential strains were derived based on this reflection. The constitutive equations, which do not contain any conventional strength constants such as tensile strength, compressive strength, angle of internal friction, etc., look like simple linear functions. One of the two stiffness coefficients, however, is a function of strain and its value decreases with expansion.

These constitutive equations can simulate the Class II as well as the Class I strain-softening behavior. Stress-strain curves of actual brittle rock can also be simulated by adjusting these constants. Numerical results which support the assumption of the stiffness coefficients will also be shown.

2 Relationship between Axial Strain, Circumferential Strain and Axial Stress

Samples, 30 mm in diameter and 60 mm in length, were prepared for Kamisunagawa sandstone. The specimens were dried at 80°C for 24 hours and were stored in a desiccator for two weeks before being tested. An Instron's screw-type 5586 loading frame, with a maximum load of 300 kN, was used for the uniaxial compression tests. The axial strain rate was set at $8.3 \times 10^{-5}\ s^{-1}$.

Instron's axial and chain-type circumferential extensometers were used to measure the core deformations. The axial extensometer was clamped onto the specimen with a pair of rubber bands, and contacted the specimen by two steel blades. For various reasons, the distance between the blades was set at 25 mm out of an entire specimen length of 60 mm. The axial strain taken by the extensometer, ε_E, can be inaccurate in the post peak region when macroscopic fractures appear near specimen ends. On the other hand, the axial strain evaluated from the displacement of the cross head, ε_D, contains errors due to an initial unknown clearance (e.g. between the platen and the end piece, between pieces of the spherical seat, etc.) and the extra inherent deformation of the loading frame, the end pieces, etc. Therefore, the axial strain was evaluated in the following way:

(1) Tangent Young's modulus at a 50% maximum load was evaluated from axial extensometer data (E_E) and from cross-head displacements (E_D);

(2) ε_D' was evaluated by subtracting the extra deformation from ε_D so that E_D equals E_E, assuming that the extra deformation is proportional to the load, namely,

$$\varepsilon_D' = \varepsilon_D - (1 / E_D - 1 / E_E)\, \sigma_A, \tag{1}$$

where σ_A is the axial stress;

(3) ε_D'' was evaluated by subtracting the initial clearance so that the ε_D'' value equals ε_E at a 60% maximum stress, namely,

$$\varepsilon_D'' = \varepsilon_D' - \varepsilon_D'(60\%) + \varepsilon_E(60\%), \text{ and;} \tag{2}$$

(4) ε_E was regarded as the axial strain where the axial load had less than a 60% maximum load. ε_D'' was regarded as the axial strain once the axial load exceeded a 60% maximum load.

Figure 1 shows the axial strain-circumferential strain-axial stress trace for Kamisunagawa sandstone. The trace looks curvilinear (Figure 1(a)). After rotating the coordinate system, however, the trace appears as a straight line except for at the initial loading stage (Figure 1(b)). Similar results were obtained for many other kinds of rock. The equation for the plane can be represented as:

$$\sigma_A = A\,(\varepsilon_A - \varepsilon_0) + B\,\varepsilon_C, \tag{3}$$

where ε_A, ε_C, and σ_A are axial strain, circumferential strain, and axial stress, respectively. A, B, and ε_0 are constants and should be equal to $\lambda + 2\mu$, λ, and zero, respectively, for an elastic medium. λ and μ are Lamé's constants. Table 1 shows the constants evaluated by the least squares method with apparent elastic constants E' and ν' evaluated by assuming A equals $\lambda + 2\mu$ and B equals λ.

3 Constitutive Equations

3.1 Derivation

Okubo[1] assumed that the axial compliance increases with axial strain. It is well known, however, that axial cracks develop during compression tests on rock (e.g. Hallbauer *et al.*[2]). This will cause a decrease in radial stiffness accompanied with radial expansion. Therefore, constitutive equations will be derived here assuming that stiffness decreases with expansion.

Replacing constant A with a function of normal strain $A'(\varepsilon)$ and ignoring constant ε_0 in equation (3),

$$\sigma_A = A'(\varepsilon_A)\,\varepsilon_A + B\,\varepsilon_C. \qquad (4)$$

By analogy, the following equation can be written for confining pressure case σ_C:

$$\sigma_C = A'(\varepsilon_C)\,\varepsilon_C + B\,\varepsilon_A. \qquad (5)$$

$A'(\varepsilon)$ should converge to a certain value as ε tends to ∞. It should decrease with expansion and converge to a certain value as ε tends to $-\infty$. The following function (Figure 2) was chosen as $A'(\varepsilon)$ to satisfy the above requirements:

$$A'(\varepsilon) = D\,[\,\tan^{-1}\{C\,(\varepsilon / \varepsilon_S + 1\,)\,\} + \pi / 2\,] + F, \qquad (6)$$

where C, D, and F are constants and ε_S (> 0) is a reference strain. The value of this function becomes A (= $D\,\pi + F$) as ε tends to ∞. It decreases as ε decreases and converges to F as ε tends to $-\infty$. The rate of decrease shows the maximum at $\varepsilon = -\varepsilon_S$. The maximum gradient increases with constant C.

Eliminating ε_C between equations (4) and (5), leads to

$$\sigma_A = [\,\{\,B^2 - A'(\varepsilon_A)\,A'(\varepsilon_C)\}\,/\,B\,]\,\varepsilon_C + \{\,A'(\varepsilon_A)\,/\,B\,\}\,\sigma_C. \qquad (7)$$

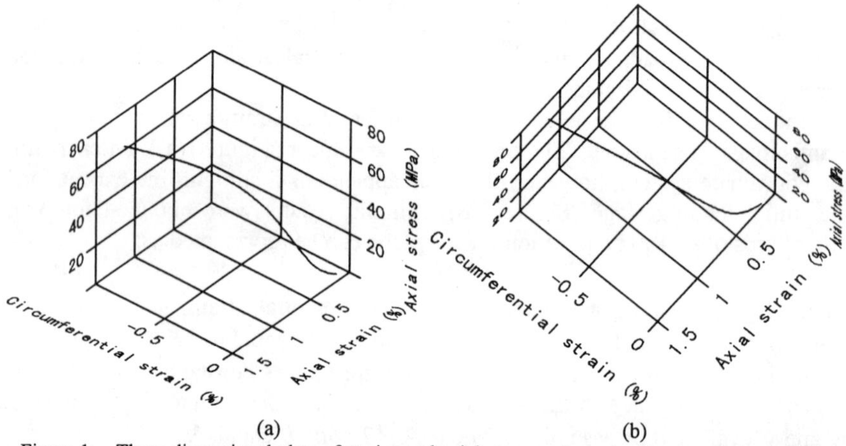

Figure 1: Three-dimensional plots of strains and axial stress under uniaxial compression; after rotating the coordinate system, the originally curvilinear trace (a) appears as a straight-line (b), except for at the initial loading stage.

Table 1: Peak strength σ_{AP}, tangent Young's modulus measured at a 50% peak load, $E_{t,50}$, Poisson's ratio measured at a 50% peak load, $\nu_{t,50}$, critical tensile strain ε_{TC} (circumferential strain at peak load), constants A, B, and ε_0 in equation (3) for uniaxial compression test; the apparent elastic constants are also shown.

σ_{AP} (MPa)	$E_{t,50}$ (GPa)	$\nu_{t,50}$	ε_{TC} ($\times 10^{-2}$)	A (GPa)	B (GPa)	ε_0 ($\times 10^{-2}$)	E' (GPa)	ν'
64.9	10.1	0.39	-0.59	8.7	1.8	0.19	9.4	0.17

The axial stress will show a constant residual strength only if the first term on the right side of equation (7) becomes zero as ε_A and ε_C tend to ∞ and $-\infty$, respectively. Therefore,

$$F = B^2 / A, \tag{8}$$

$$D = (A^2 - B^2) / (A\pi). \tag{9}$$

3.2 *Example*

A set of axial stress and axial strain values for a given circumferential strain value is calculated as follows:

(1) Evaluate A' for ε_C;

(2) Substitute the above values for A', σ_C, ε_C, and B into equation (5) so that ε_A is given;

(3) Evaluate A' for ε_A; and,

(4) Substitute the above A', ε_A, B, and ε_C into equation (4) so that σ_A is given.

Figure 3(a) to (e) show examples of triaxial and uniaxial stress-strain curves obtained with such constitutive equations. Calculations were conducted up to -1×10^{-2} of the circumferential strain. Figure 3(a) shows the effects of the confining pressure. Note that the stress and strains in the figure are total values, not differential values. Figure 3(f) shows the uniaxial tension stress-strain curves for several C values. The following similarities with actual rock are observed:

(1) The tensile strength is less than the compressive strength;

(2) The peak load increases with confining pressure;

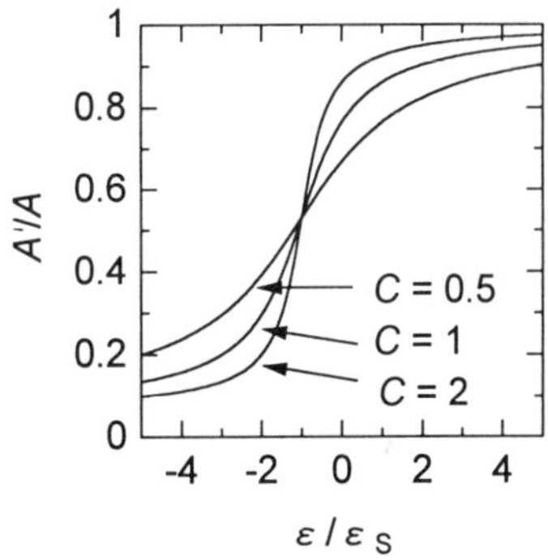

Figure 2: Relationship between A' and ε for several C; A = 20 (GPa) and B = 5 (GPa).

(3) The critical compressive strain (axial strain at peak load) increases with confining pressure; and,

Table 2: Constants used to simulate stress-strain curves.

A (GPa)	B (GPa)	C	ε_S ($\times 10^{-2}$)
13	2	0.4	0.1

(4) The critical tensile strain (circumferential strain at peak load) is not affected by the confining pressure.

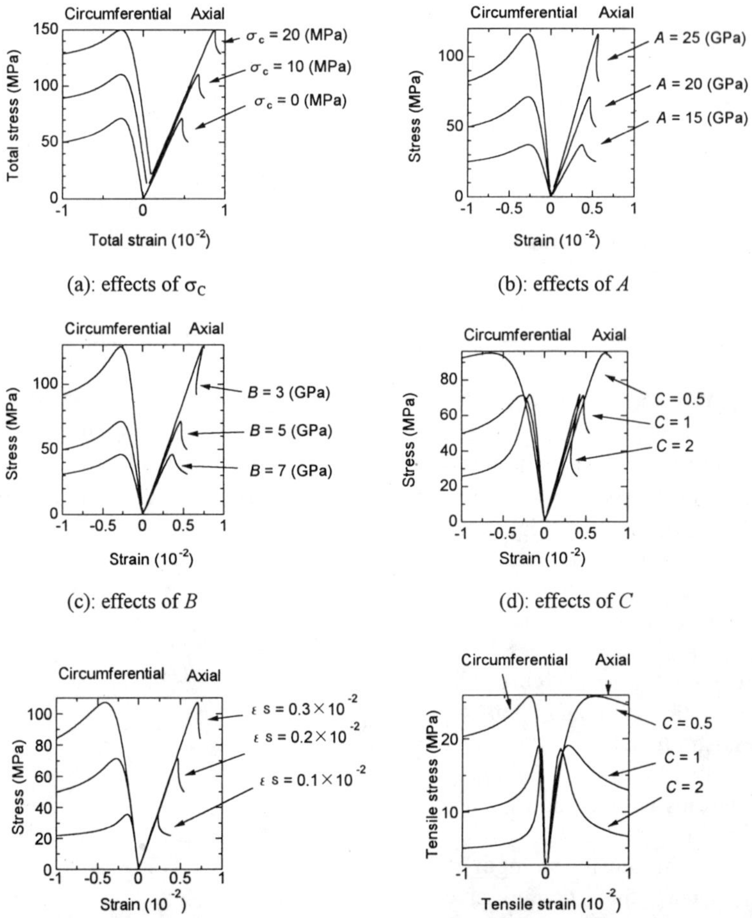

(a): effects of σ_C (b): effects of A

(c): effects of B (d): effects of C

(e): effects of ε_S (f): effects of C (uniaxial tension)

Figure 3: Stress-strain curves obtained from proposed constitutive equations; the curves for $A = 20$ (GPa), $B = 5$ (GPa), $C = 1$, $\varepsilon_S = 0.2$ ($\times 10^{-2}$), and $\sigma_C = 0$ are set as a standard; stress and strains in (a) are assumed to be zero-balanced before applying the confining pressure.

Namely, the tensile strain criterion for brittle rock, proposed by Fujii *et al.*[3], is automatically derived from these constitutive equations. However, the brittle-ductile transition phenomenon can not be represented by the constitutive equations.

Constant A mainly affects Young's modulus, strength, and the critical compressive strain (Figure 3(b)). Constant B mainly affects Poisson's ratio and the curvature of the axial stress-axial strain curve around the peak load (Figure 3(c)). Constant C significantly influences strength, the curvatures of the stress-strain curves around the peak load, and the critical strains (Figure 3(d)). Constant ε_S simply determines the strength and the critical strains (Figure 3(e)).

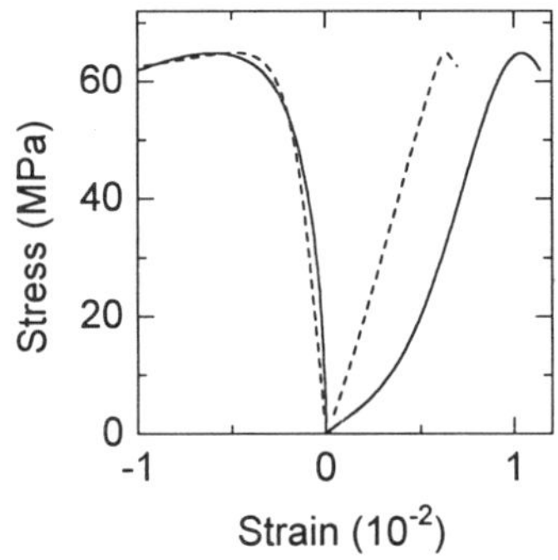

Figure 4: Approximations of the uniaxial compression stress-strain curves obtained with the proposed constitutive equations; solid and dotted lines are measured and approximated values, respectively.

3.3 Approximation of the test results

Necessary constants in the constitutive equations are A, B, C, and ε_S. σ_C is known as one of the experimental conditions. The approximation of the test results, therefore, can be carried out as follows:

(1) Adjust A so that the gradient of the axial stress-axial strain curve can be simulated;

(2) Adjust B so that the gradient of the axial stress-circumferential strain curve can be simulated;

(3) Adjust C so that the entire shape of each stress-strain curve can be simulated;

(4) Adjust ε_S so that the peak load point can be simulated; and,

(5) Iterate the above steps, if necessary.

The stress-strain curves, approximated with the constants in Table 2, are shown in Figure 4. The values of the constants are different from those evaluated by the least squares method in Table 1. The difference may be mainly caused by ignoring ε_0. A reasonable consistency between the measured and the approximated curves, however, was observed. Most of the differences between the axial stress-axial strain curves may also be caused by ignoring ε_0.

4 Consideration of the Assumptions for **A'(ε)** *and* **B** *based on the Deformations in a Cracked Plate*

Assume a plate that is 20 mm × 10 mm in size and contains an open crack 6 mm in length under plane strain conditions (Figure 5). The outer boundary was modeled by the body force element method (Nishitani[4]) and the crack by the displacement discontinuity method (Crouch and Fairhurst[5]). The following two cases were considered:

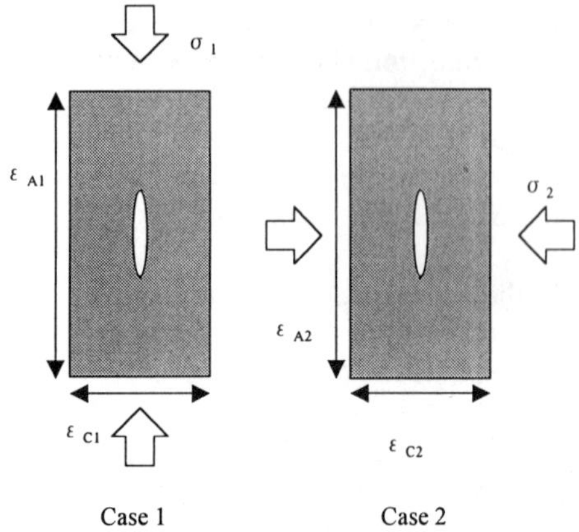

Figure 5: Model used to determine the change in stiffness due to the existence of a crack.

Case 1: σ_1 parallel to the crack is subjected;

Case 2: σ_2 perpendicular to the crack is subjected.

The average axial and lateral strains in Case 1, ε_{A1}, ε_{C1}, and those in Case 2, ε_{A2} and ε_{C2} were numerically calculated.

Constants A_1, A_2, and B were calculated using the following equations:

$$\sigma_1 = A_1\,\varepsilon_{A1} + B\,\varepsilon_{C1}, \tag{10}$$

$$0 = A_2\,\varepsilon_{C1} + B\,\varepsilon_{A1}, \tag{11}$$

$$0 = A_1\,\varepsilon_{A2} + B\,\varepsilon_{C2}, \tag{12}$$

$$\sigma_2 = A_2\,\varepsilon_{C2} + B\,\varepsilon_{A2}. \tag{13}$$

The results are as follows:

$$A_1 = 1.01\,(\lambda + 2\mu), \tag{14}$$

$$A_2 = 0.66\,(\lambda + 2\mu), \tag{15}$$

$$B = 0.69\,\lambda. \tag{16}$$

In the case of an uncracked plate, constants A_1, A_2, and B should be $\lambda + 2\mu$, $\lambda + 2\mu$, and λ, respectively. The assumption that $A'(\varepsilon)$ decreases with expansion is determined to be reasonable from these results. B was assumed to be constant based on the measured stress-strain curve. The validity of this assumption should be further investigated.

5 Conclusions

The trace of axial strain, circumferential strain and axial stress of brittle rock under uniaxial compression was located roughly on a plane in the three-dimensional space defined by these quantities. Constitutive equations for brittle rock were derived from these observations. The constitutive equations which do not contain any conventional strength constants, such as tensile strength, compressive strength, the angle of internal friction, etc., look like simple linear functions. One of the two stiffness coefficients, however, is a function of strain and its value decreases with expansion.

Stress-strain curves obtained by the newly-deduced constitutive equations simulated Class II as well as Class I strain-softening behavior and had the following similarities with actual brittle rock:

(1) The tensile strength was less than the compressive strength;

(2) The compressive strength increased with confining pressure;

(3) The critical compressive strain (axial strain at peak load point) increased with confining pressure; and,

(4) The critical tensile strain (circumferential strain at peak load point) was not affected by the confining pressure.

Stress-strain curves of brittle rock were also simulated by adjusting the constants. The physical meanings of the constants, the effects of the environments, and the function form of the stiffness coefficient should be further investigated.

Acknowledgments

Professor J-C. Roegiers of The University of Oklahoma and Mr. T. Kiyama at Mitsui Construction Co., Ltd. gave the authors valuable advice.

References

1. Okubo S., *J. MMIJ* **108**, 601 (1992 in Japanese).
2. Hallbauer, D. K., Wagner, H. and Cook, N. G. W., *Int. J. Rock Mech. Min. Sci. & Geomech. Abstr.* **10**, 713 (1973).
3. Fujii, Y. and Ishijima, Y., *Int. J. Rock Mech. & Min. Sci. & Geomech. Abstr.* **34**, 3-4, Paper No. 086 (1997).
4. Nishitani, H., *Computational and Experimental Fracture Mechanics* (Computational Mechanics Publications, Berlin, 1994).
5. Crouch, S. L. and Fairhurst, C., *U. S. B. M. Contract Report* **H0101778** (1973).

POROELASTIC STABILITY ANALYSES OF INCLINED BOREHOLES IN SATURATED FORMATIONS

L. CUI, J-C. ROEGIERS
Rock Mechanics Institute, The University of Oklahoma
Norman, OK 73019, USA
E-mail: cui@rmg.ou.edu

Y. ABOUSLEIMAN
Department of Civil Engineering, School of Engineering and Architecture,
The Lebanese American University, Byblos, Lebanon
E-mail: yasleman@lau.edu.lb

Most boreholes are drilled through fluid-saturated formations. The solid-pore fluid coupling induced by drilling as well as pumping/injection influences the stress and pore pressure distributions around boreholes significantly, hence their stability. However, conventional approaches for borehole stability usually ignore this coupling effect. The present paper introduces the recently-developed poroelastic models of inclined boreholes with considerations of solid-pore fluid coupling, and presents poroelastic stability analyses of inclined boreholes.

1 Introduction

In the energy industry, the borehole failure is a continuing problem that results in huge losses in operation time and revenue. Since inclined and horizontal borehole drilling technology has become popular, the stability analyses of far-reaching boreholes have drawn more and more attention since the late 70's. (A borehole is "inclined" if its longitudinal axis is deviated with respect to the principal axes of the in-situ stress tensor.) Early work on inclined boreholes was presented by Bradley,[1] in which the rock was assumed to be a linear elastic material, and the fluid flow into or out of the formation was not considered. Although this solution is widely used in the industry, field observations and laboratory tests showed phenomena, particularly transient ones,[2] that could not be accounted for by this linear elastic analysis.

Boreholes are usually drilled through fluid-saturated formations; hence, the stress concentration around a borehole results in a disturbance of the pore pressure field; strongly modifying the effective stresses which controls the formation failure. The theory of poroelasticity[3,4] deals with this solid deformation-pore pressure coupling mechanism. Therefore, the application of poroelasticity to the borehole stability is expected to give more accurate predictions of instabilities and more optimized designs of borehole trajectories.

In this paper, several recently-developed poroelastic models of inclined boreholes in saturated formations are presented. Using these models, the analyses of the effects of solid-fluid coupling mechanism on the borehole stability were carried out.

2 Poroelastic Models

2.1 Governing Equations

A set of governing equations of poroelasticity in a Cartesian coordinate system is briefly described below:

$$\sigma_{ij} = D_{ijkl}\epsilon_{kl} - \alpha_{ij}p \tag{1}$$

$$p = M(\zeta - \alpha_{ij}\epsilon_{ij}) \tag{2}$$

$$\sigma_{ij,j} + b_i = 0 \tag{3}$$

$$\epsilon_{ij,kl} + \epsilon_{kl,ij} - \epsilon_{ik,jl} - \epsilon_{jl,ik} = 0 \tag{4}$$

$$\frac{\partial \zeta}{\partial t} + q_{i,i} = \gamma \tag{5}$$

$$q_i = -\frac{k_{ij}}{\mu} p_{,i} \tag{6}$$

where σ_{ij} is the total stress tensor (positive for tension), ϵ_{ij} is the overall strain tensor of the porous body, p is the pore pressure, D_{ijkl} is the drained elastic modulus tensor, α_{ij} is the Biot's effective stress coefficient tensor, ζ is the volumetric variation of pore fluid content in the porous body, M is Biot's modulus, b_i is the body force vector, q_i is the pore fluid discharge vector, and γ is the source/sink density.

In the above, the materials have been assumed linear elastic; and, the quasi-static state has also been assumed. The coupling between the pore pressure/pore fluid flow and stresses/strains and the time-dependency of variables are clearly seen from Eqs. (1) to (6).

2.2 Problem Description

A schematic of an inclined borehole is presented in Fig. 1. $S_{x'}$, $S_{y'}$, and $S_{z'}$ are three principal stresses of the in-situ stress state; and, p_o is the virgin pore pressure. The deviation of the borehole is measured by the azimuth φ'_z and the inclination angle φ_y. The local coordinate system xyz, or $r\theta z$, attached to the borehole (see Fig. 1 (b)) is preferred for borehole analyses; and, S_i denotes the components of the in-situ stress state in the local coordinate system which were calculated from $S_{x'}$, $S_{y'}$, $S_{z'}$, and the coordinate transform matrix.

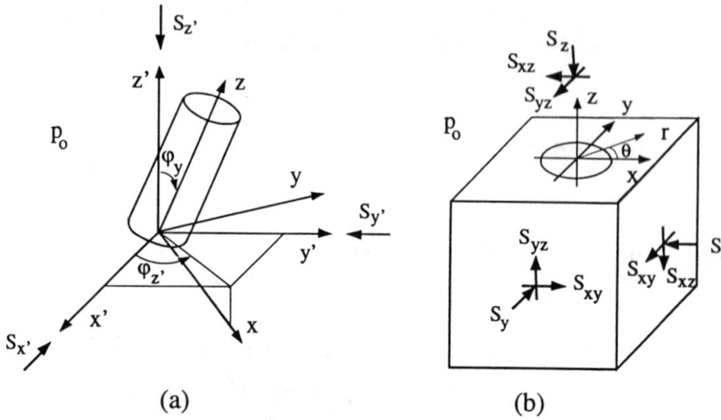

Figure 1: A schematic of the inclined borehole.

The boundary conditions at the borehole wall are usually imposed as follows: the total radial stress, σ_{rr}, is the same as the well pressure, p_w (the mud/injection fluid pressure); the pore pressure or the flow rate is a known function of time which is dependent of p_w and the degree of the permeability of the borehole wall; and, other variables are trivial. At infinity, all stresses and pore pressure remain the same as the in-situ stresses and the virgin pore pressure. It is usually assumed that the pore pressure at the borehole wall is equal to p_w for the permeable borehole wall; but, for the very poor permeable condition of the wall, it is rather assumed that no flow is allowed through the wall.

2.3 Solution

For the isotropic saturated formations, following Cui et al,[5] general expressions of stresses and pore pressure around the borehole were obtained as:

$$\sigma_{rr} = -P_o + S_o \cos 2(\theta - \theta_r) + (P_o - p_w)\frac{r^2}{R^2} + \sigma_{rr}^{(1)} + \sigma_{rr}^{(2)} \tag{7}$$

$$\sigma_{\theta\theta} = -P_o - S_o \cos 2(\theta - \theta_r) - (P_o - p_w)\frac{r^2}{R^2} + \sigma_{\theta\theta}^{(1)} + \sigma_{\theta\theta}^{(2)} \tag{8}$$

$$\sigma_{zz} = -S_z + \nu\left(\sigma_{rr}^{(1)} + \sigma_{rr}^{(2)} + \sigma_{\theta\theta}^{(1)} + \sigma_{\theta\theta}^{(2)}\right) - \alpha(1-2\nu)(p^{(1)} + p^{(2)}) \tag{9}$$

$$\tau_{r\theta} = -S_o \sin 2(\theta - \theta_r) + \sigma_{r\theta}^{(2)} \tag{10}$$

$$\tau_{rz} = -(S_{xz}\cos\theta + S_{yz}\sin\theta)\left(1 - \frac{R^2}{r^2}\right) \tag{11}$$

$$\tau_{\theta z} = (S_{xz}\sin\theta - S_{yz}\cos\theta)\left(1 + \frac{R^2}{r^2}\right) \tag{12}$$

$$p = p_o + p^{(1)} + p^{(2)} \tag{13}$$

where $2\theta_r = \tan^{-1}(2S_{xy}/(S_x - S_y))$, $S_o = \sqrt{((S_x - S_y)/2)^2 + S_{xy}^2}$, $P_o = (S_x + S_y)/2$, and:

$$\tilde{\sigma}_{rr}^{(1)} = -\left(C_1\frac{\mathrm{K}_1(\xi)}{\xi} - \frac{C_2}{r^2}\right) \tag{14}$$

$$\tilde{\sigma}_{\theta\theta}^{(1)} = C_1\left(\mathrm{K}_0(\xi) + \frac{\mathrm{K}_1(\xi)}{\xi}\right) - \frac{C_2}{r^2} \tag{15}$$

$$\tilde{p}^{(1)} = -\frac{C_1}{2\eta}\mathrm{K}_0(\xi) \tag{16}$$

$$\tilde{\sigma}_{rr}^{(2)} = \left[D_1\left(\frac{1}{\xi}\mathrm{K}_1(\xi) + \frac{6}{\xi^2}\mathrm{K}_2(\xi)\right) - \frac{D_2}{r^2} - \frac{D_3}{r^4}\right]\cos 2(\theta - \theta_r) \tag{17}$$

$$\tilde{\sigma}_{\theta\theta}^{(2)} = \left\{-D_1\left[\frac{1}{\xi}\mathrm{K}_1(\xi) + \left(1 + \frac{6}{\xi^2}\right)\mathrm{K}_2(\xi)\right] + \frac{D_3}{r^4}\right\}\cos 2(\theta - \theta_r) \tag{18}$$

$$\tilde{\sigma}_{r\theta}^{(2)} = \left[2D_1\left(\frac{1}{\xi}\mathrm{K}_1(\xi) + \frac{3}{\xi^2}\mathrm{K}_2(\xi)\right) - \frac{D_2}{2r^2} - \frac{D_3}{r^4}\right]\sin 2(\theta - \theta_r) \tag{19}$$

$$\tilde{p}^{(2)} = \frac{B(1+\nu_u)}{3}\left(\frac{1-\nu}{\nu_u - \nu}D_1\mathrm{K}_2(\xi) + \frac{D_2}{r^2}\right)\cos 2(\theta - \theta_r). \tag{20}$$

Instead of their explicit expressions in the time domain, Laplace transforms of $\sigma_{rr}^{(1)}$, $\sigma_{rr}^{(2)}$, $\sigma_{\theta\theta}^{(1)}$, $\sigma_{\theta\theta}^{(2)}$, $\tau_{r\theta}^{(2)}$, $p^{(1)}$ and $p^{(2)}$ are presented above. Their inversions can be obtained via any analytical or numerical method.[6,7] In the above, $\tilde{}$ denotes the Laplace transform; s is the transform parameter; $\xi = r\sqrt{s/c}$ in which c is the diffusivity coefficient; η is a poroelastic stress coefficient; ν is the drained Poisson's ratio; ν_u is the undrained Poisson's ratio; B is the Skempton's pore pressure coefficient; K_n is the modified Bessel function of the second kind of order n; and, D_i and C_i are the functions of s to be determined by the boundary conditions (see Refs.[5,8,9] for details).

For more general anisotropic formations or formations with material nonlinearity, finite element models[10] can be applied.

3 Poroelastic Models with Other Effects

3.1 Viscoporoelastic Model

It was reported that rheological phenomena, such as creep deformations and time-delayed borehole failures, had been observed in the field. [2] These phenomena are usually considered as resulting from the rock's viscosity. The theory of poroelasticity with such considerations (i.e., viscoporoelasticity) has been formulated. [11,12] Based on the correspondence principles between poroelasticity and the viscoporoelasticity, the viscoporoelastic solution for inclined boreholes was developed recently. [13] For the borehole problems, the viscoporoelastic formulation has almost the same form as the poroelastic one, but with the correspondent viscoporoelastic parameters.

3.2 Thermoporoelastic Model

Aforementioned poroelastic models were developed under the assumption of isothermal conditions. In reality, the temperature of drilling/injection fluids is usually quite different from the one of the formation; therefore, heat exchange takes place through the borehole wall, influencing stresses and pore pressure around the borehole. The thermal effect on the borehole problem depends on the temperature variation at the borehole wall as well as the formation's thermal properties; and this effect may not be ignored for many cases. [2]

For formations with low permeability, such as shales, a fully-coupled poroelastic model of inclined boreholes with thermal coupling (thermoporoelasticity) was developed recently. [14] The solution showed that thermal effects, especially on the pore pressure magnitude, is quite significant, [14] since the thermal expansion of the pore fluid is usually much greater than the one of the solid matrix.

4 Stability Analyses

It was demonstrated that for many cases, especially for rocks with lower permeability, poroelastic results were quite different from the elastic ones in several aspects. [5,9,17] Based on developed poroelastic models, the stability analyses of inclined boreholes for several cases are presented below. The failure analysis discussed herein is based on two failure criteria: the minimum principal stress criterion for tensile failures such as fracturing and spalling; and, the Drucker-Prager criterion [15] for shear failure.

The following specific geometry and material properties were used in the ensuing analyses. The radius of the borehole is $R = 0.1$ m. The rock is a Gulf of Mexico shale with the following properties: $G = 7.6 \times 10^2$ MPa; $\nu = 0.219$;

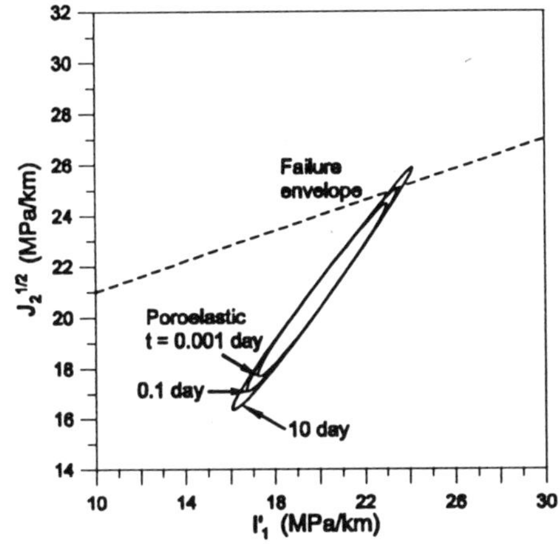

Figure 2: Stress cloud expands with the increase of time for an inclined borehole.

$\nu_u = 0.461$; $B = 0.915$; and, $k = 1 \times 10^{-7}$ darcy with $\mu = 0.001$ Pa·s. Two typical in-situ stress states were adopted: Set 1 [16]: $S_{x'} = 29$ kPa/m; $S_{y'} = 20$ kPa/m; $S_{z'} = 25$ kPa/m; and, $p_o = 9.8$ kPa/m. Set 2 [1]: $S_{x'} = 18$ kPa/m; $S_{y'} = 14$ kPa/m; $S_{z'} = 22.6$ kPa/m; and, $p_o = 10.4$ kPa/m.

4.1 Time-dependent Failure

Since stresses and pore pressure are time-dependent due to the coupling of the pore fluid flow, it is expected that the borehole failure in saturated formations will also be time-dependent.

Fig. 2 presents the expansion of the stress cloud [1] in the $(\sqrt{J_2}\text{-}I_1')$ space at the borehole wall with the increase of time for a penetrating model. In the figure, if any part of the stress cloud (closed curve) is above the (Drucker-Prager) failure envelope (straight line), the borehole fails; otherwise, the borehole is stable. In this case, $\varphi_{z'} = 30°$; $\varphi_y = 60°$; set 1 of in-situ conditions was applied; and, $p_w = 0$. The figure shows that at the borehole wall, the stress cloud at $t = 0.001$ day is totally below the envelope, but part of the cloud at $t = 10$ days is above the envelope. This suggests that the failure potential increases as time progresses; i.e., indicating a time-delayed shear failure for low well pressures.

In view of the time-delayed failure, the range of the mud pressure to maintain a stable well is also dependent of time. Fig. 3 presents the safe range of the mud weight varying with time and inclination. In this case, $\varphi_{z'} = 0$; and,

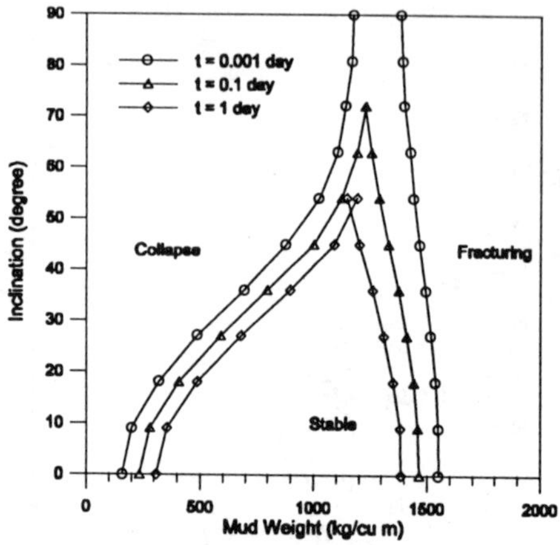

Figure 3: The safe range of the mud weight at various inclinations varies with time.

set 2 of in-situ conditions was applied. It is observed that the range becomes smaller with increasing time. (However, it remains constant after a certain time duration.[17]) This indicates that with a certain mud pressure an initially stable borehole may fail later on if the mud pressure remains constant.

It should be noted that the time-dependency of the borehole failure is affected by the in-situ stresses and formations properties. Generally speaking, the higher the difference between principal in-plane far-field stresses, the more significant the time effect; and, the higher the rock's permeability, the faster the dissipation of the time effect.[17]

4.2 Effect of Relative Wall Permeability

The permeable conditions of the wall influence the pore fluid flow around the borehole significantly. There are two extreme cases: fully permeable walls (penetrating model); and, impermeable walls (impermeable model). In conventional models, it is simply assumed that the pore pressure around the borehole is the same as the well pressure for the penetrating model, or the virgin pore pressure for the impermeable model. However, these assumptions are not correct for the coupled cases since the pore pressure field is disturbed during drilling or pumping/injection, especially in the region of abnormal tectonic stresses or for horizontal wells.

For the penetrating model or the cases of very permeable walls, the pore pressure in a thin layer around the borehole wall reaches the well pressure very

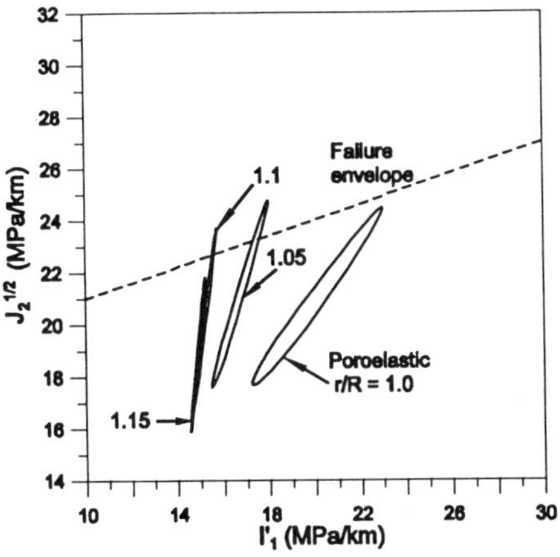

Figure 4: Stress clouds at various radial distances.

quickly. With a low well pressure applied, and a short period after drilling, the pore pressure inside the borehole wall may be significantly higher than the well pressure due to Skempton's effect.[18,9] Hence, the peak values of the pore pressure may occur at a short distance inside the borehole wall. On the other hand, for the case of impermeable walls, the maximum variation of the pore pressure around the borehole always occurs at the borehole wall. Therefore, the borehole failure may initiate inside the borehole wall for permeable walls, but it always starts at the borehole wall for impermeable walls

Fig. 4 presents shifting of the stress cloud at $t = 0.001$ day in the ($\sqrt{J_2}$-I_1') space with the change of the ratio r/R for a penetrating model. The conditions applied here are the same as the ones for the former case discussed in the previous section. It is observed that: at the borehole wall ($r/R = 1$) or distant from the borehole ($r/R > 1.15$) elasticity prevails; while the regions around $r/R = 1.05$, and 1.1 fails, indicating that shear failure may initiate at a distance inside the borehole wall at earlier times. This also suggests that a highly permeable wall may decrease the potential of shear failure around the borehole wall at small times.

Fig. 5 shows the effective fracturing stresses, $\sigma_{fr} = \min_{0 \leq \theta \leq 2\pi} \{-(\sigma_{\theta\theta} + p)\}$, varying with the mud weight. Since the tensile strength was assumed to be zero, $\sigma_{fr} > 0$ indicates a stable well, but $\sigma_{fr} < 0$ indicates fracturing. In this case, $\varphi_{z'} = \varphi_y = 0$; and, set 2 of in-situ conditions was applied. In this figure, comparison between the penetrating model and the impermeable model is presented. It is seen that the critical mud weight for fracturing for the pene-

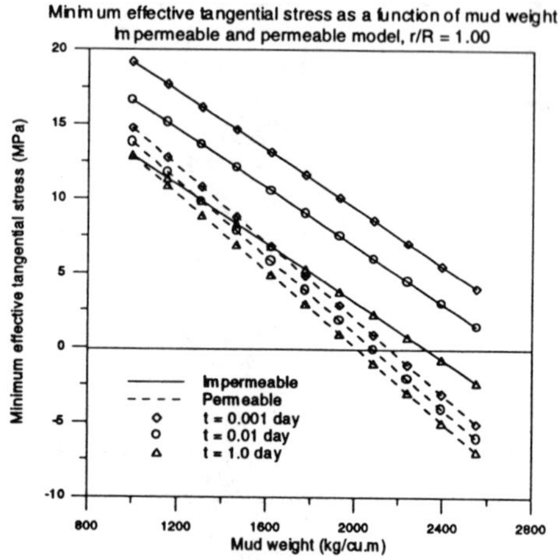

Figure 5: Effective fracturing stress varying with the mud weight for both permeable and impermeable models.

trating model is much lower than for the impermeable model. In other words, the more permeable the borehole walls, the higher potential for borehole fracturing. This is attributed to the fact that permeable walls may increase the total tangential stress ("back stress"),[19] if the mud pressure is higher than the virgin pore pressure.[17]

5 Conclusions

Several recently-derived poroelastic models of inclined boreholes were described in this paper. Based on these models, stability analyses of boreholes in fluid-saturated formations were conducted. The analyses showed that: (i) coupling of the pore fluid flow in saturated formations is a direct factor for the time-dependent borehole failure; (ii) at an earlier stage, highly permeable walls may reduce the failure potential around the borehole wall for a low mud pressure, indicating borehole failure may initiate inside the borehole wall; and, (iii) for the case of over-balance ($p_w > p_o$), however, an impermeable wall may reduce the potential of fracturing.

Acknowledgments

This work is supported by a National Science Foundation grant to the Rock Mechanics Research Center, the Oklahoma Center for the Advancement of Science and Technology, and the O.U. Rock Mechanics Consortium.

References

1. W.B. Bradley, *J. Energy Resour. Tech., ASME.*, **101**, 232-239 (1979).
2. E. Fjær, R.M.. Holt, P. Horsrud, A.M. Raaen, and R. Risnes, *Petroleum Related Rock Mechanics* (Elsevier Science, 1992).
3. M.A. Biot, *J. Appl. Phys.*, **12**, 155-164 (1941).
4. M.A. Biot, *J. Appl. Phys.*, **26**, 182-185 (1955).
5. L. Cui, A.H-D. Cheng and Y. Abousleiman, *J. Appl. Mech., ASME*, **64**, 32-38 (1997).
6. H.S. Carslaw and J.C. Jaeger, *Conduction of Heat in Solids, 2nd ed.* (Oxford Univ. Press, London, 1959).
7. H. Stehfest, *Comm. ACM.*, **13**, 47-49 and 624 (1970).
8. E. Detournay and A.H-D Cheng, *Int. J. Rock Mech. Min. Sci. & Geomech. Abstr.*, **25**, 171-182 (1988).
9. L. Cui, S. Ekbote, Y. Abousleiman, M. Zaman, and J.-C. Roegiers, to appear in *Proc. of 3rd North American Rock Mechanics Symposium* (Cancun, Quintana Roo, Mexico, June 3-5, 1998).
10. L. Cui, Y. Abousleiman, A.H-D. Cheng, V.N. Kaliakin and J-C. Roegiers, *Int. J. Numer. Anal. Methods Geomech.* **20**, 381-401 (1996).
11. M.A. Biot, *J. Appl. Phys.*, **27**, 459-467 (1956).
12. Y. Abousleiman, A.H-D. Cheng, C. Jiang and J-C. Roegiers, *Acta Mechanica*, **119**, 119-219 (1996).
13. Y. Abousleiman, L. Cui, S. Ekbote and J-C. Roegiers (manuscript under preparation).
14. X. Li, L. Cui and J.-C. Roegiers, to appear in *Proc. of Eurock'98* (Trondheim, Norway, June 8-10, 1998).
15. C.S. Desai and H.J. Siriwardane, *Constitutive Laws for Engineering Materials, with Emphasis on Geologic Materials* (Prentice-Hall, 1984).
16. D.C. Woodland, *SPE Drill. Eng.*, **5**, 27-33 (1990).
17. L. Cui, Y. Abousleiman, A.H-D. Cheng and J-C. Roegiers, submitted to *J. Energy. Res. Tech.*
18. A.W. Skempton, *Géotechnique*, **4**, 143-147 (1954).
19. A.H-D. Cheng, Y. Abousleiman and J-C. Roegiers, *Int. J. Rock Mech. Min. Sci. & Geomech. Abstr.*, **30**, 1119-1126 (1993).

ADVANCES IN THE CREEP COMPLIANCE THEORY OF VISCOELASTIC SURROUNDING ROCK MASSES

LIN XUE , RONG LUO
Qingdao Institute of Architectural Engineering
Qingdao 266033 P.R.China

This paper studied the regularity and creep limit of generalized creep compliance for all viscoelastic models. The reduction formula of the basic formula of creep displacement in viscoelastic plane question is discovered. A new method of investigation viscoelastic question can be found out by the creep compliance theory, and meanwhile, the quantitative relation of results of field and laboratory creep tests has been established. The new theory and testing method will play an important part in the structured media of rock masses.

1 Guidclincs

1.1 *Viscoelastic Displacement Theory Bring Forward*

In the past two years, authors successfully contributed the displacement theory of viscoelastic rock mass mechanics[4]. The basis of new theory is the viscoelastic continuum mechanics. Generally speaking, the viscoelastic problem can be solved in two different ways. The first is direct solution method, the second is worked out with elastic-viscoelastic correspondence principle. When the elastic solution had been known, choosing second method is simple. For a various of the elastic plane problem in the rock mass engineering, the elastic solutions had been known. Thus, the engineers usually choose the second method.

In elastic-viscoelastic correspondence principle, the Laplace transform method is used for the different problem of many viscoelastic models. But, for example, the solving viscoelastic problems of Burgers model are difficult to do because of a complex computation for Laplace transform. When the mechanical models of rock mass are simple and the shape of hole-room is complex, the computations of Laplace transform is complex or difficult, too. Thus, before then engineers can only do Laplace transform to find out the theoretical solution of circular hole-room for simple viscoelastic mechanical models.

When solving viscoelastic problem, new method need not Laplace transform with elastic-viscoelastic correspondence principle. The elastic solution can be directly applied to the viscoelastic displacement formula. As a result, solving the viscoelastic problem reduced to the solving corresponding elastic problem at plane state. Because the elastic solutions in many engineering problems have been worked out, new method of solving viscoelastic problem is simplest. It is more important for viscoelastic analysis to find out a new numerical method. Laplace transform is not taken into account when solving viscoelastic plane problem by the numerical method. Thus, these answers are well adapted to all viscoelastic models in the plane problem with the any shape pore. Where, Burgers model is added to. It is more import development of the analysis and back analysis for viscoelastic displacement in rock mass underground engineering.

The displacement theory of viscoelastic rock mass is divided into two parts, the first one is the creep compliance theory and its reversion theory of viscoelastic continuum mechanics for surrounding rock masses, the other one is the generalized stress and strength theory of plane deformation state for surrounding rock masses[1~7].

Because of space, this paper studied new advances in the creep compliance theory. It is often spoken of a center of the displacement theory for viscoelastic rock mass.

1.2 *Basic Concepts and Formula of Viscoelastic Displacement*

In the light of assumption of continuum, the basic concepts employed of displacement formula are creep compliance, generalized creep compliance and two generalized stresses. They are new physical quantities in viscoelastic mechanics.

The complex representation of viscoelastic displacement in plane problem is expressed as follows[1,4]

$$u(t) + iv(t) = \chi_1(z)J_1(t) + \chi_2(z)J_2(t) \tag{1}$$

Where, $u(t)$ and $v(t)$ are the components of viscoelastic displacements.

$J_1(t)$ is creep compliance and $J_2(t)$ is generalized creep compliance of viscoelastic rock masses.

The creep compliance $J_1(t)$ and generalized creep compliance $J_2(t)$ may be expressed by using the constitutive equations of the five viscoelastic models. They are time-dependent functions.

The complex variable functions $\chi_1(z)$ and $\chi_2(z)$ are called the analysis functions.

There, $\chi_1(z) = \varphi(z) - z\overline{\varphi'(z)} - \overline{\Psi(z)}$

$\chi_2(z) = 6\varphi(z)$

Where, $\varphi(z)$ and $\Psi(z)$ are complex potential functions in the correspondence elastic problem.

The analysis functions can be solved by the method of complex variable function in elastic plane problem, and by the numerical method for elastic plane problem[6].

There is the basic formula of viscoelastic displacement for plane problem can be expressed as follows[5]:

$$\vec{U}(z,t) = \vec{P}_1(z)J_1(t) + \vec{P}_2(z)J_2(t) \tag{2}$$

Where, $U(z,t)$ is viscoelastic displacement vector of plane state.

It is vector function of time and positional coordinate in plane field.

$\vec{P}_1(z)$ and $\vec{P}_2(z)$ are generalized stresses of surrounding rock masses.

The generalized stresses are the vector functions of positional coordinate of the point z in plane field, they are the medium-independents, time-independents, are independents of each other.

1.3 *The Creep and Generalized Creep Compliance in Plane State*

In fact, in the formula(1), $J_1(t)$ is referred to the creep compliance of rock masses for plane state. It was deduced from the viscoelastic displacement formula in plane problem. But then, this creep compliance are the same functional expressions as the creep compliance $J(t)$. And, $J(t)$ is the creep compliance of corresponding viscoelastic constitutive models, namely $J_1(t) = J(t)$. Generally, the creep compliance $J_1(t)$ of corresponding five viscoelastic models in common use is expressed as follows[1,3]

Maxwell model: $J(t) = J_1(t) = \frac{1}{E} + \frac{t}{\eta}$

Kelvin model: $J(t) = J_1(t) = \frac{1}{E}[1 - exp(-\frac{t}{\tau_d})]$

Kelvin-Voigt model and Poynting-Thomson model:

$$J(t) = J_1(t) = \frac{1}{E_L} + (\frac{1}{E_e} - \frac{1}{E_L})exp(-\frac{t}{\tau_d})$$

Where, E_e is instantaneous elasticity modulus. E_L is long-term elasticity modulus.

Burgers model $J(t) = J_1(t) = \frac{1}{E_1} + \frac{t}{\eta_1} + \frac{1}{E_2}[1 - exp(-\frac{E_2}{\eta_2}t)]$

The generalized creep compliance $J_2(t)$ of viscoelastic rock masses is different in the state of plane stress and the state of plane strain. In the state of plane stress, the generalized creep compliance $J_2(t)$ is a time function.

The generalized creep compliance $J_2(t)$ of corresponding above five viscoelastic models may be illustrated as follows[1,3]

Maxwell model: $J_2(t) = \frac{1-2\mu}{2E}[1 - \upsilon exp(-\beta\frac{t}{\tau_r})]$

Where, τ_r is relaxation time.

β and υ are algebraic expression of Poisson ratio μ , namely,

$\beta = 2/(3-2\mu), \upsilon = (1-2\mu)/(3-2\mu)$

Kelvin model: $J_2(t) = \frac{\upsilon}{E}[1 - exp(-\frac{t}{\upsilon\tau_d})]$

Where, τ_d is delayed time.

Kelvin – Voigt model and Poynting-Thomson model:

$$J_2(t) = \frac{\upsilon}{\beta E_e + \upsilon E_L}[1 + \upsilon(\frac{E_L}{E_e} - 1)exp(-\frac{\beta t}{\tau_t} - \frac{\upsilon t}{\tau_d})]$$

Burgers model:

$$J_2(t) = \frac{1-2\mu}{2E_1}\{\frac{1}{\tau_p - \tau_Q}[A_Q exp(-\frac{t}{\tau_Q}) - A_p exp(-\frac{t}{\tau_p})] + 1\}$$

Where, $A_Q = \frac{\beta\tau_b}{\tau_Q} + \tau_Q - \tau_m$, $A_p = \frac{\beta\tau_b}{\tau_p} + \tau_p - \tau_m$

$$\tau_p = \tau_a + \sqrt{\tau_a^2 - \tau_b} , \quad \tau_Q = \tau_a - \sqrt{\tau_a^2 - \tau_b}$$

$$\tau_a = \frac{1}{2}(\frac{\eta_1}{\beta E_1} + \frac{\eta_2}{E_2} + \frac{\eta_1}{E_2}) , \quad \tau_b = \frac{\eta_1\eta_2}{\beta E_1 E_2}, \tau_m = \frac{\eta_1}{E_1} + \frac{\eta_2}{E_2} + \frac{\eta_1}{E_2}$$

In the tunnel with any shape, the analysis functions above can be solved by the method of complex variable function in elastic plane problem, and by the numerical method for elastic plane problem[6]. According to the formulae of the creep compliance and generalized creep compliance above, providing correspondence elastic solution is known, the viscoelastic displacement solutions of all viscoelastic models are worked out with Eq. (1) at the same time.

1.4 *The Generalized Stresses of the Point in Plane Field*[4,5]

The magnitude and direction of the generalized stresses are closely related with stress condition and geometrical condition of boundary in plane field. Most important, the generalized stresses are the vector functions of positional coordinates of the point in plane field. They have nothing to do with physical property of surrounding rock masses. Thus, the two generalized stresses are be calculated on the principle of elastic mechanics. The considerable simplification in solving the viscoelastic questions can be achicved.

First of all, the values of analysis functions $\chi_1(z)$ and $\chi_2(z)$ about a point z in the plane field are solved. After the values of two analysis functions are worked out, the two generalized stress components can be determined by the simple operation for complex number.

As a result, can be found the viscoelastic displacement of the surrounding rock masses about all models, providing the creep compliance, the generalized creep compliance and the generalized stresses are known. Where, the unit of the generalized stresses is Mpa. m, the unit of the creep compliance and generalized creep compliance is 1/Mpa, the unit of the viscoelastic displacement is m.

Generally, the generalized stress $P_1(z)$ is about the size of the generalized stress $P_2(z)$.

1.5 *The Viscoelastic Principal Displacement*[4]

In the formula(2), let

$$\vec{U}_1(z,t) = \vec{P}_1(z)J_1(t) \tag{3}$$

Where, $\vec{U}_1(z,t)$ is referred to the principal displacement in viscoelastic plane state. $\vec{P}_1(z)$ is referred to the principal generalized stress for plane problem.

This problem is very easy. The magnitude of the principal displacement vary as the value of the creep compliance $J_1(t)$ at time t, and vary as the magnitude of the generalized stress $\vec{P}_1(z)$ at point z. The viscoelastic principal displacement is in the same direction as the principal generalized stress. The direction of the principal generalized stress is called the principal direction of the creep displacement in viscoelastic plane problem.

2 Initial Value and Limit Value of the Generalized Creep Compliance[1,7]

The most important of which are the models of standard viscoelastic solid and Burgers model. The initial value and limit value of generalized creep compliance of the standard viscoelastic solid are expressed as follows

With $t = 0$, initial value is $J_2(0) = \upsilon/E_e$; with $t \to \infty$, limit value is

$J_2(\infty) = \upsilon/(\beta E_e + \upsilon E_L)$. There are initial value $J_1(0) = 1/E_e$, and limit value $J_1(\infty) = 1/E_L$, and $E_e > E_L$.

The creep compliance is different from the generalized creep compliance in the incremental quantities with the creep time. Generally, in plane strain state, the increment of generalized creep compliance is very small. Let $J_2(\infty) - J_2(0)$ is the incremental quantity of the generalized creep compliance from time $t = 0$ to $t \to \infty$. $J_1(\infty) - J_1(0)$ is the incremental quantity of the creep compliance from time $t = 0$ to $t \to \infty$. From analytical computation, $J_2(\infty) - J_2(0)$ is almost 0.1% ~ 3% at numerical value than $J_1(\infty) - J_1(0)$. Let the generalized creep compliance be also equal to a constant in plane strain state, and the approximate value of this constant is the initial value, that is $J_2(0) = \upsilon/E_e$.

The initial value and limit value of creep compliance and generalized creep compliance for Burgers body are expressed as follows

$$J_1(0) = 1/E_1 \qquad J_2(0) = \upsilon/E_1$$
$$J_1(\infty) = \infty \qquad J_2(\infty) = \upsilon/(\beta E_1)$$

At all time of creep deformation,

$$J_1(t) - J_1(0) >> J_2(t) - J_2(0)$$

therefore the creep effects for the generalized creep compliance can be neglected.

3 The Reduction Formula of Creep Displacement

For ease of the study out the creep law, it is necessary to subtract viscoelastic displacement vector from the initial creep displacement vector. And, from suppose $J_2(t) = J_2(0)$, the creep displacement vector of surrounding rock masses is expected as follows

$$\vec{U}_c(z,t) = \vec{U}_1(z,t) = \vec{P}_1(z) J_1^C(t) \tag{4}$$

Where, $\vec{U}_c(z,t)$ is the creep displacement vector of surrounding rock masses.

$J_1^C(t) = J_1(t) - J_1(0)$, it is the increasing part function of the creep compliance. The argument t is creep time, and $t > 0$.

The formula (4) is called reduced formula of the viscoelastic displacement in plane problem.

4 The Similarity of the Creep Displacement-time Curve

The viscoelastic displacement-time curve for the surrounding rock masses is subdivide into three ranges: transient creep, transition creep and stationary creep. In the transient creep range, the viscoelastic displacement-time curve is rectilinear creep. In fact, the rectilinear creep agrees with elastic law. In the transition creep range, the principal creep responses are the structured media and dislocation creep for structural rearrangement. The transition creep is composition creep. In stationary creep range, the viscoelastic displacement-time curve is curve creep. Its creep law is accorded with the ideal viscoelastic body.

If ratio between one function and other function is constant at any given value in determinate region, the one curve of one function is similar to other. From Eq. (4), at any given point z, the ratio between magnitude of displacement vector and the increasing part function are constant with all creep time. It goes without saying, displacement-time curve is similar to time-function curve of increasing part of creep compliance. Where, irrespective of initial value of transient creep, the creep displacement vector done is equal to the product of the increment of the creep compliance and principal generalized stress vector. The magnitude of the creep displacement vector always increases with the magnitude of the principal generalized stress vector.

As a whole, for the surrounding rock masses with the complex structured media, the displacement-time curve in the stationary state is accorded with the stationary creep curve of the ideal vicoelastic body. In the transition range, the curve surge is bigger.

From the displaccment-time curve in the stationary state, the Eq. (4) can be applicable to the surrounding rock masses with the complex structured media.

5 The Quantitative Relation of Results of Field and Laboratory Creep Tests[5]

5.1 *The Physical Meaning of Generalized Stresses*[5]

What do the generalized stresses mean? They are the action functions in boundary stresses of plane inferior field. These action functions can affect the deformation of surrounding rock masses in three ways: magnitude, direction and application point. The functions are described as the property of stress boundary condition and geometrical boundary condition in physical plane. They are said as the action of mechanics. The action functions are called the generalized stresses. Where, $P_1(z)$ is the principal generalized stress at point z in plane field. $P_2(z)$ is the 2nd generalized stress at the same point z in plane field. They are independent of each other.

The action of 2nd generalized stress also relates to cross creep property. In other words, it also relates to generalized creep compliance. Assume the generalized creep compliance is constant. Then taking creep property at the point in plane field, the action of 2nd generalized stress may be omitted.

In Eq. (4), the value of the creep displacement increases with the value of principal generalized stress. With respect the creep displacement, it is the same orientation of the principal generalized stress. As a result, the principal generalized stress is a measure of creep response in plane field. Assume it is the generalized force on a unit of area.

5.2 *The Generalized Stress in the State of Uniaxial Stress.*

In the state of uniaxial stress, the viscoelastic displacement of the rock body in test is expressed as follows[5]

$$u(x,t) = x\varepsilon(t) = \sigma x J(t) = P(x)J(t) \tag{5}$$

Where, $P(x) = \sigma x$, it is called generalized stress in the state of uniaxial stress. $J(t)$ is creep compliance in the state of uniaxial stress. σ is the stress of loading. x is the axial length of the rock body in test.

5.3 *The Quantitative Relation of Results of Field and Laboratory Creep Tests*

The problem of field and laboratory test must be solved. Above all, the field test in soft surrounding rock masses is always accompanied with some creep test of laboratory. The rock bodies in laboratory test are selected according to rock masses in engineering field. These rock bodies and rock masses are alike in creep physical property of materials. Suppose rock masses are the isotropic solid and continuum.

In the field and laboratory test, let $J_1(t) = J(t)$ in some physical condition, and then,[5] let $P(x) = P_1(z)$. The equivalence relation of creep displacement can be described as follows

$$U(x,t) = P(x)J(t) = P_1(z)J_1(t) = U_1(z,t) \tag{6}$$

Where, $P_1(z)$ is the principal generalized stress at any point z in plane field of surrounding rock masses.

In the laboratory test, the stress of loading σ may be selected as follows[5]

$$\sigma = \frac{P(x)}{x} = \frac{P_1(z)}{x} \tag{7}$$

In the field and laboratory tests, the effect of the generalized creep compliance is negligible in view of the desired precision of the final result. Hence, in two tests, the relation of viscoelastic displacement is obtained that can be described as follows

$$U(x,t) = U_1(z,t) = U_c(z,t) \tag{8}$$

With the Eq. (7) and Eq. (8), the quantitative relation of field and laboratory creep test can be found.

In Eq. (8), the axial length x measures is too small that the stress of loading σ is very large. The rheological response at which viscoelastic body depends on the pressure. When the change of pressure is very large, the constitutive relationship of rock can change. At this time, Eq. (8) could not be used. So, make use of the following method. The similarity of the creep displacement curves is the basis of the quantitative relation in field and laboratory tests. From the above description, the creep displacement-time curve of laboratory is contrast with the creep displacement-time curve at the fixed point in field test. Getting a new hypothesis, the curve of creep displacement-time at any fixed point in field tests is similar to that in laboratory tests. Many facts can prove the above description. Let $J_1(t) = J(t)$, the displacement ratio between $U_c(z,t)$ and $U(x,t)$ in field and laboratory tests is expressed as follows

$$\gamma_d = \frac{U_c(z,t)}{U(x,t)} = \frac{P_1(z)J_1(t)}{P(x)J(t)} = \frac{P_1(z)}{P(x)} \tag{9}$$

The displacement ratio γ_d is the constant for given point. From the above formula, the displacement ratio γ_d at any fixed point can be found providing its generalized stress $P_1(z)$ and generalized stress $P(x)$ of the rock bodies in uniaxial tests are known.

The creep displacement at this point in the surrounding rock masses is expressed as follows

$$U_c(z,t) = \gamma_d U(x,t) = \gamma_d P(x)J(t) \tag{10}$$

Where, the physical conditions of the environment in laboratory are important for the experimental results accord with the theoretical calculations.

For example, the circuit tunnel, radius is $2m$. q is vertical stress in infinite boundary. $q = 9Mpa$. Lateral pressure coefficient is 0.5. At the apex point, the generalized stress is $P_1(a, \frac{\pi}{2}) = qa = 18Mpa.m$. At lateral point, the generalized stress is $P_1(a, 0) = \frac{1}{2}qa = 9Mpa.m$. In laboratory test, the rock body is cube. The axial length is 10 cm. For apex point in the tunnel, if $\gamma_d = 10$, let the stress of loading is $\sigma = 18$Mpa. If $\gamma_d = 20$, let the stress of loading $\sigma = 9$Mpa.

As a result, the creep displacement at the apex point in surrounding rock masses is $18J(t)$. In the same way, let stress of loading be $\sigma = 18Mpa$, or $9Mpa$. As a result, $\gamma_d = 5$, or $\gamma_d = 10$. The creep displacement at lateral point is $9J(t)$. Where, the unit of displacement is m. The unit of creep compliance is $1/Mpa$.. The value of $J(t)$ at all time is achieved by the back analysis solving method for displacement in laboratory rock mechanics test.

6 Conclusion

This paper is working out two conclusions. First, the solution of viscoelastic plane problem is reduced very much. According to the creep compliance and generalized creep compliance of viscoelastic surrounding rock masses, the viscoelastic plane problems with all viscoelastic models is solved by means of elastic mechanics. Second, the quantitative relation of results of field and laboratory creep tests has been achieved. But, new method mus be greatly improved, and will be applied to rock engineering.

References

1. Theorem of Identification for Viscoelastic Model of the Rock Mass and its. Applications Lin Xue (Chinese Journal of Geotechnical Engineering, P. R. China, 1994).
2. Criteria for Mechanical Model of Viscoelastic Body and Their Applications in Rock Tests Lin Xue (Rock and Soil Mechanics, P. R. China, 1996).
3. Study on Viscoclastic Meehanical Model Rccognition of Rock Mass and Methods of Parameter Inversion Analysis Lin Xue (Journal of Engineering Geology, P. R. China, 1995).
4. Displacement Theory of Rock Mass Mechanics and Their Engineering Applications Lin Xue (Pock Mechanics and Engineering for 21 Century, P. R. China, 1996).
5. The Theory of Generalized Stress and Application in Rock Test Lin Xue (Mechanics and Practices, P. R. China, 1996).
6. Applications of Boundary Elcment Mcthod of Displacement Back Analysis of Tunnels in Viscoelastic Rock Mass Lin Xue and Rong Luo (The First Rock Mechanics Symposium, Korea, 1997).
7. Gencralized Earth Stress and Its Theory of Strength Lin Xue (Rock and Soil Mechanics, P. R. China, 1997).

Stability of Rock Engineering

STABILITY OF UNDERGROUND STRATIFIED HARD ROCK ROOFS

A. I. SOFIANOS
National Technical University,
157 80 Athens, Greece
E-mail: sofianos@metal.ntua.gr

A. P. KAPENIS
Edafomichaniki Ltd., 5A Delfon Street,
151 25 Maroussi, Greece
E-mail : mmepak@central.ntua.gr

Analytical formulae are developed for the evaluation of the mechanical response of an individual voussoir hard rock roof beam with cracks at midspan and the abutments. Further, a non-dimensional chart is given for the prediction of failure against the modes of buckling, crushing and sliding. This, relates the geometrical shape of the roof rock beam to the deformational and strength parameters of the roof rock. Data obtained from observed stable and failed underground stratified roofs are used to feed the stability chart with actual cases. The precompression necessary to inhibit interlayer sliding of bolted voussoir beam roofs is calculated for various strata configurations and lower bound formulae are given. Finally, the effect of partial precompression on the mechanical behaviour of bolted strata is investigated.

1 Introduction

The competence of the roof structure of an underground opening in mining and civil engineering operations determines the mining method, the dimensions of the opening, the choice of the support and more importantly the degree of safety. Roofs of such openings are commonly of stratified hard rock. Fayol[1] conceived the mechanical behaviour of such roof rock beds as similar to a stack of simply supported beams and he noted experimentally that the lowest stratum is not loaded by the upper ones; the gravitational load of the uppermost beams is clearly transferred laterally to the supports, a phenomenon called arching.

Initial attempts to analyse the behaviour of such an undermined stratum, include its analysis in bending as a linear elastic beam or slab with an allowable tensile strength. The deflection at midspan and the strain at the abutment, for such a clamped at both ends beam, is given by :

$$\delta_n = \frac{\delta}{t} = \frac{Q_n \cdot s_n^3}{32}, \qquad \varepsilon_{ax} = \frac{\sigma_{ax}}{E} = \frac{Q_n \cdot s_n}{2} \tag{1}$$

$$Q_n = \frac{Q}{E \cdot t} = k_q \cdot \frac{\gamma \cdot s}{E}, \qquad s_n = \frac{s}{t} \tag{2}$$

where,
E = Young's modulus of the rock
k_q = surcharge factor
Q = total load on the rock beam
s = clear span of the rock beam
t = thickness of the rock beam
γ = unit weight of the rock
δ = vertical deflection of the rock beam at midspan
ε_{ax} , σ_{ax} = extreme fibre normal strain and stress respectively at the abutment

The behaviour of multiple such roof beams tied together with rock bolts has been investigated by Panek[2]. He distinguished between suspension and reinforcement action of the bolts. Suspended strata and supporting strata may be evaluated as individual ones, by using an equivalent unit weight γ_{ieq} for each stratum i given by :

$$\gamma_{ieq} = \gamma \cdot \frac{E_i \cdot t_i^2 \cdot h}{\sum_k E_k \cdot t_k^3}, \qquad \Delta q_i = (\gamma_{ieq} - \gamma) \cdot t_i \tag{3}$$

Tensioned rock bolts are commonly used to reinforce the strata, by increasing the frictional shear strength within the interlayers. The precompression pressure required to inhibit slip within an interlayer with friction angle φ_j, is given by Goodman[3] :

$$p_b = \left[\frac{3 \cdot \gamma \cdot s}{4 \cdot \tan \phi_j} + \Delta q \right] \tag{4}$$

where Δq is the interlayer distributed load defined in eq. 3.

However, the tensile strength of a roof beam is common to be exceeded and tensile cracks may be induced in the lower fibre at the midspan zone and in the upper fibre at the abutments. Such a beam ceases to behave further as a continuous one, and its response may further be simulated as a discontinuous structure composed of individual blocks named voussoirs. Evans[4], in his memorable paper, advanced the theory of the voussoir beam to analyse the stability of roof strata that are far too broken and cracked transversely to act as simple beams. Failure of this beam was considered to be achieved in three modes, i.e. crushing of the rock at midspan or the abutment, buckling of the beam, and sliding of the blocks at the abutments. By making arbitrary assumptions in order to solve his indetermined model, he finally produced diagrams, that related the lowest stable thickness of the beam to the span, for three elastic moduli and three values of the uniaxial strength of the rock. Wright[5] developed also a design procedure for the control of layered undermined roofs. This procedure was based on a large number of finite element model computer runs and on laboratory physical model tests; these evaluated a contact length at the joints of an individual beam of the order of 20% of its thickness. Sterling[6] captured the salient features of the work of Evans and other researchers and by experimenting on his apparatus for testing the load-deflection behaviour of laboratory size rock beams, provided a coherent picture of the deformation and failure modes of roof rock. He similarly, evaluated a contact length of about 20% of the thickness of the beam, when in the elastic range, which increased before ultimate load was reached.

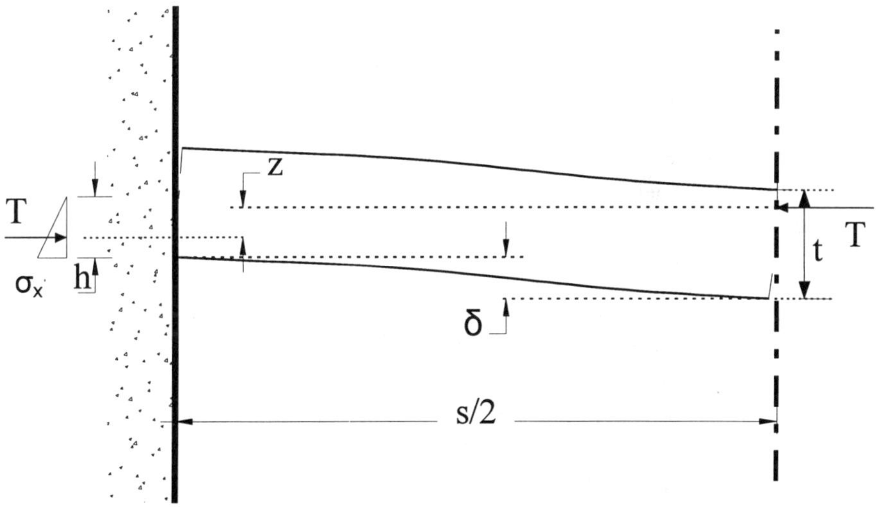

Figure 1. Left part of symmetric voussoir beam model

2 Individual voussoir beam roofs

The initial attempt, to evaluate numerically the stability of a limited number of individual voussoir beam configurations as discontinua with significant displacements, was performed by modelling them as distinct blocks (Sofianos et al.[7]) and using the discontinuum computer code UDEC™. An individual voussoir beam model, such as shown in Figure 1, is an indetermined structure. By applying static equilibrium, kinematic compatibility, and Hooke's law as the constitutive equation, and by assuming a linear distribution of the strain at the critical sections, i.e. at the abutment and the midspan, the deflection at midspan and the strain at the critical sections may be evaluated (Sofianos[8]) by :

$$\delta_{zo} = \delta_z \cdot (1-\delta_z/2)\cdot(1-\delta_z) = \frac{Q_n \cdot s_z}{16\cdot k_1 \cdot \chi}\cdot\left[s_z^2 + k_1\right] \leq \frac{1}{3\cdot\sqrt{3}} \approx 0.192 \tag{5}$$

$$\varepsilon_x = \frac{\sigma_x}{E} = \lambda\cdot\varepsilon_{ax} = \frac{Q_n \cdot s_z}{4\cdot n\cdot(1-\delta_z)} \tag{6}$$

$$\delta_z = \frac{\delta_n}{z_{on}},\quad s_z = \frac{s_n}{z_{on}},\quad z_{on} = \frac{z_o}{t} = 1-\frac{2}{3}\cdot n,\quad z_o = z+\delta,\quad n = \frac{h}{t} \tag{7}$$

where,
h = equivalent mean contact length at the critical sections
z = lever arm of the horizontal thrust couple
λ = abutment extreme strain ratio; $\lambda\in[0.3,0.8]$; $\lambda_{mean}\approx 0.6$
ε_x , σ_x = equivalent extreme fibre normal strain and stress respectively
χ = normalised (divided by t) equivalent thickness of the voussoir arch
k_1 = parameter characterising the shape of the thrust line; $k_1\approx 8/3$

The indeterminate parameters χ and n have been evaluated statistically, from numerical simulations on a variety of tested models (Wright[5], Sofianos et al.[7]), for roof rock beams for which the following assumptions are made :

- Rock is homogeneous, isotropic and elastic
- The beam is horizontal and symmetric, uniformly loaded, and has only three vertical joints, one at the midspan and one at each of the abutments
- The abutment supports are rigid and the joints are very stiff
- The rock Poisson's ratio is zero
- There is no lateral stress confining the beam prior to its deflection.

For such beams, the indetermined parameters may be obtained by :

$$k_1\cdot\chi = 1$$
$$n = \frac{1}{0.22\cdot s_n + 2.7};\quad n\in[0.1,\ 0.3] \tag{8}$$

Values of n however do scatter from the values defined by eq. 8.

If the deflection attains its maximum stable value defined by its upper limit given in the right of eq. 5, the state of incipient buckling is reached; i.e. max δ_z=0.42, max δ_n<0.40. This defines in the chart of Figure 2 the buckling line "FS_b=1" which separates, in the s_n, Q_n space, safe regions in buckling below it, to unsafe ones above it. Another buckling line "FS_b=2" is drawn in the chart for a factor of safety of 2.

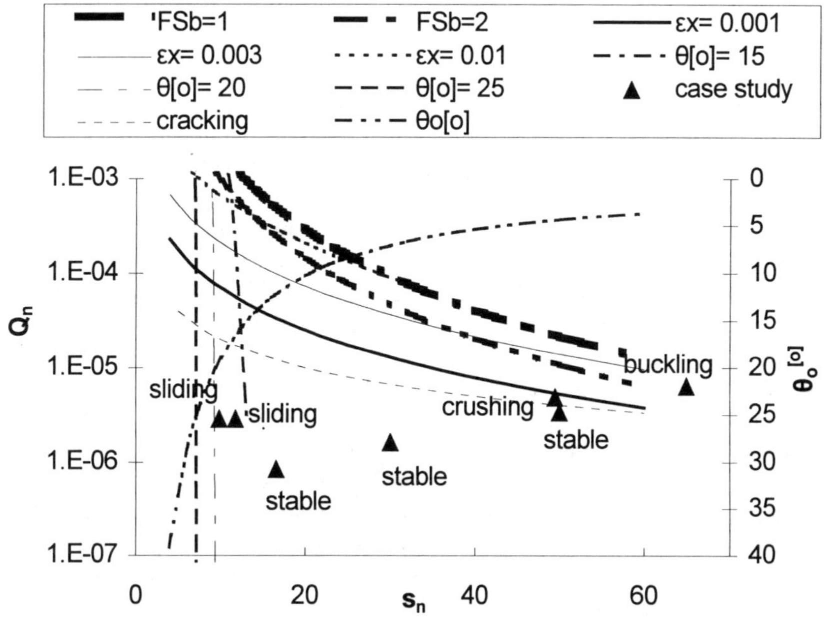

Figure 2. Design chart

In the same chart, by using eqns. 6, 5 and 8, contours are drawn for three constant values of the strain ε_x. Each contour delineates above it an unsafe region in crushing due to exceeding a maximum allowable equivalent strain, to a safe one below it. Further, the condition of incipient cracking of the continuum beam is represented in the chart by the "cracking" line, based on equation 1, for an allowable tensile strain ε_{ax} of 0.0001.

The third mode of failure is slip along a vertical joint at an abutment. This may happen if the shear force there becomes larger than the shear strength of the joint. If the strength of the joint is purely frictional, then a safe against sliding inclination angle θ, of the reaction at the abutment to the horizontal, has to be smaller than the friction angle φ_j of the joint. The inclination angle θ and its limits θ_o and θ_b, which correspond to zero deflection and at buckling respectively, are given (Sofianos[8]) by:

$$\tan\theta_o = \frac{4}{s_z} \geq \tan\theta = \frac{Q_n}{\varepsilon_x \cdot n} \geq \frac{2.32}{s_z} = \tan\theta_b \qquad \tan\theta \leq \frac{\tan\phi_j}{FS_s} \tag{9}$$

where, FS_s is a factor of safety against sliding. In the chart of figure 2 contours are drawn for three values of θ, whereas the $s_n=s_n(\theta_o)$ function, defined in eq. 9, is also included. These subvertical contour lines of θ, do separate, safe in the right to unsafe in the left, regions in the s_n, Q_n space of the design chart.

Thus, the buckling line, and the pertinent crushing strain and slip lines may define an area in the s_n, Q_n space within which it is safe against the three mentioned modes of failure. This chart, nevertheless, does not allow for any compliance at the abutments. From a number of model tests, it has been found (Sofianos and Kapenis[9]) that as the abutment width is increased, i.e. its compliance is increased, the deflection increases and the factor of safety against buckling decreases, and both tend asymptotically to a limit. In order to account for this behaviour, an equivalent larger span or a lower elastic modulus may be assumed. From the models considered, a reduction in the factor of safety against buckling of the order of no more than 10% has been evaluated.

Validation of the chart in-situ has started in a Greek bauxite mine exploited with the room and pillar method. Roof rock there consists of layered limestone; its deformability modulus ranges from 55 to 80GPa, whereas its crushing strain may be estimated to range from 0.001 to 0.002. Room spans range from 2.5 to 13m, and strata thicknesses from 0.16 to 0.86m. A case study (Kotinis et al,[10]) made there on the response of the roofs of the excavated rooms, provided the data shown in the chart of Figure 2. Stable conditions are drawn within the safe region enclosed by the pertinent failure lines, whereas the observed roof fall cases, due to buckling, crushing and sliding, are drawn in the fringes of that region.

3 Reinforced multilayered voussoir beam roofs

If the individual voussoir rock beams, treated in the previous section, are unable to stand up alone, reinforcement of multiple strata with tensioned rockbolts may be used to increase the shear strength within interlayers and hence increase their joined bearing capacity. The average precompression pressure p_b in an interlayer, required to inhibit slip between two strata may be estimated by :

$$p_b = \frac{2 \cdot T}{s \cdot \tan\phi_j} = \frac{Q_n \cdot E}{4 \cdot z_{on} \cdot (1-\delta_z) \cdot \tan\phi_j} \approx \frac{Q_n \cdot E}{3.5 \cdot \tan\phi_j} \tag{10}$$

where, T is the lateral thrust acting on the bolted strata at the abutment. The right part of the equation holds for small deflections only. For a two bed beam, the average precompression pressure p_{bb} for conditions close to buckling, is given (Sofianos and Kapenis[11]) by :

$$p_{bb} \approx \frac{E}{s_n^3 \cdot \tan\phi_j} \geq p_b \tag{11}$$

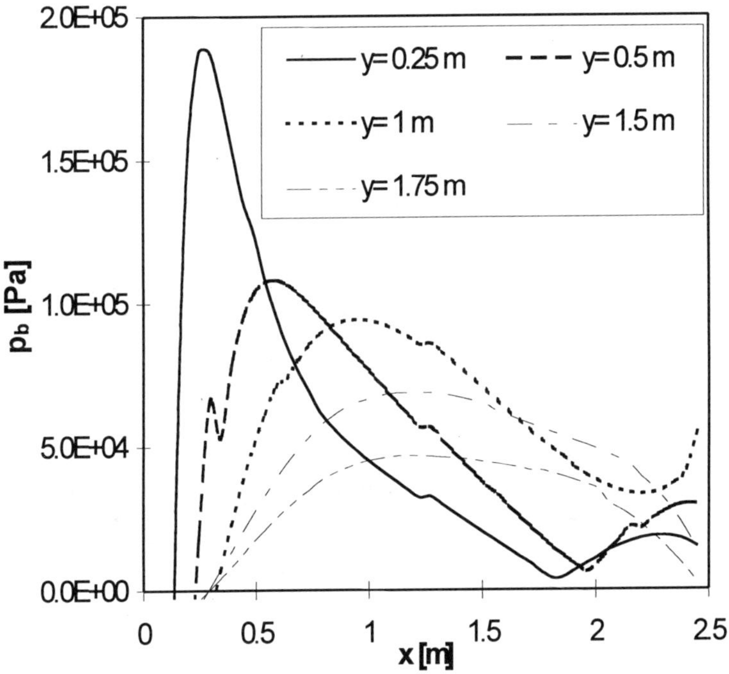

Figure 3. Required precompression in order to inhibit slip
s=5m, t=2m, γ=30kN/m^3, E=8GPa, φ_j=30°

These precompression pressures are very large and may not be used in common practice. Further, the required precompression evaluated by the formulae above is an average value of its distributed value along the bolted strata. In Figure 3 the necessary precompression, to prevent slip between interlayers which possess friction angle φ_j=30° and no cohesion, is drawn along the half length of a 5m long and 2m thick voussoir beam at various interlayers located at distance y from the lower fibre. It may be seen that the maximum required precompression close to the abutment is much larger than the average of the distributed ones along each layer; across the beam the envelope, of the required precompression distributed along each interlayer, has to be satisfied. Further, it may be observed that even for a 0.5m thick lower stratum, the necessary precompression pressure close to the abutment is ~110kPa, which is close to an upper limit of the ability of the rock bolts to provide it. In table 1, the precompression required to inhibit slip for three span configurations, is evaluated numerically at the level of two interlayers for two values of their friction angle φ_j; a precompression lower bound is also calculated according to eq.10. It may be observed that complete precompression of these strata requires large pressures, which may practically be not feasible.

Table 1. Precompression p_b[kPa] required to inhibit slip; t=2m, E=8GPa,γ=30kN/m^3

s [m]=	5		10		20	
y [m]	φ_j=30°	φ_j=40°	φ_j=30°	φ_j=40°	φ_j=30°	φ_j=40°
0.25	190	90	450	260	1000	520
0.50	110	70	310	170	640	410
eq.10	74	51	148	102	297	204

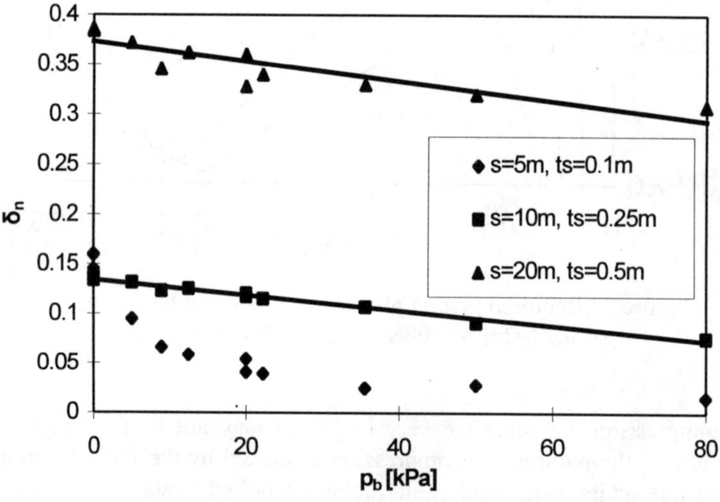

Figure 4. Reduction in deflection vs. precompression
t=2m, E=8GPa, γ=30kN/m^3, φ_j=30°

However, by partially inhibiting sliding, due to partial precompression, an improved behaviour may be achieved. In Figure 4, the partial bolt pretension applied to three multilayered voussoir beam configurations 2m thick each, with span s, 5, 10 and 20 m and strata thickness t_s, 0.1, 0.25 and 0.5 m respectively (Sofianos and Kapenis[12]), is found to reduce significantly the deflection of the beam, albeit sliding between interlayers has occurred. However, for the particular studied cases, the maximum stress at the midspan and at the abutment is not particularly affected. Finally, for such beams, the deflection and consequently the factor of safety against buckling is found (Sofianos and Kapenis[13]) to depend highly on the thickness t_s of the individual strata.

4 Conclusions

Closed form solutions are developed for the evaluation of the behaviour of an individual hard rock voussoir roof beam, bearing on certain assumptions. Based on these solutions, design charts are drawn for the visual estimation of the stability of an individual voussoir beam; these do separate the space of the non dimensional quantities s_n, expressing the geometry of the beam, and Q_n, expressing the load on the beam, into safe and unsafe regions in buckling, crushing and sliding. These, do provide very good estimates for the deflection of the roof and its factor of safety against buckling. Safety against crushing is established in terms of a maximum allowable equivalent strain, which may range practically from 0.001 to 0.01. However, further research is under development aiming to an improvement in the evaluation of the extreme strain and the contact length at the critical sections. It has to be noted here that beams crossed by many flexible joints, such as joints with infilling, will exhibit larger contact length at the abutments and the midspan, and may not be treated adequately by the proposed method. For such beams an analysis, such as developed by Beer and Meek[14] or Brady and Brown[15] might be appropriate; there, equations 5 and 6 may still be used, but with appropriate values for χ and n.

The chart is then fed with data obtained by observing the behaviour of some of the bedded limestone roofs at the Helikon bauxite mines; these are found to fit well in the pertinent regions of the chart. Collection of data is progressing nowadays in a larger scale at the same mine (e.g. Kotinis et al.[16]). Moreover, the chart is used also, in parallel to a conventional procedure, to apply an observational safe span evaluation approach in a pilot underground marble quarry (Sofianos, et al.[17]).

Multilayered voussoir beam models are used to examine the effect of the precompression on the increase in the bearing capacity of reinforced, with tensioned bolts, strata due to the increase of the shear strength of the interlayers. However, a large precompression force is required to totally inhibit slip within these interlayers, which for greater openings may practically be not feasible with bolts. Studies, on discontinuum models of partially precompressed bolted strata, illustrated the dependence of roof deflection on the precompression. Finally, lower bound formulae are given for an initial assessment of the necessary reinforcement system.

References

1. Fayol M. "Sur les movements de terain provoques par l'exploitation des mines. *Bull. Soc. l'Industrie Minerale,* **14**, 818 (1885).
2. Panek L.A. "Design for bolting stratified roof", *Trans. Soc. Min. Eng., AIME*, vol. **229**, pp. 113-119 (1964).
3. Goodman R.E. "*Introduction to rock mechanics*". Wiley, (1980).
4. Evans W.H. "The strength of undermined strata". *Trans. Am. Inst. Min. Metall.* **L(50),** 475-500 (1941).
5. Wright F.D. "Design of roof bolt patterns for jointed rock", *Bureau of Mines*, N.T.I.S., US Department of Commerce (1974).
6. Sterling R.L. "The ultimate load behaviour of laterally constrained rock beams", *21st US Symposium on Rock Mechanics*, pp.533-542 (1980)
7. Sofianos A.I., Kapenis A.P. and Tsoutrelis C.E. "Underground voussoir rock beams with large deflections", (*MJFR-2*), ed. Rossmanith, 639-643, (1995).
8. Sofianos A.I., "Analysis and design of an underground hard rock voussoir beam roof", *Int. J. Rock Mech.Min.Sci. & Geomech. Abstracts*., **33**, 2 (1996).
9. Sofianos A.I. & Kapenis A.P. "Effect of abutment compliance on the stability of an underground bedded roof formation". *Eurock'96*, 833-837.
10. Kotinis D.H., Sofianos A.I. and Economopoulos I.N. "Flat arch response of the bedded limestone roofs of the underground bauxite exploitations of Helikon". *3rd Hel. Conf. on Geot. Eng.,* **2**, 185-191, Patra (1997), in Greek.
11. Sofianos A.I. & Kapenis A.P. "Bolted voussoir rock roof beds with large deflections", *MPES 95*, Calgary, 1069-1073, Balkema (1995).
12. Sofianos A.I. and Kapenis A.P. "Bolt pretension of an idealised underground bedded roof formation", *Proc. Int. Conf.,* Wollongong, Australia, (1998).
13. Sofianos A.I. & Kapenis A.P. "Effect of strata thickness on the stability of an idealised bolted underground roof". *MPES 96*, 275-279, Balkema (1996).
14. Beer G. and Meek J.L. "Design curves for roofs and hanging walls in bedded rock based on voussoir beam and plate solutions", Trans. I.M.M., 91 (1982).
15. Brady B.H.G. and Brown E.T. "Rock Mechanics for underground mining", George Allen & Unwin, London (1985).
16. Kotinis D.E., Sofianos A.I. and Kapenis A.P. "Analysis of a roof failure in a bauxite mine". *MPES 97*, 347-352, Chech Republic, Balkema (1997).
17. Sofianos A.I., Kapenis A.P. and Rogakis C. "Voussoir beam analysis of an underground marble quarry", *MJFR-3*, Vienna, Austria, April 4-9, (1998).

DEFORMATION AND FRACTURING IN UNDERGROUND MINING

A.VERVOORT

Department of Civil Engineering, KULeuven, W.de Croylaan 2, B-3001 Heverlee, Belgium
E-mail: andre.vervoort@bwk.kuleuven.ac.be

Around mining excavations, various types of induced fractures are observed. These fractures can be classified based on different criteria: on the extent of the fractured zone, on the type of fracture or on the cause of the stresses responsible for creating a fracture. The cause of an increase in the stresses can either be attributed to a stress redistribution or to a movement or at least a tendency of a part of the rock mass to move. Apart from the mining induced fractures, natural fractures are also present. The combination of all these fractures has a negative effect on the stability of the excavations, but it can have a positive effect on the mining process and hence on the global behaviour.

1 Introduction

For non-mining applications, the main aim of rock engineering is to limit the deformation and the fracturing of the rock surrounding the excavations as much as possible. This has a direct and positive effect on the stability, the safety and the support requirements. For mining excavations, the picture is much more complex. The presence of induced fractures or large deformations is not desirable from the stability and safety point of view. This is not necessarily so from the mining point of view. For several mining methods (the so called caving methods like for example longwall, sub level caving (SLC) and block caving), collapses must be induced by the redistributed stresses for either support or production purposes. These collapses should occur in a controlled way, which means that they should happen at regular and frequent intervals, so that no large sudden collapses take place. As fractures and deformation are required to induce these collapses, they affect at the same time, but in a negative way, the stability of the excavations, which have to remain stable and safe, such as access tunnels and ore passes. For the mining methods that do not make use of caving techniques, the allowable deformations and the number of induced fractures are generally also larger than for non-mining applications.

The cause and the importance of fractures and deformation around mining excavations will be discussed in the paper and this will be illustrated by several examples from coal and metal mines. The choice of the examples is based on the personal experience of the author. It is not the aim of the paper to make an exhaustive list of mining examples, but rather to describe and illustrate the induced fractures (e.g. orientation and type) for well selected cases.

2 Different types of fractures

In non-mining applications, the natural fractures normally dominate stability studies; around mining excavations, however, the induced fractures are often playing the most important role.

2.1 Natural fractures

The natural occurring fractures are described by intrinsic and non-intrinsic characteristics, namely characteristics which are respectively independent and dependant on the measuring and sampling method. If large surfaces are exposed, the various fracture patterns can be mapped. On the other hand, if one has only access via boreholes to a rock mass, the available information is limited. Geophysical borehole techniques can give some additional information on fracture orientation and dip, but a limited amount of core samples is often the only hard data available. To get as much out of these samples as possible, one should try to measure as many parameters as possible. The Computerized X-ray Tomography (CT) is a non-destructive technique, which is very suitable to analyse samples before any destructive tests are carried out. In a CT scanner the object under investigation is placed in a circular tube at the centre of a revolving X-ray source and the X-rays impinge at different angles. CT-scans give a three-dimensional visualization of the core material and they allow as such to detect, classify and quantify various kinds of structural variations. Images generated by CT are based predominantly on the variation in density within the sample. In Figure 1, scans through a coal sample is presented. The CT-numbers, related to the density, are coded in grey shades. The darker the shade, the lower the density of the material; the lighter, the denser the material. On Figure 1, only natural fractures are present. The black regular fracture pattern corresponds to an open cleat system (filled with air), while the white fractures correspond to cemented fractures. Some small inclusions of anorganic material are also observed (high density). The white circumference observed is due to an artefact (F.Verhelst et.al.[1]). J.Caers et.al.[2] showed that by a detailed petrographic study combined with CT-scans, the large spread in rock mechanical test results could be explained. With the CT-images, one could distinguish massive, carbonate-cemented fractures from kaolinite-/carbonate-cemented fractures. Both groups have a significant different rock mechanical behaviour.

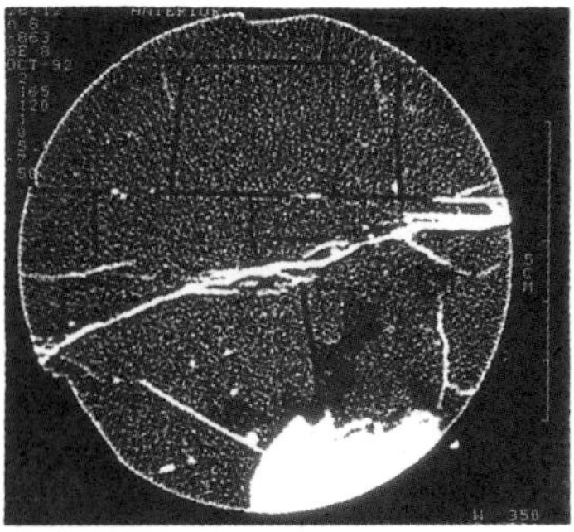

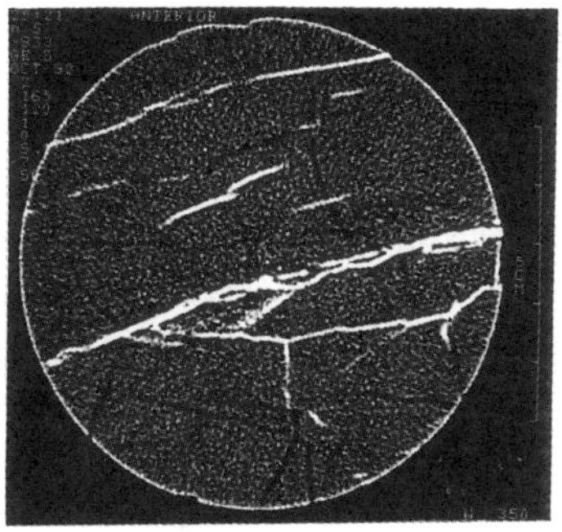

Figure 1: Computerized Tomography X-ray scans through a coal sample [diameter core sample: 80 mm].

2.2 Induced fractures

One can look at induced fractures from different angles.

First of all, there are two different stress conditions which induce fractures, namely fractures induced by tensile stresses and fractures which are formed by shear stresses. The first group consists of the direct tensile fractures, like observed in a bending test, and the extension or cleavage fractures, like for example observed in the Brazilian tensile test.

Another approach in studying induced fractures is the extent of the fractured zone. Either individual or single fractures are observed, or entire zones are completely fractured with dimensions of several decimeters or meters. If only a single fracture is observed, this means that the tensile or shear stress conditions were such that the rock material failed, but that due to the presence of this fracture, associated with or without movement, the redistributed stresses were not large enough to induce more fractures. However, if large fractured zones are observed, it either means that in the entire zone small fissures were induced, which grew with time, or it means that an individual fracture was induced, but that the redistributed stresses remained large enough to induce additional fractures. The distinction between the latter two is not always that clear, as the time effect is often obscured. First of all, time plays an important role in the damage and fracture flow rules, and secondly the prevailing stresses can vary with time for example due to a moving face. A supplementary factor playing a role at the moment of fracturing is the response of the rock, namely if one observes a brittle or ductile behaviour.

A third way of looking at fractures is the cause of the stresses, that are large enough to induce fractures. Firstly, due to an excavation, the original forces present inside the excavation have to be carried by the surrounding rock mass. Before and after excavating a volume, there must be an equilibrium between the total forces acting on any section through the rock mass and the excavation. Normally, this creates a significant increase in the compressive stresses acting in the rock mass surrounding the excavation. Secondly, large stresses can be caused by movement of the rock or at least the tendency to move. This creates mainly tensile stresses.

In the next paragraphs, regular reference will be made to the various approaches, mentioned above.

3 Fractures around longwall panels

The stress distribution and rock movement, including the development of the goaf, around longwall total extraction panels is complex. The mining method (see e.g. B.Brady and E.T.Brown[3] for a more detailed description) is typically applied in a tabular orebody, which is subdivided in rectangular panels with dimensions of for example 200 m by 1000 m. Starting along its short side of the panel, the ore is excavated parallel to this side in successive slices. As such, the face moves forwards until the entire panel is excavated. Around the excavated zone, stress peaks occur. When no backfilling is used, the peaks of the redistributed stresses reach after a certain advance (e.g. 150 m), such levels that the rock under these peaks fails. Not only the rock in the orebody fails, but also above and below the orebody. Hence, an extensive net of induced fractures are present in the roof and a regular goaf occurs when the roof is exposed in one of the successive excavation steps. Due to the presence of the goafed area, an important part of the excavated area carries part of the overburden and the pressure on the rock mass ahead of the face remains limited. After an advance of typically 200 m, one has reached a stationary state (stress redistribution, creation of fractures, goaf, etc.). With each mining step, the stress peak and the extent of the fractured zone move forward. The horizontal stresses in the roof close to the mined out area are mostly much smaller, as the goaf applies no or a limited pressure on the strata above the face. Ahead of the face, typical extension type of fractures are induced (Gramberg J.[4]), due to the large compressive stresses in a vertical or sub-vertical direction and the limited compressive stresses in the perpendicular direction. These fractures are parallel to the major principal stress and hence they are either vertical or inclined towards the face advance direction. They are always parallel to the face.

Between the non-excavated orebody and the goaf, a zone of 5 to 10 m has to be protected, so that the ore can be excavated under safe operating conditions. For this reason, powered roof support is installed along the longwall face. A second type of fracture is induced in the roof in the area close to the face. It is a shear type fracture, parallel to the face, whereby the inclination can be either vertical, inclined towards the face advance direction or inclined towards the goaf (Vervoort A.[5]). Although shear stresses are the cause of this type of fractures, they are mainly caused by a tendancy of the roof to move towards the goaf area. This movement is enhanced if weak support is installed along the face (Vervoort A.[6]).

Due to the presence of both types of fractures in the roof, the roof above and in front of the powered longwall support is entirely fractured and composed of relatively small blocks. The dimensions of these blocks are of order of magnitude of centimeters to decimeters. In Figure 2, a typical example of such a block is presented. It was taken out of the roof in a Belgian coal longwall at a depth of around 800 m. In the middle, a general view of the block is presented, while at left a closer view of the extension type fracture is presented and at right the shear type of fracture is presented. The surface of the shear fractures is characterized by white dust, caused by the relative movement. The surface, if touched, feels silky or talcous. Even though the surface is not necessarily plane, the surface variation is very regular (smooth). The surface of the extension fracture is much more irregular (rough); normally, no relative movement has occurred, retaining the original fracture surface. On the latter, fan-shaped striations are sometimes observed.

In Figure 3, an example of a block, taken in the roof of a salt longwall operation, is presented. The left side is a straight clean cleavage fracture, while the other side that is pictured is more irregular and not as clean due to the shear movement.

Figure 2: Picture of a roof block taken in a coal longwall [at left, extension type fracture; in the middle, general view of block; at right, shear type fracture].

4 Fractures in room and pillar excavations

In comparison to longwall mining, the movement and stress redistribution in and around room and pillar panels is much smaller. The mining method is relatively simple. On a mostly regular mesh, square or rectangular pillars are left; the area around the pillars is mined (see e.g. B.Brady and E.T.Brown[3] for a more detailed description).

Figure 3: View of a roof block taken in a salt longwall.

In the discussion, a clear distinction must be made between typical coal mine and metal mine room and pillar panels. In the coal mine room and pillar excavations, the maximum room width is firstly determined (e.g. 6 m wide), followed by the calculation of the minimum pillar width to carry the overburden. In typical metal mines, the approach is opposite: first, the smallest possible pillar width is determined which can be excavated in a safe way, followed by the maximum room width for such a pillar. If this width does not correspond with the maximum room width from a roof stability point of view, the pillar width can be increased to accommodate for a larger excavated room width. In general, coal pillars are wider than the rooms, while in metal mines the rooms are often wider than the pillars. The reason for this difference in approach is that the roof and the orebody is much stronger in metal mines, compared to coal mines.

The stress redistribution around the pillars is relatively easy to calculate. While a first impression could be that the state of stress is uniaxial, this is certainly not the case for wide pillars. At the circumference of the pillars, an uniaxial stress state is present and the vertical stress component is there also largest. Both factors increase the probability of creating stress induced fractures. Most of these fractures are shear type fractures, however, cleavage and fracturing due to buckling cannot be excluded. If fractures are induced, the stress peak moves further inward, away from the edge. Deeper into the pillar a triaxial stress state exists, larger vertical stresses can therefore be accommodated without creating additional fractures and an equilibrium is reached (Vervoort A.[7]). Even an entirely fractured zone surrounding the pillar helps in creating the triaxial stress state. In Figure 4, two typical examples of coal pillar corners are presented, showing various degrees of fracturing (Vervoort A. and Jack B.[8]), and one example (at right) taken in a copper mine, showing a similar behaviour.

In the roof of the rooms, tensile type fractures can be observed in the middle of the span, due to the deflection of the beam (Vervoort A.[9]). Above the pillar corners, a shear type fracture can occur, especially when the pillar corner forms a rigid transition between the pillar and the room (Vervoort A.[9]). If the pillar corner is softened by fracturing the probability that such fractures occur, decreases.

Figure 4: Some views of pillar corner behaviour [at left and in the middle: coal mine; at right: copper mine].

5 Fractures in cut and fill operations

The cut and fill mining method is applied in vertical or sub-vertical orebodies. Successive horizontal slices are mined and immediately filled. One starts at the bottom of a panel and moves upwards (see e.g. B.Brady and E.T.Brown[3] for a more detailed description). The aim of the backfill is to limit the horizontal movement of the hanging- and footwall, and the horizontal stress peaks above the excavation. However, as backfill is a passive type of support, some horizontal movement must occur before any stresses are induced in the backfill and before it starts to play a significant role in assisting to reach an equilibrium between the various parts (e.g. hanging- and footwall, orebody, room and backfill). The failure of the hanging- and footwall can be due to a horizontal movement over the entire height of the panel. As the first layers around the orebody are closing towards the excavated zone by deflection and are compressing the backfill, their length should increase. Hence, they pull at the rock above the last excavated slice, creating fractures in the orebody and in the hanging- and footwall. The behaviour of the rock can be deteriorated by additional deflection of these layers and possible buckling. In Figure 5, a detailed picture is presented of a sidewall, showing its large deformation and fracturing. Horizontal movement of up to 25 mm over a height of 3 m is being observed. The picture is taken in a metal mine (pyrite-chalcopyrite orebody) at a depth of nearly 1000 m. Grouted bolts and cables are installed and steel fibre reinforced shotcrete is applied on the sidewalls and roof. The latter has the advantage of providing a good areal coverage of the exposed rock and to be more ductile than plain shotcrete. Hence, the energy which can be absorbed is much larger.

Figure 5: View of a sidewall in a cut and fill mining operation.

6 Discussion of findings

In Table 1, an overview is given of all fractures discussed above. Two criteria are used for their classification: type of fracture (shear vs. tensile or extension) and the cause of the stress levels, that are high enough to induce one or more fractures.

Table 1: Overview of fractures observed, classified as a function of fracture type and cause of large stresses.

CAUSE OF HIGH STRESS	FRACTURE TYPE	
	Shear	Tensile/Extension
Stress redistribution	Edge of pillar	Longwall: ahead of face
Movement	Longwall: around face	Center of roof span (room) Sidewall in CAF

The Table shows the complexity of the behaviour of rock masses around mining excavations. Except for the fractures in the roof of room and pillar workings, the other fractures discussed have in a mining environment at least also some beneficial effect. Some are even essential in the entire mining process. To the latter category belong without doubt the fractures induced ahead of and around of the face of a longwall. They facilitate a regular goaf process and decrease the probability of having large hang-ups, resulting in large, sudden and dangerous collapses. The lateral movement, associated with the sidewalls in cut and fill operations and creating sidewall fracturing, induces much quicker the beneficial support pressure in the backfill. The fracturing of the extremities of a pillar softens the pillar circumference, which is beneficial for the roof stability. Such a softening associated with a more ductile behaviour of the entire pillar is also beneficial. To demonstrate this, one has only to make the comparison with the brittle and violent collapse of a rigid rock sample in a soft testing equipment, while carrying out an uniaxial compression test.

Although most of the fractures around mining excavations do have some beneficial effect, their negative effect from a local stability point of view remains. In the mining industry, one has to try to operate as safe as possible under these conditions, which is often not that easy.

Acknowledgements

The roof blocks taken in coal longwalls were collected at the Belgian coal mines (K.S.), while the author was preparing his Ph.D. (F.P.Ms.). The pictures of the coal room and pillar panels were taken, when the author was employed by the South African Chamber of Mines (COMRO), Coal Mining Division. The pictures of the various metal mines were taken on single visits to these mines. The CT-images were produced at the university hospital 'Gasthuisberg'. KULeuven. The author acknowledges the assistance and comments by colleagues researchers and supervisors, while working at various institutes (F.P.Ms. and KULeuven, Belgium, and COMRO, South Africa).

References

1. F.Verhelst, A.Vervoort, Ph.De Bosscher and G.Marchall, X-ray computerized tomography, determination of heterogeneities in rock samples, Proc. of the 8th Int. congress on Rock Mechanics, A.A.Balkema, p.105-108 (1995).
2. J.Caers, R.Swennen and A.Vervoort, Petrography and X-ray computerized tomography applied as an integral part of a rock mechanics investigation of discontinuities, Transaction of the IMM (Section B, Applied earth sciences), v.106, p.B1-B58 (1997).
3. B.Brady and E.T.Brown, Rock mechanics for underground mining, George Allen & Unwin, London (1993).

4. J.Gramberg, Analyse des cassures, mouvements et contraintes aux alentours d'une voie de chantier. ECC, Luxemburg, (1969).
5. A.Vervoort, Mining induced fractures in the roof of longwalls in deep coal mines, Proc. of the Int. Conf. on Mechanics of jointed and faulted rock, A.A.Balkema, p.733-739 (1990).
6. A.Vervoort, Geostatistical analysis of longwall face roof convergence data, Int.J.of Mining and Geological Engineering, v.7, p.227-237 (1989).
7. A.Vervoort, Monitoring and simulation of sidewall fracturing in coal mines, Proc. of the Int.Conf.Eurock92, A.A.Balkema, 6 p. (1992).
8. A.Vervoort and B.Jack, Analysis of sidewall fracturing and spalling during coal-pillar extraction, J.of the SAIMM, v.92, n 1,p.3-11 (1992).
9. A.Vervoort, Improved roof support in South African coal mines, Proc. of the Int.Conf. MASSMIN92, SAIMM, p.395-401 (1992).

A NEW METHOD FOR STABILITY ANALYSIS OF ROCK SLOPE

D.Y.ZHU, Z.S.ZHOU
Dept. of Civil Engineering, Nanjing Engineering Institute
No. 1 Haifuxiang Guanghuamen, Nanjing 210007, PR China
E-mail: KFZXB@public.hf.ah.cn

A new method is presented in this paper for assessing simultaneously the local and overall stability of rock slopes, which can easily deal with complex geometrical structure and discontinuities in rocks. On the basis of principle of limit equilibrium, along with method of slices, slope slip fields including dangerous slip field (DSF), critical slip field (CSF), locally critical slip field (LCSF) and globally critical slip field (GCSF) are defined. Based on the principle of optimality, the principle of maximum thrust and the principle of maximum generalized-thrust are proposed for simulating the above fields numerically. The critical slip surface of arbitrary shape, as well as other locally critical slip surfaces, can be numerically calculated with great accuracy and efficiency, rather than searched in conventional methods. No initial trial location is needed and the solution is always converged with few numerical difficulties encountered. Illustrative and case examples are presented, results of which indicate the effectiveness and applicability of the new method.

1 Introduction

Rock slope mass contains various discontinuities, such as bedding planes, joints, or faults, which make strength properties of rock heterogeneous, anisotropic, and discontinuous and frustrate geotechnical engineers or researchers in assessing slope stability, especially in locating reasonable critical slip surface corresponding to the lowest factor of safety. In the absence of discontinuities, the stability analysis of rock slope is similar to that of soil slope, which can be performed by means of all variety of conventional methods (Chowdhury[1]). When the stability of rock slope is wholly dictated by the well-defined discontinuities in the rock, the position of the potential failure surface may be determined, with ease to some extent, by virtue of experience and try-error method (Hoek and Bray[2]).

In most cases, potential failure surface in the rock slope may pass in one portion along a few of discontinuities and shear the intact rock mass in the other, thus that surface is usually of composite shape. It is difficult and almost impossible to locate reasonable critical slip surface, in such complicated situation, by use of commonly used search methods such as optimization technique (Chen and Shao[3]), random search methods (Boutrup and Lovell[4]) and dynamic programming (Baker[5]), which are feasible in the case of soil slope. The factor of safety associated with any unreasonable critical slip surface is always greater than the actual one and leads to evaluation of slope stability on the unsafe side.

Critical Slip Field Method (Zhu[6]) was first proposed only for searching the critical slip surface of soil slope with simple assumption about the interslice force as used in the simplified Janbu's method[7]. It was shown that this method was efficient and accurate in finding the critical slip surface of general shape and the result for simple homogeneous soil is in good agreement with that obtained by variational calculus (Revilla and Castillo[8]) which can be taken as the correct one. In this paper, the writers improve this method by incorporating more general assumptions regarding the interslice forces and more general concepts of slop slip fields are proposed.

2 Generalized Formulation of Method of Slices

Method of slices within the framework of limit equilibrium has been used for stability analysis of slopes in soil or rock since the early years of this century. With the failure surface defined, the sliding mass is divided into several slices, vertical in most situations, then assumptions about the interslice forces are made to render the indeterminate problem to be determinate. By satisfying the condition of force or/and moment equilibrium, the factor of safety against failure is derived. Different assumptions result in different methods, thus leading to different values of factors of safety. Comparison of these methods has be given by Fredlund and Krahn[9]. Detailed review of various methods are beyond the scope of this paper, however, brief discussion must be included for the derivation of the writers' method and generalized formulation of method of slices is presented in the following.

As shown in Figure 1, the potential sliding mass above the failure surface is divided into n slices with vertical boundaries. The forces acting on a typical slice is shown in Figure 2. For the sake of simplicity, only force equilibrium is taken into account, however, the writers' method presented in subsequent sections may be easily extended with consideration of moment, or both force and moment, equilibrium. In many analysis methods , the inclination of interslice force is assumed to be a constant or a variable related to the geometry of slope mass. Some of commonly used assumptions are presented in table 1, which can be categorized into two groups: Type A, the inclination of interslice force is independent of the base of slice, as used in the simplified Janbu's[7], the US Army Corps'[10] and the Sarma's[11] methods; Type B, the inclination of interslice force is related to the inclination of base of slice, as used in the Lowe and Karafiath's[12] and the residual thrust method[13]. Among these assumptions, the one corresponding to the residual thrust method is most applicable to stability analysis of rock slopes according to the investigation by the writers. In the residual thrust method, which is widely used in China and adopted by Chinese Codes of foundation design[13], the inclination of the interslice force is assumed equal to that of the base of associated upper slice. According to Morgersten and Price[14], the equilibrium condition in respect of moment is automatically satisfied with this assumption, although unreasonable for soil slopes. In the case of circular slip surface, the calculated factor of safety by the residual thrust method is very close to that by the simplified Bishop's method[15], which is widely accepted as a standard method. With this connection, the assumption of residual thrust method is mainly taken in this paper.

The starting value of factor of safety is assumed to be F_s and the shear strength of rock for each slice is reduced by F_s .

$$\overline{\varphi}_i = \tan^{-1}\left(\tan\varphi_i / F_s\right) \quad ; \quad \overline{c}_i = c_i / F_s \tag{1}$$

where

c_i = cohesion of the base of slice i;

φ_i = angle of internal friction of the base of slice i.

Resolving the forces acting on the slice as shown in Figure 2 and introducing the Mohr-Coulomb failure criteria give a recursive expression for the interslice forces:

$$P_i = \frac{\cos(\alpha_i - \theta_{i-1} - \overline{\varphi}_i)P_{i-1} + \sin(\alpha_i - \overline{\varphi}_i)W_i + \cos(\alpha_i - \overline{\varphi}_i)K_c W_i}{\cos(\alpha_i - \theta_i - \overline{\varphi}_i)} + \frac{\sin(\alpha_i - \omega_{i1} - \overline{\varphi}_i)Q_i + u_i l_i \sin\overline{\varphi}_i - \overline{c}_i l_i \cos\overline{\varphi}_i}{\cos(\alpha_i - \theta_i - \overline{\varphi}_i)} \tag{2}$$

where

P_i = the interslice force between slice i and slice i+1;

P_{i-1} = the interslice force between slice i and slice i -1;

W_i = the weight of slice i;

K_c = coefficient of seismic force which is assumed to be horizontal, the value is

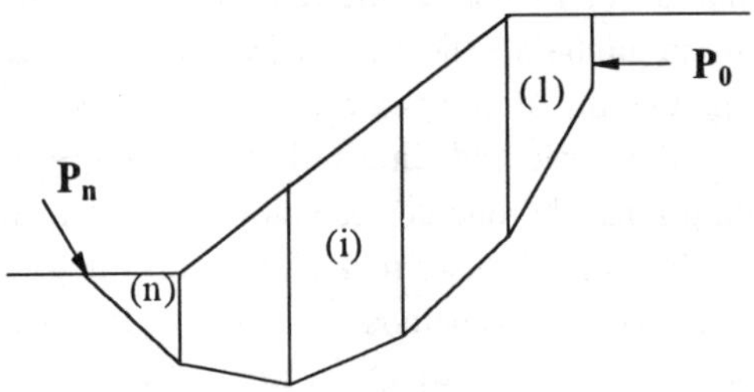

Figure 1: Division of sliding mass

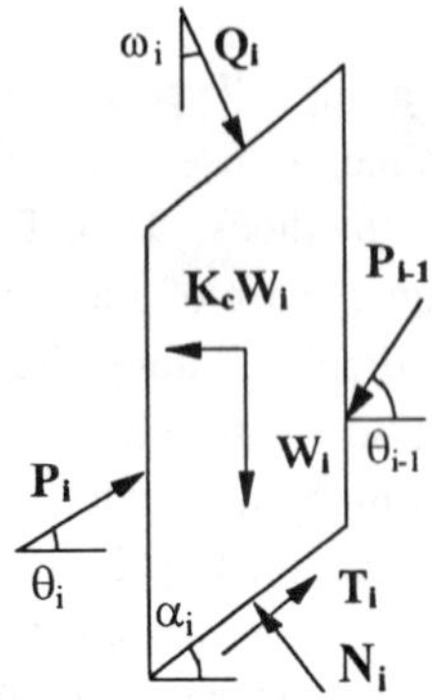

Figure 2: Forces on a typical slice

Table 1: Assumptions regarding interslice forces

Type	Method	Assumption	Equilibrium condition
A	the simplified Janbu's	$\theta=0$	force only
	the US Army Corps'	$\theta=\beta$	force only
	the Sarma's	$\theta=\varphi$	force only
B	the Lowe and Karafiath's	$\tan\theta=(\tan\alpha+\tan\beta)/2$	force only
	the residual thrust	$\theta=\alpha$	force and moment

notations:
θ= the inclination of interslice force
α= the inclination of the base of slice
β= the inclination of the surface of slope
φ= the angle of internal friction of the interface with reduction by the factor of safety
The effect of cohesion along the interface is neglected in the Sarma's method herein.

ratio of seismic acceleration to that of gravity;
Q_i = external force acting on slice i;
u_i = water pressure at the base of slice i;
l_i = length of the base of slice i;
α_i = the inclination the base of slice i;
θ_i and θ_{i-1} denote the inclination of P_i and P_{i-1} respectively;
ω_i = the angel between Q_i and vertical direction.

Commencing at the first slice at the upper end with prescribed external thrust P_0, usually equal to zero, and proceeding using Equation (2) to the last slice at the slip exit, the final thrust P_n is obtained, which is required to act on the last slice to satisfy the condition of force equilibrium. P_n may be defined as residual thrust. When P_n is greater than zero, the assumed value F_s must be greater than the actual one; and vice versa. The factor of safety F_s is modified repeatedly until the residual thrust P_n is much close to zero and the difference between the values of F_s at the last two stages is small enough. This indirect procedure for calculating the factor of safety is more simple and efficient than other direct ones in which numerical difficulties are often encountered. The generalized formulation will be employed for constructing slip fields in the following.

3 Definitions of Slope Slip Fields

In analyzing stability of a slope that has not failed or even not formed before cutting, search is required for the critical slip surface associated with the lowest factor of safety. Some available approaches for this purpose are mentioned in Introduction. Most currently used methods in practice and academic researches attempt to find the critical slip surface among a large number of candidates which are generated by some optimization techniques. Nevertheless, problems of local optimum, solution-instability and numerical difficulties often arise in using these methods, and the complicated characters of rock masses make them more inefficient, and furthermore, even futile.

3.1 Dangerous Slip Surface

In fact, the factor of safety is a complicated functional of the location of slip surface, with slope configuration and rock properties known; and the critical slip surface is the extremal of that functional which is difficult to express in a mathematical equation. From the preceding section, it can been seen that to locate the failure surface with minimum factor of safety is equivalent to find the same surface with maximum residual thrust whose value is approximately equal to zero. For a specified value factor, it may be easier to locate the surface with maximum residual thrust due to its explicit expression. This surface is defined as dangerous slip surface in contrast to critical slip surface. If the approach for locating dangerous slip surface is found, it is not difficult to locate the critical slip surface by modifying the trial value of safety factor to render residual thrust to be zero. It is evident that the determination of dangerous slip surface is crucial to the solution of problem.

3.2 Dangerous Slip Field, DSF

It is apparent that the above-mentioned dangerous slip surface is in overall sense. In fact, for any slip exit, there also exists a corresponding dangerous slip surface with maximum residual thrust at that exit point. This dangerous slip surface is in local sense. All of dangerous slip surfaces for all slip exits constitute a slip field which is defined as dangerous slip field, abbreviated to DSF.

3.3 Critical Slip Field, CSF

There is one residual thrust for one dangerous slip surface in DSF, the value of which is maximum among all potential slip surfaces through corresponding exit. When the maximum of these residual thrusts is equal to zero, that dangerous slip surface is then the critical slip surface, thus in this situation the DSF is defined as the critical slip field, abbreviated to CSF. The factor of safety corresponding CSF is just the lowest factor of safety to be found, and the critical slip surface associated with that exits having zero residual thrust is included in CSF. With exception of the critical slip surface, other dangerous slip surfaces can be also used for indicting potential sliding areas with residual thrusts as an indictor of degree of stability.

3.4 Locally Critical Slip Surface and Locally Critical Slip Field, LCSF

For any specified slip exit, the residual thrust can become zero by employing a special factor of safety. The dangerous slip surface for that exit is defined as locally critical slip surface, and such DSF is locally critical slip field, abbreviated to LCSF. Local critical slip surface is useful in analyzing stability of slope in local zone. This situation is often encountered in analysis and design of high rock slope such as surface mining slope.

3.5 Globally Critical Slip Field, GCSF

All of locally critical slip surfaces constitute a field that is defined as globally critical slip field, abbreviated to GCSF. The stability of slope in overall and local sense is indicated in GCSF with numerous locally critical slip surfaces and associated factors of safety. Thus, GCSF will be mainly stressed in this paper.

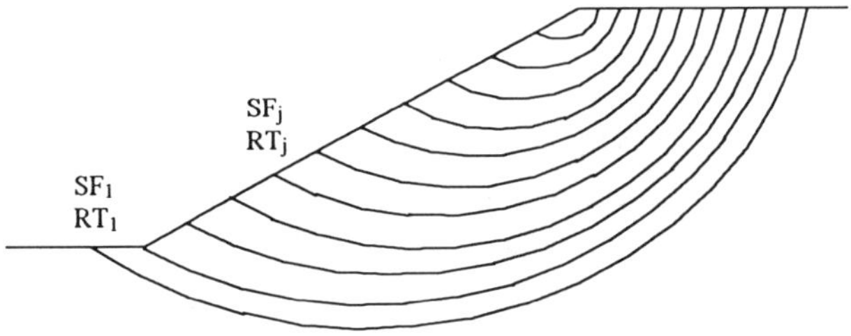

Figure 3. Schematic configuration of slope slip field

DSF, CSF, LCSF and GCSF are in general called Slope Slip Fields, a schematic configuration of slope slip field is shown in Figure 3.

Assume RT_1 , RT_2, ⋯, RT_m to be residual thrusts for those dangerous slip surfaces in any of SSF, and SF_1 , SF_2 , ⋯ST_m to be their specified (not final) factor of safety respectively. By definitions in the above, the relationship among these fields is summarized as follows.

SSP = DSF , if $SF_1 = SF_2 = \cdots = SF_m$

DSF = CSF , if $\mathrm{Max}[RT_1 , RT_2 ,\cdots RT_m] = 0$

DSF = LCSF, if $RT_j = 0$ j is the number of slip exit specified

SSP = GCSF , if $RT_1 = RT_2 = \cdots = RT_m = 0$

4 Principles of maximum thrust and maximum generalized-thrust

For the sake of conciseness, a single dangerous slip surface with fixed exit is chosen for investigation as shown in Figure 4. Any potential or actual slip surface is unnecessary to be smooth, and may be approximated by a broken line with connection points on the interfaces of slices if the number of slices is enough large. Now assume that the position of one point, point U in Figure 4 , on this dangerous slip surface is known beforehand, several potential curves between slip entrance and point U, i.e. UA_1, UA_2, ···, are specified using some methods, among which the optimum one for selection is such that the resulting residual thrust is maximized.

Type A assumptions are first made regarding interslice forces that the inclination of the interslice force is independent of the position of the base of slice. From Equation (2), it is shown that the process of calculating interslice forces constitutes a serial structure which can be analyzed by the principle of optimality, stated by Bellman[14]: An optimal policy has the property that whatever the initial state and initial decision are, the remaining decisions must constitute an optimal policy with regard to the state resulting from the first decision. This means that if curve EUA in Figure 4 is optimum one, curve UA is optimum too. In other words, if curve EUA is a dangerous slip surface giving maximum residual thrust at exit E, then curve UA is also a dangerous slip surface giving maximum interslice force at point U, which could be regarded as residual thrust if U taken as a slip exit. Interslice forces may be called thrust. Because the point U is arbitrarily chosen, thus the principle of maximum thrust holds as follows:

With assumption that the inclination of interslice force is independent of the base of slice, thrust between any two adjacent slices is maximum for a dangerous slip surface.

However, if Type B assumptions are taken, the above principle is invalid in that the process of calculating thrusts, in this situation, violates the principle of optimality. In fact, the residual thrust at E is not only dependent upon the value of thrust P_U , but also upon its orientation. In the case of type A assumption, the orientation of P_U is constant irrespective of variance in the base of slice, the residual thrust at E is only related to thrust P_U. But, it is not true for the case of type B assumption. By mechanics intuition, the horizontal thrust is liable to cause the

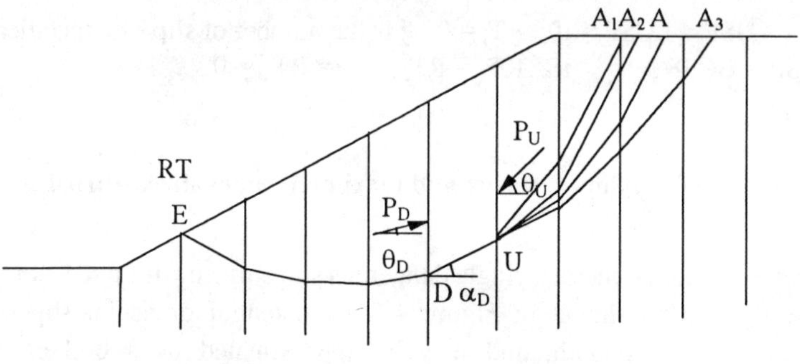

Figure 4. Investigation of dangerous slip surface

residual thrust to increase, while the vertical thrust is of opposite effect. However, this statement is not rigorous in theory.

Assume that the inclination of base of slice to the left of U , line DU as shown in Figure 4, is known to be α , the accurate position of curve EU still remains unknown. Now, select optimum curve EA in order for maximization of RT, i.e. residual thrust, at E. Since the inclination of thrust at D, θ_D, is fixed at current stage, RT is only dependent upon the varying value of P_D. The optimum curve EA among potential ones is such that the value of P_D is maximum. Simplifying Equation (2) gives:

$$P_D = \frac{\cos(\alpha - \theta_D - \overline{\varphi})}{\cos(\alpha - \theta_U - \overline{\varphi})} P_U + G \tag{3}$$

where $\overline{\varphi}$ is the friction angel of line DU, G is a term unrelated to the position of curve EA as well as P_U and θ_U. The essential and sufficient condition for maximization of P_D is that the first term on the right of Equation (3) attains to maximum. This term, $\frac{\cos(\alpha - \theta_D - \overline{\varphi})}{\cos(\alpha - \theta_U - \overline{\varphi})} P_U$, is defined as generalized-thrust. The principle of maximum generalized thrust is stated as follows:

With general assumptions about interslice forces, thrust between any two adjacent slices must be such that the corresponding generalized-thrust is maximum for a dangerous slip surface.

It is apparent that the principle of maximum generalized-thrust is an extension of that of maximum thrust, and, the latter is a special case of the former. The two principles are theoretical bases on which slope slip field is constructed.

5 Numerical Technique for Construction of Slope Slip Field

There are innumerable dangerous slip surfaces in a slope slip field and a few of them representing the whole is necessary to be located for engineering practice. From conventional viewpoint, every type of slope slip field can be constructed by methods currently used for searching the critical slip field, but the same difficulties still exist and alternative numerical procedure should be developed. It is the starting point that the field is to be constructed, not individual surface.

5.1 Discretization of Slope Mass

To the beginning, the analysis area of slope mass is chosen by margin lines beyond which the potential slip surface is unlike to pass through. This is easy for any engineering practitioner. Then the analysis area is divided into n vertical slices in Figure 5, as done in other methods based on the principle of limit equilibrium. These slices are of equal width, b, in order for ease of implementation of computer program , without causing intolerable error if the number of slices, n, is large enough. The slice boundaries are further divided by a number of points, called state points, with equal distance, d. The bottom endpoint of any slice boundary must be added to state points. The surface of the slope is divided into segments of slip entrance and slip exit as shown in Figure 5.

For any state point, such as i,j , there are one state variable, thrust $P_{i,j}$, and one decision variable, inclination of slip surface, $\alpha_{i,j}$, which are nomenclatures in dynamic programming. If all potential slip surfaces are assumed to be connected by state points, dynamic programming approach, along with principles proposed in preceding section, may be used for constructing slope slip field. However, as to a little more complex problem, the writers found out, the number of state points should be so large that computation time was intolerable even if the required space of computer was available. Now in this method, dangerous slip surface may pass

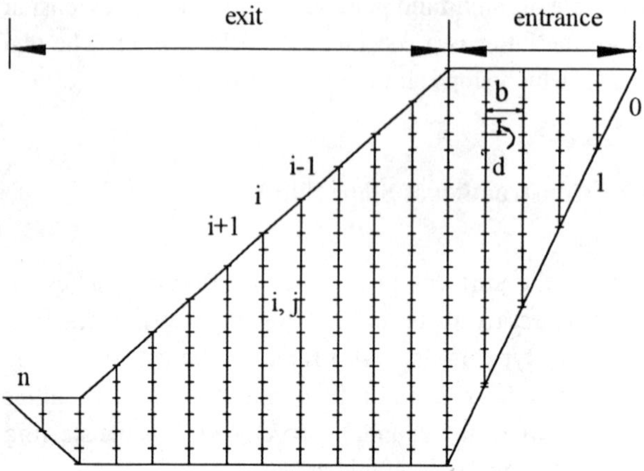

Figure 5. Discretization of slope mass

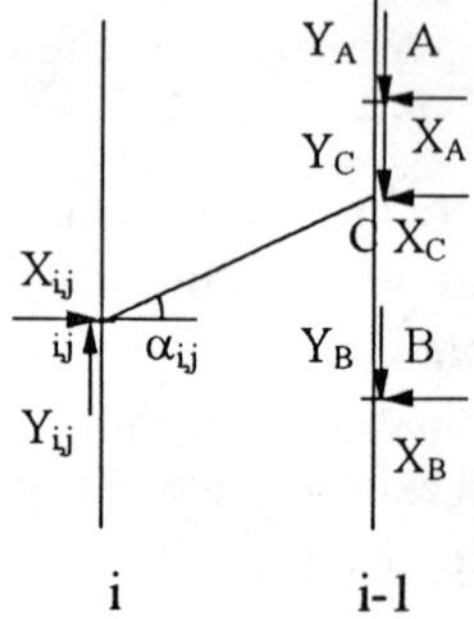

Figure 6. Linear interpolation of thrust

freely through any location between two state points. $P_{i,j}$ and $\alpha_{i,j}$ are redefined as dangerous thrust and dangerous slip direction respectively, both of which also called state values in general. This two state values are not independent of each other, and their combination should follow the principle of maximum generalized-thrust

5.2 Procedure for Constructing DSF

The factor of safety is specified equal to F_s , all strengths of rocks are reduced by Equation (1). From the first slice boundary to the last , and from the uppermost to the bottom on individual slice boundary, dangerous thrusts and dangerous slip directions of all state points are calculated in a recursive manner.

Taking Type A assumption (Type B will be discussed next) about the interslice forces, assume that the state values along i-1st slice boundary have been obtained, now calculate state values along ith slice boundary. Consider point i,j , in Figure 6, the dangerous slip direction, $\alpha_{i,j}$, is searched for by trial-error approach in order that the thrust is also dangerous one according to the principle of maximum thrust. When calculating a value of thrust, usually denoted by its components, $X_{i,j}$, $Y_{i,j}$, for a trial $\alpha_{i,j}$, the thrust from upper slice, X_c , Y_c in figure 6, is required to be determined. The slip surface with inclination of $\alpha_{i,j}$ intersects the upper slice boundary at point C, which is situated between state points of A and B, whose components, X_A , Y_A, X_B, Y_B, are predetermined. X_c , Y_c can be computed by linear interpolation as follows:

$$X_c = \frac{\overline{CA}}{\overline{AB}} X_A + \frac{\overline{CB}}{\overline{AB}} X_B \qquad (4)$$
$$Y_c = \frac{\overline{CA}}{\overline{AB}} Y_A + \frac{\overline{CB}}{\overline{AB}} Y_B$$

The state values for subsequent state points can be calculated step-by-step in the same way.

It should be noted that negative thrust in entrance segment is not permitted and may be eliminated by introducing tension crack near the crest. However, in exit segment, negative thrust is admissible artificially for computation of the residual thrusts, the values of which may be positive or negative. Under all circumstances, the effective force acting on the base of slice should not be equal to, or less than, zero. In order to avoid numerical difficulties, the suggestion of Whitman and Bailey[15] is adopted for choosing trial dangerous slip directions:

$$m_\alpha = \cos\alpha + \sin\alpha \tan\overline{\varphi} > 0.2 \quad (5)$$

The whole of dangerous slip directions form a field as shown in Figure 7. Each dangerous slip surface is able to be traced out in such field by interpolating dangerous slip directions between those of state points using linear interpolation function similar to Equation (4). The tracing commences from the exit, proceeding upwards, and terminates at the bottom of tension crack.

When using Type B assumptions about interslice forces, the principle of maximum generalized-thrust should be followed in determining dangerous slip direction for every state points. However, the slip direction through current state point in the lower slice is not known, that is, α_D and $\overline{\varphi}$ in Equation (3) are unknowns, so generalized thrust cannot be obtained. This problem can be resolved by iterative approach. For the first iteration, the horizontal component of thrust substitutes the generalized thrust. In other words, α_D and $\overline{\varphi}$ in Equation (3) are assumed to be zero. A initial field consisting of dangerous slip directions is constructed at the end of this iteration. Then the second iteration is to be set out, the generalized thrust can be calculated approximately, with values of α_D and $\overline{\varphi}$ in Equation (3) interpolated linearly from the previous field of dangerous slip directions. The above iteration is performed repeatedly until converged dangerous slip directions are obtained, or the difference in residual thrusts between two successive stages is within prescribed limits. Generally, only 3~5 iterations are sufficient for obtaining convergent dangerous slip field.

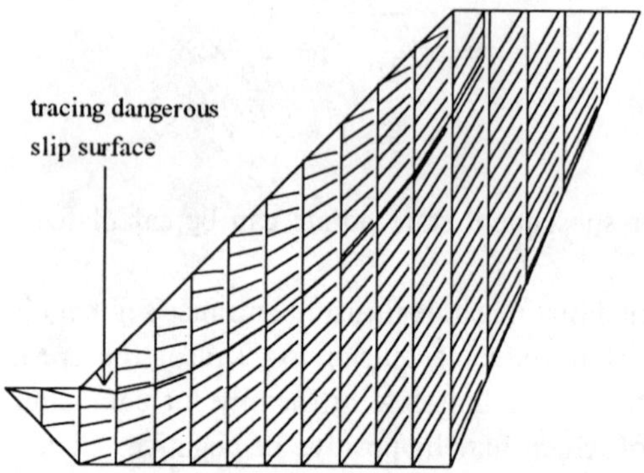

Figure 7. Field of dangerous slip directions

5.3 *Procedure for Constructing CSF, LCSF and GLCSF*

CSF, LCSF and GLCSF can be numerically simulated by improving the above porcedure for constructing DSF. The procedure of simulating CSF is as follows:

Step 1 Assume initial factor of safety F_{s0} and construct field of dangerous slip directions for F_{s0}. It is unnecessary for this initial field to be converged when using Type B assumption regarding interslice forces, thus no iteration is required.

Step 2 Modify the factor of safety according to the maximum value of residual thrusts. At this stage, the generalized thrusts are able to be calculated approximately from preceding field of dangerous slip directions. Field of dangerous slip directions for this modified factor of safety is constructed.

Step 3 Repeat Step 2 and terminate when the maximum value of residual thrusts is close to zero, at the same time, the factor of safety is converged.

Step 4 CSF is simulated from the final field of dangerous slip directions.

Step 5 Re-calculate factor of safety of slope, using conventional method of slices, according to the critical slip surface in CSF. From viewpoint of mathematics, the obtained factor of safety in Step 3 is a numerical solution by discretization and linear interpolation and does not correspond exactly to the critical slip surface. The less the difference between the re-calculated value of factor of safety and that in Step 3, the better the solution. The re-calculated value is recommended for representing the stability of slope.

LCSF can be simulated exactly the same way as CSF except that the maximum of residual thrusts in Step 3 is replaced by residual thrust at specified exit, corresponding to which the dangerous slip surface is locally critical surface.

GCSF consists of all locally critical surfaces that can be located by constructing all of LCSF, further presentation is unnecessary.

Example 1: A slope, with height of 100m and inclination of 45°, is in homogeneous rock. The properties of rock are: unit weight $\gamma = 260$ KM/m^3, c = 100 KPa, $\varphi = 32°$.

The CSF and GCSF, using assumption of simplified Janbu's method are presented in Figure 8. The factors of safety in GCSF have been corrected according to geometric ratio of thickness to length of corresponding sliding mass, Janbu et al[7]. The slip surfaces in GCSF are approximately like circle. Figure 9 presents three critical slip surfaces:

a. one from CSF using assumption of simplified Janbu's method;

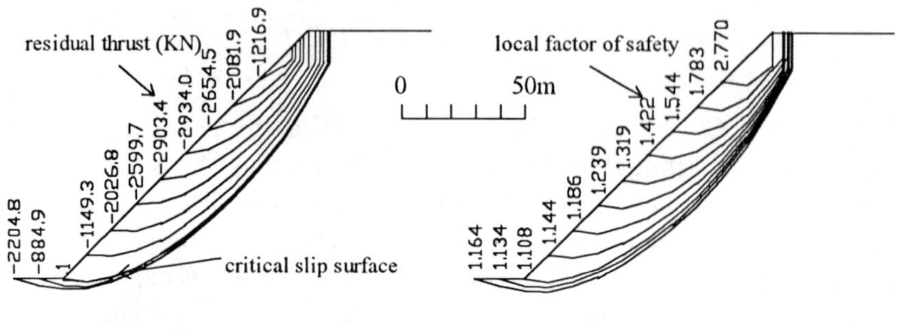

(a) CSF

(b) GCSF

Figure 8. Example 1: CSF and GCSF

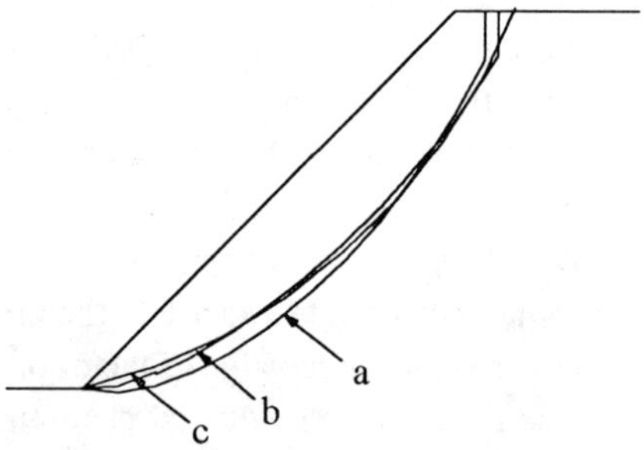

Figure 9. Example 1: Comparison of critical slip surfaces

b. one from CSF using assumption of residual thrust method;

c. critical circle by means of method of slip circle.

In this simple example, it is found that the locations of critical slip circles searched out with factor of safety defined by the simplified Bishop's method and by residual thrust method are almost identical, so are values of their factors of safety. For the three slip surfaces , a, b and c, re-calculate the factor of safety, with results: F_a = 1.160; F_b =1.139; F_c =1.161. F_b is the least as expected, but the difference between them may be neglected in practice. At the same time, this validates slip circle method for homogenous rock or soil. However, in the case of non-homogenous rock slope, the actual critical slip surface may largely deviate from a circle. In addition, it can be concluded that the location of critical slip surface is also related to the adopted assumption about interslice forces. The assumption used for constructing CSF should remain unchanged when re-calculating safety factor.

5.4 Considerations of Rock Discontinuities

When rock mass contains discontinuities such as joints, bedding planes or faults, special considerations are needed. Every individual rock may have several sets of joints including the bedding plane, which should be assumed to be closely distributed. When the trial slip direction coincides with one set of joint, or the difference between their inclinations is less than a small prescribed value, the shear strength of relevant joint is used in computation.

Example 2: A rock slope contains a set of bedding planes having a dip of 50°. Slope geometrical parameters: height = 500m, inclination = 45°. Rock strength properties : for intact rock mass, γ = 27KN/m^3, c = 360KPa, φ = 40°; for bedding plane, c = 120KPa, φ = 36°.

GCSF of this slope is shown in Figure 10 using assumption of residual thrust method. The active portion of the locally critical slip surfaces including the critical one is passing along the bedding plane, as expected by practical experience, while passive portion of curved shape crossing rock mass.

Example 3. The same rock slope as in Example 2, with two sets of joints added, is to be analyzed as shown in Figure 11. The cohesion and frictional angle of joints are: c = 0, φ = 37°. One set of joints, i.e. Joint 1 in Figure 11, is facing to slope surface with dip of 15°; the other, Joint 2, dips to the opposite direction at angle of 30°.

GCSF in Figure 11. shows that the exit portion of slip surface is along Joint 1. Joint 1 decreases considerably the stability of slope, while Joint 2 has no effect. This is consistent with experience. It is interesting that one locally critical slip surface

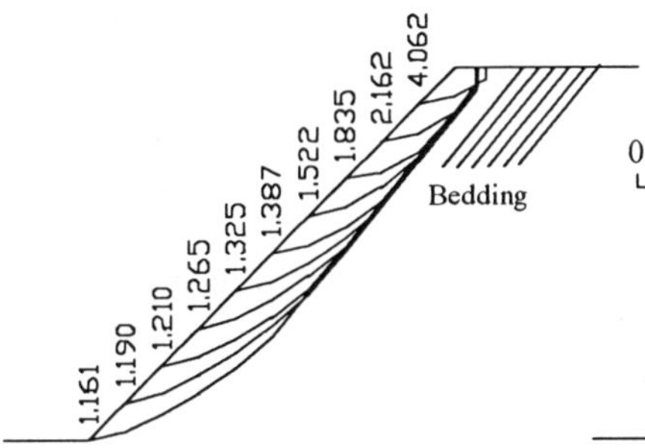

Figure 10. Example 2: GCSF
(with bedding plane)

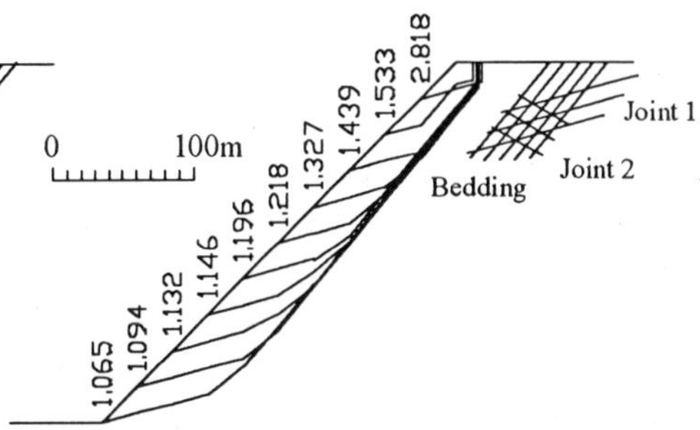

Figure 11. Example 3: GCSF
(with bedding plane and joints)

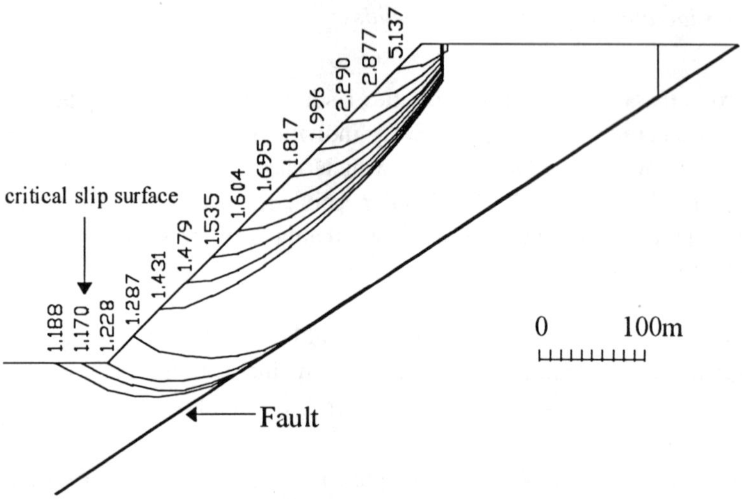

Figure 12. Example 4: GCSF (with fault)

near the crest has a bench-like shape with Joint 1 occurring as slip surface at two places. This unexpected slip surface, though of little importance to overall evaluation of stability, reveals the effectiveness and efficiency of the method herein.

In the presence of faults, special treatment should be involved in discritizing the slope mass. The fault with filling may be considered as a usual rock, while those with filling should be assumed to be filled with a artificial material of small thickness and the strength parameters of filling is those of fault plane. At least one or more state points must enter the fault. If the thickness of fault is less than the vertical distance between two state points, additional one in the fault is required.

Example 4. A slope, with height of 300m and inclination of 45°, contains a fault with dip of 33^0 and thickness negligible. The rock mass parameters are: γ = 27KN/m^3, c = 300KPa, φ = 39°; the shear strength of the fault: c = 50KPa, φ = 20°.

The calculated GCSF for this slope is presented in Figure 12. It can be seen from the GCSF that the fault effects the position of several locally critical slip surfaces exits of which are near the toe of slope. The critical slip surface is passing beyond the toe, the passive portion of which has approximately circular shape. If the toe of slope were chosen for the critical exit as in routine analysis, the factor of safety for overall slope would be enlarged by 5%. The overall and local stabilities of the slope are clearly exhibited in the GCSF.

6. Application

Case example: A surface mining slope in construction in North China has been studied using the writers' method. The designed height of slope ranges from 300m to 600m. The rock masses are mainly composed of a variety of schists containing quartz, mecca, and chlorite and folded into complex structure due to tectonic forces. The depth of weathering is about 60m. A large number of field and laboratory tests have been performed on rocks and discontinuities. Mechanical parameters entering into stability analysis are presented in Table 2. Note that layers of rock is denoted by symbols, with full names omitted. These layers of rock can be grouped into two classes: weak rock, Ma, and hard rocks, the rest.

Table 2. Mechanical parameters of rocks and their bedding planes

No.	rock	unit weight γ (KN/m^3)	mass shear strength c (KPa)	mass shear strength φ (°)	bedding shear strength c (KPa)	bedding shear strength φ (°)
1	Ma	28.2	293	34.7	32	32.6
2	Ma(weathered)	26.7	179	31.7	22	30.4
3	Mg	27.3	365	40.3	35	38.9
4	Mg(weathered)	26.9	230	37.6	25.	36.1
5	δ	30.0	303	40.8		
6	Fep, Fec	33.4	420	40.5	57	38.9
7	Sn	31.6	401	40.3	53	38.4
8	Ln	30.6	383	40.3	45	37.9
9	FAULT	20.0	78	29.5		

There are more than ten sections of slope are to be analyzed, only two of them will be presented here, Section A and B as shown in Figure 13. The coefficient of seismic force Kc=0.04, the phreatic surface is below the slope toe due to the future drainage.

The slope in Section A, with height of about 570 meters, has a fault dipping towards cutting face. The fault, of 5m thick, is somewhat convex. All of the 25 locally critical surfaces in GCSF do not pass beyond the fault and half of them in upper portion lie some distance away from the fault. The critical slip surface, with the lowest factor of safety, 1.019, is not through the slope toe. It is interesting that there are 8 surfaces with factor of safety less than 1.05. These surfaces are nearly critical and must have been missed by other routine methods.

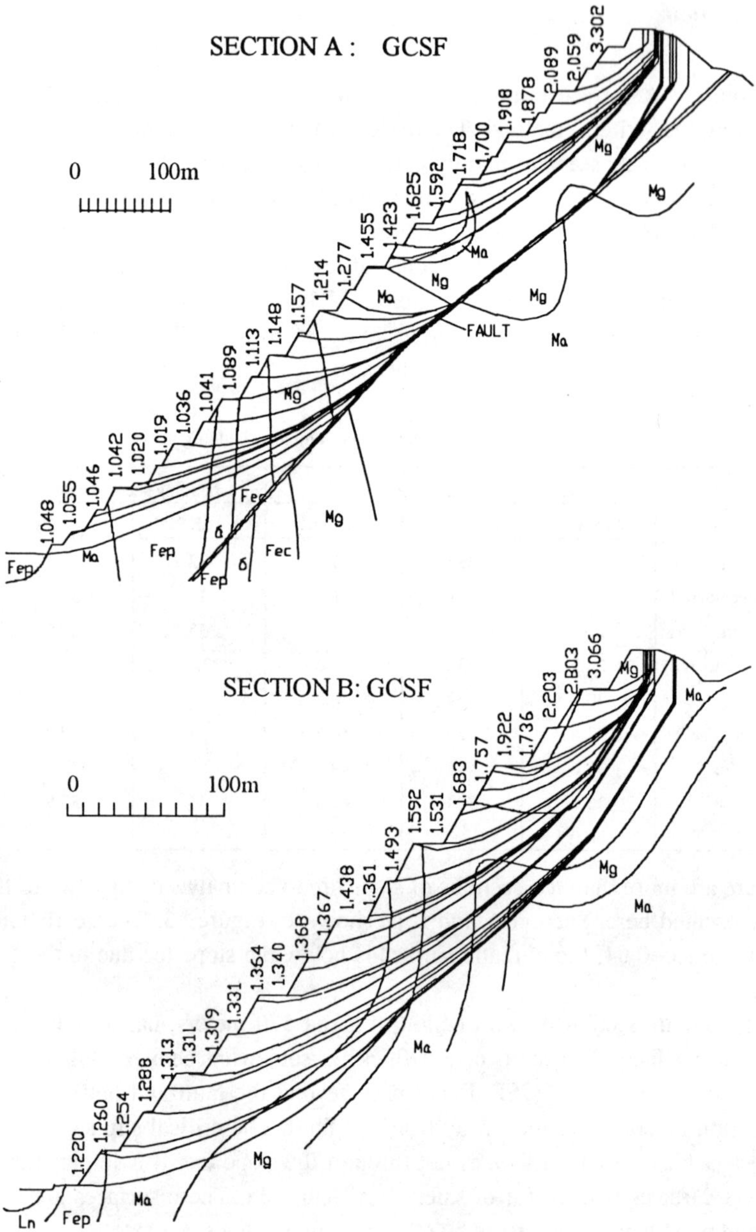

Figure 13. Case example : GCSF

The slope mass in Section B is composed of two interbedded layers of rock, the weak one, Ma, and the hard, Mg. The critical slip surface, through the toe of slope, may divided into four segments. The first one is passing along the bedding plane of rock Ma, the second crossing the center of an anticline, then the third segment passing along the bedding plane of rock Mg, in the end, the last crossing rock mass. It's shape is somewhat convex upwards at the intersection of the second and third segments. The critical slip surface of the slope of complex configuration like this one cannot be reasonably located using conventional methods.

7. Conclusion

The method presented in this paper is a powerful tool for analyzing slope stability. Compared with currently used method, it has many advantages.

The critical slip surface can be located accurately and efficiently by means of this new method. In using this method, no restrictions need to be placed on the shape of slip surface. The potential slip surface is approximated by a broken line of as many as 100 segments, sufficient for any extremely complicated shape. Discontinuities in rock slope are able to be treated with great ease, and very complex geometry and structure can be accommodated. The location of the critical slip surface is a numerical approximation to the correct one, such as solution of variational calculus if available.

The slope slip fields including DSF,CSF,LCSF and GCSF are new concepts in slope stability analysis. They can be used for evaluating stability of slopes for different purposes, and GCSF, applicable to every situation, is stressed in this papers. GCSF is able to exhibit the stability of slope everywhere by locally critical slip surface and locally minimum factor of safety. The studies of GCSF's of many slopes in rocks or soils have showen that single critical slip surface probably conceals other surfaces having factors of safety a little larger than, or even equal to, the lowest. GCSF may be substituted for the conventional stability-indicting mode consisting of the critical slip surface and the lowest factor of safety.

The program has been implemented incorporating numerical method for simulating every type of slope slip field and applied to some of engineering projects. At the first stage, slope slip filed is on the background of limit equilibrium method. More rigorous field, considering reasonable stress distribution available now by finite element method, is to be developed.

References

1. R.N. Chowdhury, *Slope Analysis.* Elsevier Scientific, Amsterdam, (1978).
2. E. Hoek and J.W. Bray, *Rock Slope Engineering*, Revised Second Edition, Inst. of Min. and Mat. London. (1977).
3. Z. Chen and C. Shao, Evaluation of minimum factor of safety in slope stability analysis. *Can. Geotech, J.*, 25(4), 735-748. (1988).
4. E. Boutrup and C.W. Lovell, Search technique in slope stability analysis. *Engrg. Geol.*, 16(1), 51-61. (1980).
5. R. Baker, Determination of critical slip surface in slope stability computations. *Int. J. for Numer. and Analytical Meth. in Geomech.*, 4, 333-359. (1980).
6. D.Y. Zhu, Critical slip field and its numerical simulation. *Chinese J. of Geo. Engrg.*, 19(1), 63-69. (1997).
7. N. Janbu, *et al*, Soil mechanics applied to some engineering problems, Norwegian Geotech. Institute, *Publication*. No. 16. (1956).
8. J. Revilla and E. Castillo, The calculus of variation applied to stability problems, *Geotechnique* 27(1), 1-11. (1977).
9. D.G. Fredlund and J. Krahn, Comparison of slope stability methods of analysis. *Can. Geotech, J.*, 14(3), 429-439. (1977).
10. U.S. Army Corps of Engineers, Engineering and design — stability of earth and rock fill dams. *Engr. Manual EM* 1110-2-1902. (1970).
11. S.K. Sarma, Stability analysis of embankments and slopes. *J. Geotech. Engrg.*, ASCE, 110(2), 231-154. (1979).
12. J. Lowe and E. Karafiath, Stability of earth dams upon drawdown. *Proc. 1st Pan-Am. Conf. on Soil Mech. and Found. Engrg.*, Mexico City, 2, 537-552. (1960).
13. Chinese Building Codes, Dsign Codes of Building Foundation. GBJ7-89, Chinese Building Publication Corp. Beijing. (1989).
14. N.R. Morgenstern and V.E. Price, The analysis of the stability of general slip surfaces. *Geotechnique*, London, 15(1), 79-93. (1965).
15. A.W. Bishop, The use of the slip circle in the stability analysis of slopes. *Geotechnique*, London, 5(1), 7-17. (1955).
16. R. Bellman, Dynamic Programming, Princeton University Press, Princeton, N.J., (1957).
17. R.V. Whitman and W.A. Bailey, Use of computers for slope stability analysis. *J. Soil Mech. and Found. Div.*, ASCE, 93(4),475-498. (1967).

LOAD APPROACH ON CIRCULAR TUNNELS IN ROCK CHARACTERIZED BY MAJOR JOINT SETS

CH. N. MARANGOS

Department of Civil Engineering, University of Thessaloniki, Greece

e-mail: cmaragos@ccf.auth.gr

In the paper a simple method for approaching the unstable areas and the loads acting on circular tunnels in rock characterized by sets of discontinuities is presented. The method can be applied on rock characterized by one or more major joint sets (MJS) or other discontinuity sets whose strikes are parallel to the tunnel axis: for example on rock masses in which the properties are controlled by a single set of discontinuities such as bedding planes or on highly schistose rocks. The method is applied for conditions usual in tunnel praxis for which graphs are presented aiming to help the engineer to approach quickly and without calculations the loads acting on the temporary supporting media and to plan the necessary supporting works.

1 Introduction

The design of underground opening supporting works is based on the approach of the unstable areas which will be developed around the excavation. In rock characterized by MJS, the joint orientation, their spacing and their angle of friction are the sizes which together with the initial stress state and the excavation geometry will determine the position and the extent of the unstable areas.

Tools disposed at the present time for approach the loads acting on tunnels in mechanical anisotropic rock masses are empirical classification systems and complicated numerical analysis methods.

The proposed method is simple: The unstable areas and the loads acting on the tunnel lining are determined by using the strength factor theory.

2 Principles of the analysis method

The opening of an underground excavation results in a strong stress redistribution in the narrow area surrounding the excavation; the stress state becomes worse. In jointed hard rock, the problem of setting boundaries to the unstable areas is focused on the determination of the areas in which the shear stress exceeds the joint shear strength.

The principles of the proposed method are presented in Figure 1. Figure 1a illustrates the static system: A vertical disc located transversally to the tunnel axis is loaded by the initial geostatic stresses $\sigma_v, \sigma_h, \sigma_n$ under plane strain conditions. σ_v,

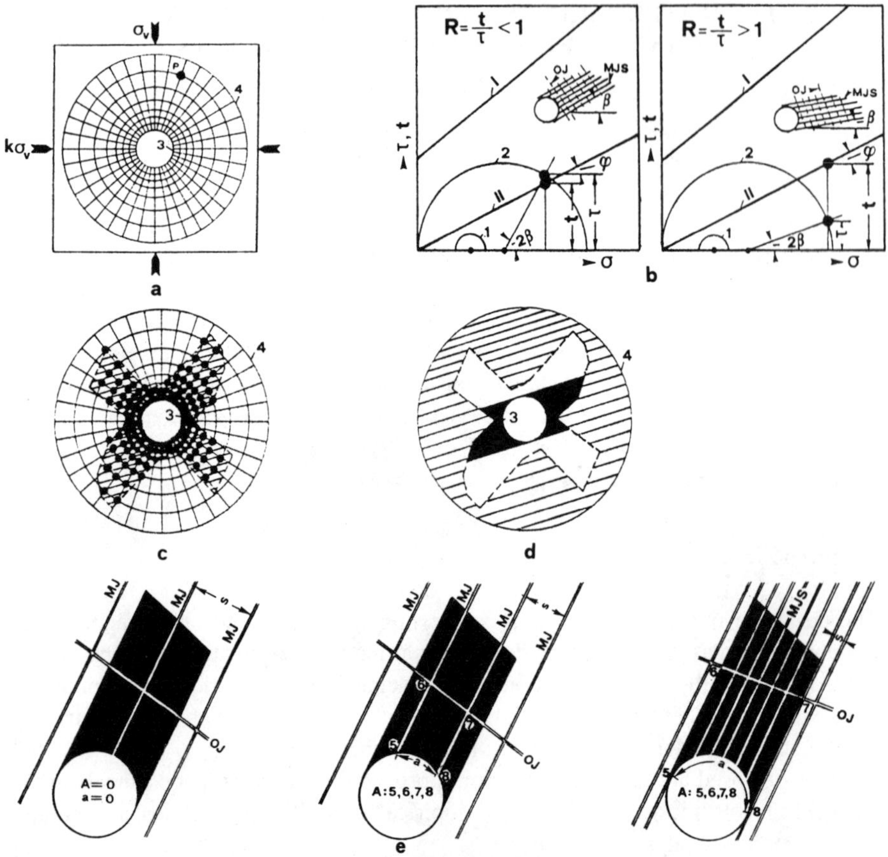

Figure 1: Method principles. a) Static system: vertical disc loaded by the initial geostatic stresses; polar grid situated on the tunnel surrounding area. 3= The tunnel section. 4= The external limit of the examined surrounding area. b) Failure criteria; left: R<1, right: R>1. c) Strength factors R and boundary setting of the failure zones, •=R≤1. d) Boundary setting of the critical zones. e) Determination from the joint spacing s and from the location of other joints (OJ) of the unstable area A (= area 5, 6, 7, 8) and of the width a of the supporting media part on which the pressure will act.

■= critical zones ▨= areas where R>1; the parallel lines in Fig. 1d are identified to the MJS dip □= shear failure areas

$\sigma_h = k \cdot \sigma_v$ are the maximum and the minimum principal stress respectively; σ_n is the intermediate principal stress. In Figure 1b, the applied failure criteria are given: the primary and the secondary stress field (circles 1 and 2 respectively; these circles are referred to the points in the middle of the tunnel walls for k= 0.5) and the straight lines I and II characterizing the shear strength of the intact rock and the joint shear strength of MJS respectively. The ratio t/τ determines the strength factor, $R = t/\tau$, where t = the joint shear strength, $t = \sigma \cdot tg\varphi$, τ = the shear stress and φ the joint friction angle. The only difference between the two diagrams of Figure 1b is the MJS dip β; due to the different dip β, $R = t/\tau < 1$ on the left diagram and $R>1$ on the right diagram.

The approach of the unstable areas is carried out as follows: A polar grid (Fig. 1a) is placed on the disc; the grid covers the area which is laying between the tunnel walls (circle 3) and the circle with radius equal to four till six times the tunnel radius (circle 4). This area is the area in which stress redistribution is appeared; in the analysis the self-weight of this area is ignored.

Each MJS is examined separately. Applying the equations of elastic stresses, the joint shear strength t and the shear stress τ acting on the joint planes of the examined MJS are determined on each grid point. That is at first is assumed that a joint of the examined MJS could pass from each grid point. On grid points where $R<1$, the shear strength will be exceeded. These points are characterized by small black circles in Fig. 1c. Consequently, based on the values of R, it is possible to set the boundaries of zones where if there are joints in these they could be activated. In Figure 1c the setting of boundaries of these zones (failure zones, $R \leq 1$) is illustrated schematically. Stability problems will appear only in the parts of these failure zones where rock body slidings are possible. These critical zones (characterized in black in Figure 1d) are determined by drawing from the excavation boundaries, the two tangential lines which are parallel to the dip line of the joints (Fig. 1d). The stability of the rock mass included in these critical zones will depend on MJS spacing s and on the presence, in these zones, of other joints with different orientation or of joints of another set which therefore will permit the sliding towards the excavation of the whole or part of the critical zones (Fig. 1e). The parts of these critical zones will consist the unstable areas. In rocks with more than one MJS the critical zones are the sum of the critical zones of all MJS.

The approach of the loads acting on temporary supporting media can be carried out as follows: For the part of supporting media which is in contact to the unstable area whose self-weight forwards its sliding towards the excavation, the pressure acting on this will result from the difference between the following components of the sliding body self-weight: the component which is parallel to the sliding direction (= dip line) and the component of friction force developed on the sliding surface (= joint surface):

$$p_i = \frac{A\gamma(\sin\beta - \cos\beta \, tg\varphi)}{a} \qquad [MN/m^2]$$

p_i = the pressure acting on this part of the supporting media, γ = the rock unit weight in MN/m^3, A = the unstable area in m^2, a = the unstable area width in m. A and a depend on joint spacing s. Figure 1e illustrates the determination of A and a

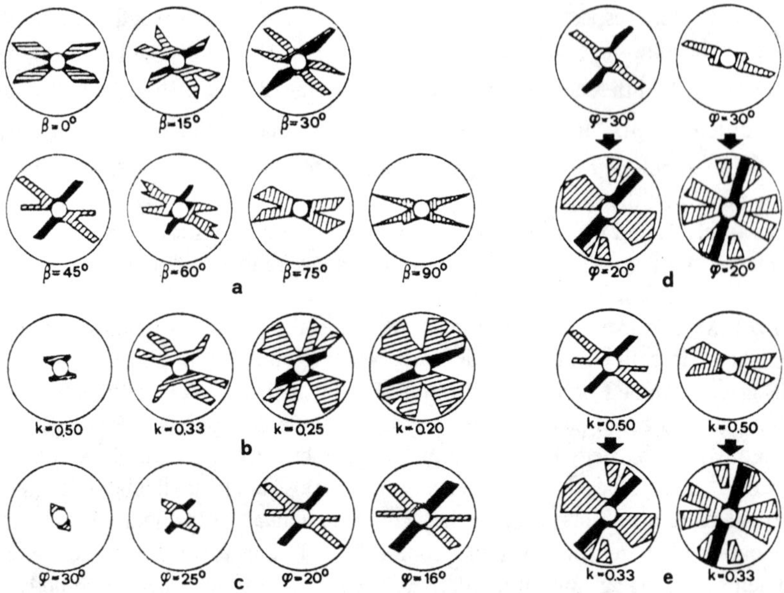

Figure 2: Influence on the extension of the critical zones of MJS dip β, of k factor and of friction angle φ. a) Influence of MJS dip β; case k = 0.5, φ = 20°. b) Influence of k; case β =15°, φ = 30°. c) Influence of friction angle φ; case β = 45°, k = 0.5. d) Elimination of the dip influence with reduction of φ value from 30° to 20°. Left: case β = 45°, right: case β = 75°. e) Elimination of the dip influence with reduction of k value from 0.5 to 0.33. Left: case β = 45°, right: case β = 75°.

■= critical zones □= areas where R>1 ▨= shear failure areas; the lines in the graphs are identified to the MJS dip

from the joint spacing s and from the location in the critical zone of other joints, of micro-joints or of joints of another MJS. When the location of the other joints can not be estimated, the design must be made by the maximum expected pressure; in this case A is the product of the penetration of the critical zone and a is its width determined according to Figure 1e.

The dimensioning of supporting media by the above equation presupposes flexible temporary supporting media able to permit failure zone development. Notice that the failure zones are referred to limit static equilibrium conditions and that in the failure zone determination, their self-weight was ignored.

Critical zones whose self-weight contradicts their sliding towards the excavation do not need to be supported; if not permissible rock intrusion towards the excavation is observed, the pressure can be determined by Kastner method (Kastner[2]).

In Figure 2 the influence on the critical zone location and extension of the joint dip, of the angle of friction and of the k factor is presented.

Table 1: Critical zones for MJS having dip β = 0° in dependence on friction angle φ and k factor.

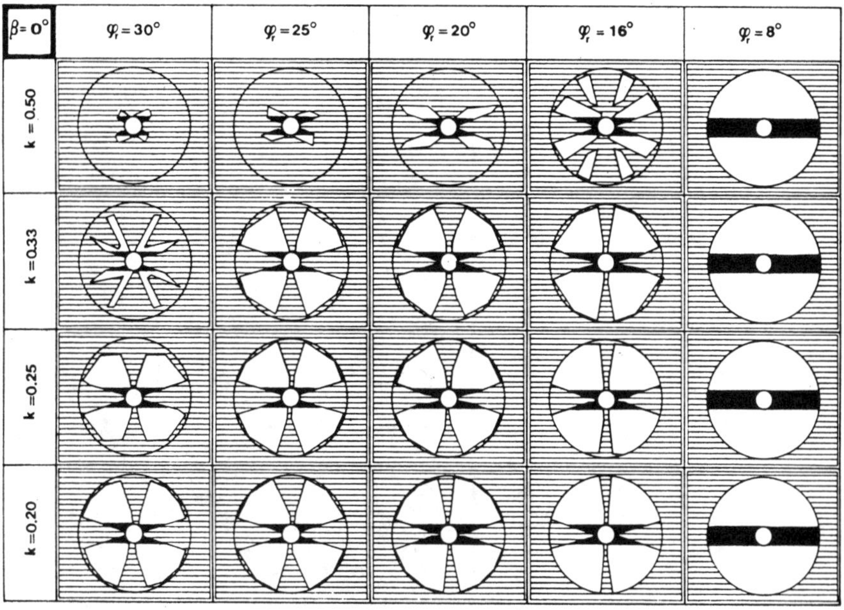

Table 2: Critical zones for MJS having dip β = 15° in dependence on friction angle φ and k factor.

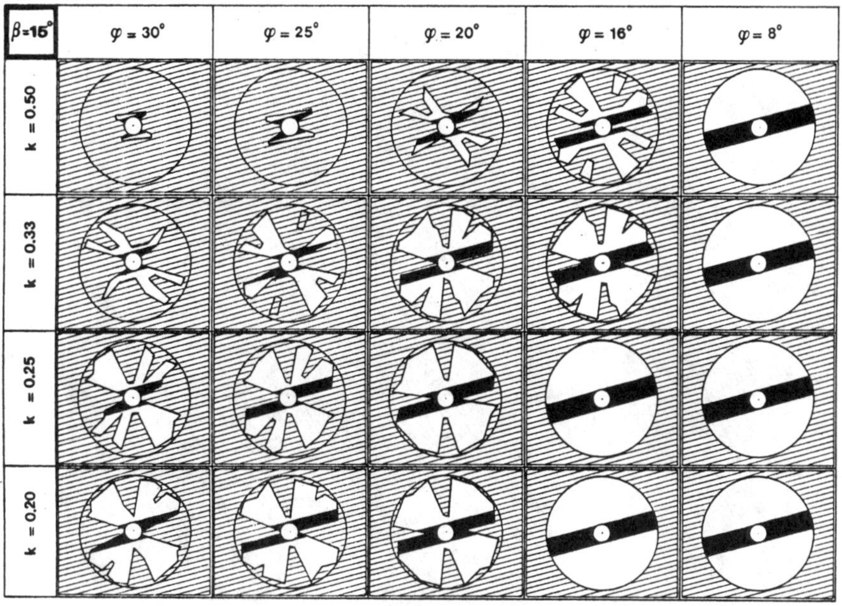

Table 3: Critical zones for MJS having dip β = 45° in dependence on friction angle φ and k factor.

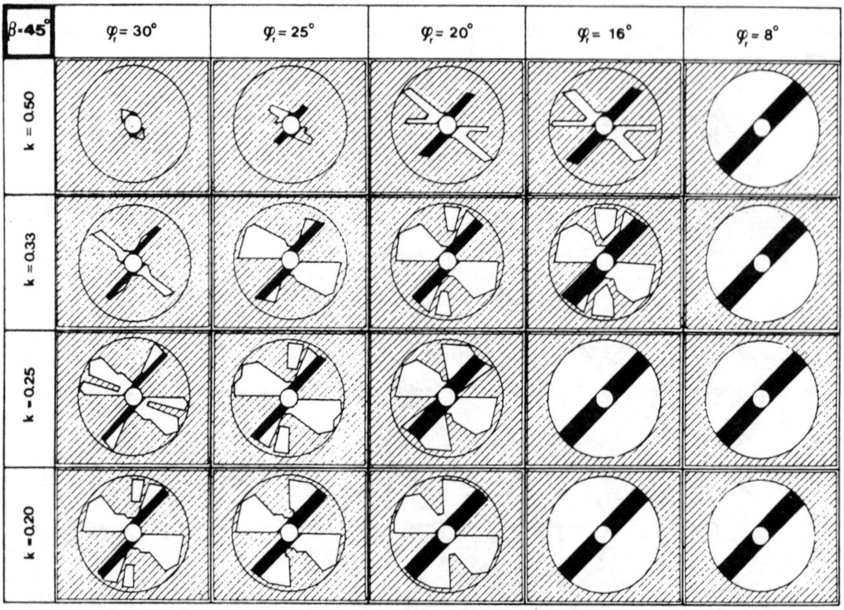

Table 4: Critical zones for MJS having dip β = 60° in dependence on friction angle φ and k factor.

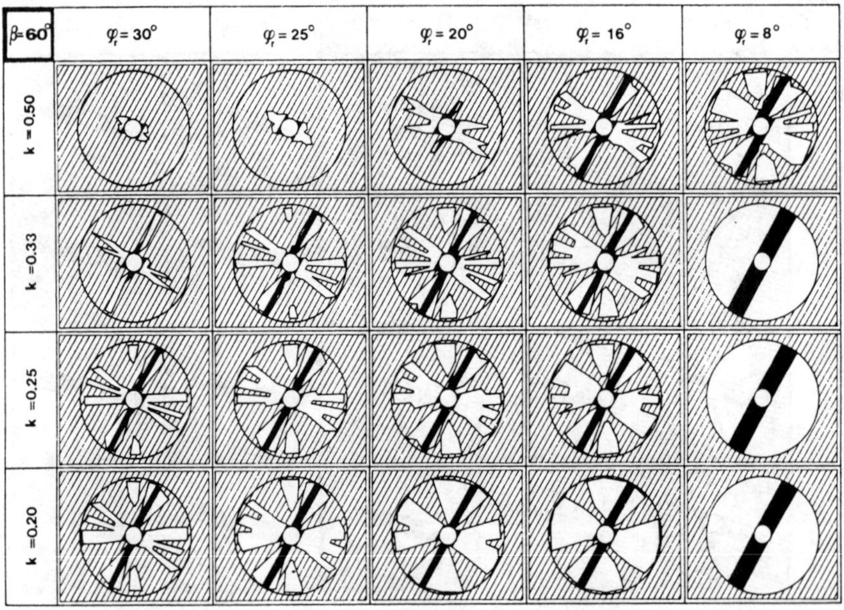

Table 5: Critical zones for MJS having dip $\beta = 75^{\circ}$ in dependence on friction angle φ and k factor.

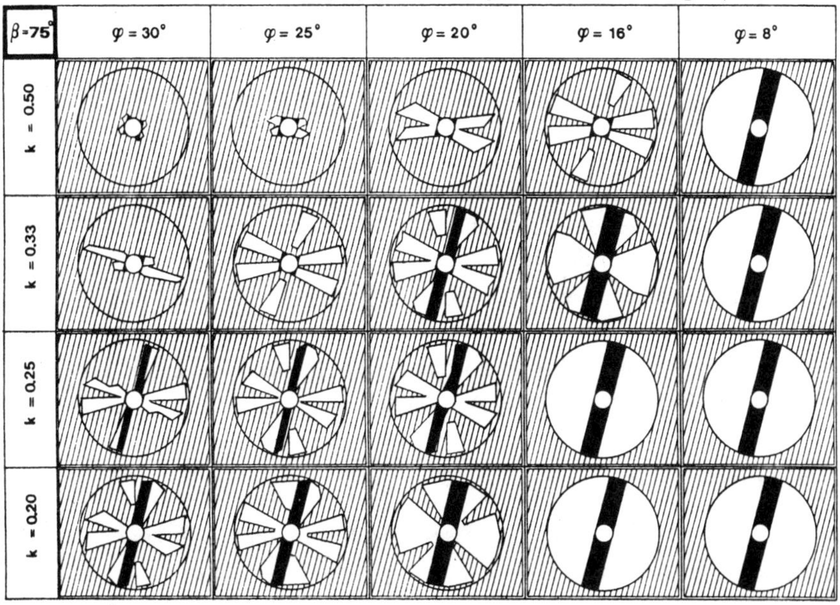

Table 6: Critical zones for MJS having dip $\beta = 90^{\circ}$ in dependence on friction angle φ and k factor.

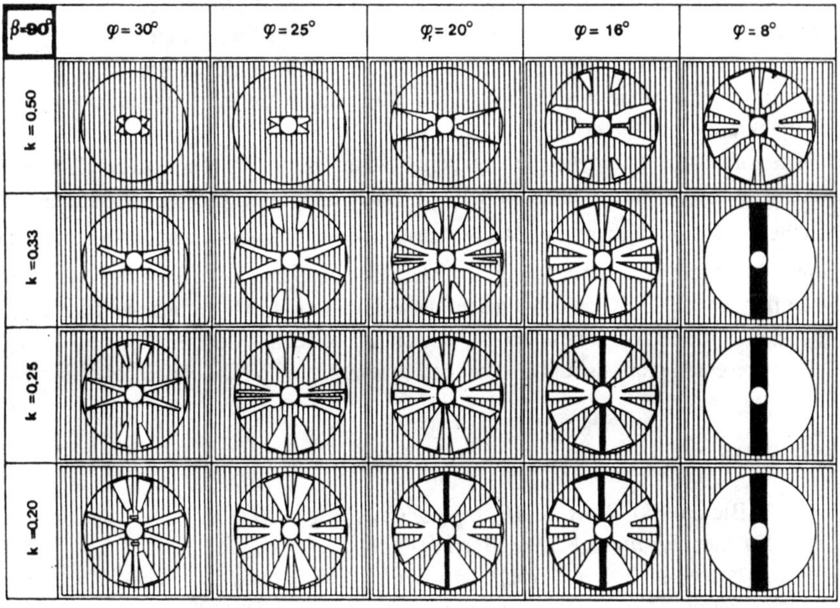

In Figure 2a, the influence of MJS dip is illustrated. MJS dips of the order of 30°-45° are the most unfavourable. As shown in this Figure, a great part of the failure zones extends towards the joint direction with consequence deep penetration of the critical zones, or in other words moving possibilities of sliding bodies in a great distance from the opening. On the contrary, in case of vertical MJS as well as those having dip >60°, the failure zones cut the joints obliquely or almost transversally. Hence, the length in which each joint will be activated, is small.

For given joint set dip, the failure figure is considerably influenced both by the value of k and by the value of the angle of friction φ (Fig. 2b, 2c).

The dip influence on the critical zones depends on the values of k (Fig. 2d) and φ (Fig. 2e). With the reduction of the values of these parameters, the failure zones cover more and more the surrounding area, with consequence the dip influence on the extension of the critical areas to be eliminated considerably.

3 Table references

In the paper, six Tables concerning cases of MJS with dips $\beta = 0^\circ, 15^\circ, 45^\circ, 60^\circ, 75^\circ, 90^\circ$ are illustrated. Each Table is referred to a specific dip β and includes cases when the angle of friction φ of the joints assumes the values $\varphi = 30^\circ, 25^\circ, 20^\circ, 16^\circ, 8^\circ$ and the k factor the values k=0.5, 0.33, 0.25, 0.20. The influence of the spacing s is not coestimated in the graphs; its coestimation must be made according to Figure 1e.

In contradiction to Figure 2, in the Tables 1-6 the areas having lines parallel to MJS dip are the areas in which R>1, the white areas are the shear failure zones and the black areas are the critical zones.

The Tables can be applied to any tunnel radius and to any tunnel depth. The only requirements are the measurement of MJS dip β and of its spacing s, the measurement of OJ and the evaluation (if there are not any measurements) of k factor and of friction angle φ. However the graphs can not be applied when the joints of the MJS are not continuous or when they are filled by cohesive materials with cohesion $c \neq 0$ because in these cases the extension of the critical zones depends also on tunnel depth (Marangos[3]).

The presented graphs could also be used to improve empirical classification systems.

References

1. Z. T. Bieniawski, *Tans. S. Afr. Inst. of Civil Eng.,* (1974).
2. H. Kastner, *Springer Verlag,* Berlin (1971).
3. Ch. Marangos, *Proc. Int. Symp. on Unique Underground Structures, U.S. Bureau of Reclamation, Denver, Colorado, Vol. 1, Chapt. 4, p. 1-18,* (1990).

Rock Reinforcement

A NEW APPROACH TO CONVERGENCE CONFINEMENT METHOD

NGUYEN MINH D., GUO C. [a]

Laboratoire de Mécanique des Solides. Ecole Polytechnique. 91128 Palaiseau Cedex

E_mail: duc@athena.polytechnique.fr

Determining the instant for support activation in plane strain tunnel calculations is an essential step, as for Convergence Confinement Method. Recent researches on that subject lead to an original development of that Method which allows for an accurate determination of that instant. The proposed methodology results from "experimental numerical laws" learned from comparison between plane strain approach and direct three dimensional one. For this purpose, a reference three dimensional geometry is considered: a constant rate advancing circular tunnel in anisotropic geostatic stress state, with one principal direction parallel to tunnel axis. The isotropic state of stress is first considered. It is shown that two fundamental principles allow to treat completely the plane strain approach. It is then shown how these principles can be extended to the anisotropic stress state, using a "mean convergence curve concept Finally, it is proposed to apply such a method to three dimensional tunnels in steady state advancement. With such improvements, the Convergence Confinement Method appears as a self-contained plane strain approach to the three dimensional advancing tunnel problem, without much additional intricacy to classical approaches.

1 Introduction

Support is fundamental to the stability of deep tunnels. The three dimensional nature of this problem involves an intricate interplay between the construction phases, rock mass and support behaviour. The framework of continuum mechanics is considered herein. Although three dimensional calculation has achieved constant progress in tunnel modelling, it remains heavy to carry out, and approximate plane strain analyses are still preferred and currently used in practice.

Among other methods, Convergence Confinement (CV-CF) is the most popular, using a variable fictitious stress state at the tunnel wall in order to reproduce the effects of face advance on convergence of a tunnel section.. However, up to now, efficiency of these approaches have been seriously impeded by the drawback of undetermined instant for support activation, an essential parameter which controls rock mass-support interaction.

This paper presents recent researches on this subject which allow to eliminate now satisfactorily such undetermination, without much modifying the nature of calculations. The method presented herein is intended to be self contained, independent of the particular behaviour of the rock mass and the support system.

---

[a] Now in *GETEC, 69 rue d'Aguesseau 92771 Boulogne Billancourt. France.*

2 Defining a reference tunnel problem

The method for determining the support activation instant is based on general rules learned from comparison between the direct three-dimensional problem and the corresponding plane strain approach. To demonstrate this, let us consider a simplified "reference tunnel problem", which contains the main ingredients of a three dimensional tunnel problem: a circular tunnel of unit radius (Figure 1), with anisotropic geostatic stress state, one principal stress direction being parallel to tunnel axis. This tunnel and its support placement are advancing at constant strain rate. Remark that steady state hypothesis is understood in CV-CF plane strain approach.

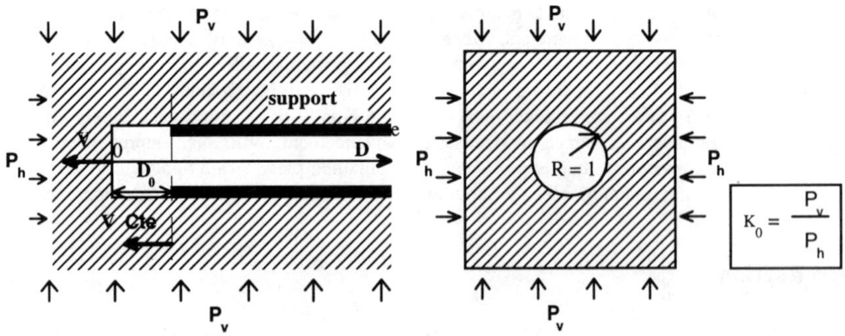

Figure 1: Reference tunnel problem

As intrinsic rules are expected to be found out, details of constitutive laws are less important, so a simple elastoplastic incompressible Mises model has been mainly used:

$$\sqrt{J_2} - C = 0$$

This problem is defined by the following parameters

$$\begin{array}{ll} \text{Stress anisotropy factor ....} & K_0 = P_h/P_v \\ \text{Mean geostatic pressure} & P = (\sigma_h + \sigma_v)/2: \\ \text{Loading factor ....} & N_s = P/C \\ \text{Support laying distance to the face ......} & D_0 \\ \text{Relative support stiffness ....} & K^* = K_n/K_m \end{array} \qquad (1)$$

K_m is the elastic ground stiffness, and K_c support normal stiffness

Let us also define K_b the support bending stiffness

For direct calculations (excavation, support placement), specific softwares have been built up, taking advantage of the peculiar hypotheses of the reference problem. There results in a versatile Finite Element programs, based on the steady state algorithm (Nguyen Q.S & al. [8], Maitournam [12], Corbetta [4], Guo [3]), which allows to treat the steady state problem on a fixed geometry, instead of using a classical sequential procedure. In case of anisotropic stress field, Fourier's development of solutions in the angular direction θ has been used, along with the steady state algorithm (Braham et al. [13], Maitournam et al.[11]), which replaces the three dimensional problem by a succession of two dimensional problems.

3 Statement of the problem

The CV-CF Method is illustrated on the generic case of a circular tunnel in rotational symmetry $K_0 = 1$ (Figure 2), with a fictitious pressure Pi applied at the wall, or conversely, a deconfining ratio λ by relationship Pi=1-λ (Pi is normalized by P).

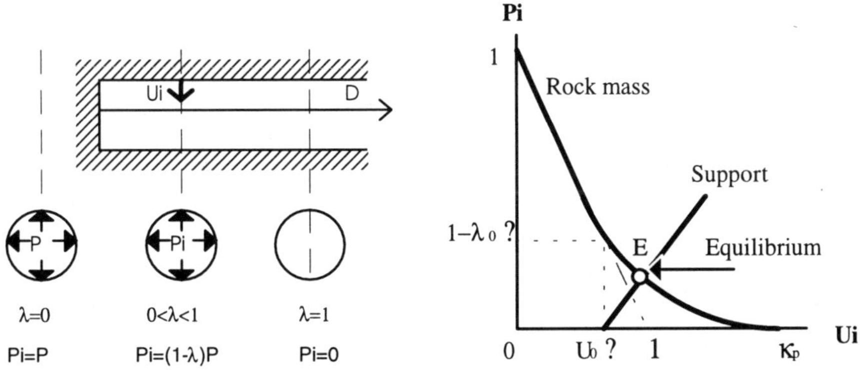

Figure 2: Fictitious pressure concept in plane strain approach and CV-CF diagram

Equilibrium of the supported tunnel appears on the non dimensional diagram Pi vs Ui (convergence normalised by free elastic one) as intersection between rock mass convergence curve, and support confinement reaction curve. This latter curve is easily obtained, but its abscissa at the origin, U_0 (or, conversely, the deconfining ratio λ_0), is undetermined. Indeed, the actual time for activation is more likely represented by distance to tunnel face D_0 for support placement, so the basic problem is to find out the relationship between this parameter D_0 and U_0.

This CV-CF method is currently extended to three dimensional cases, using plane strain FEM analyses, but of course, the undetermination of the support activation instant still remains.

Up to now, no systematic method allowed to resolve such undetermination (AFTES [1]). Different studies on this subject, initiated since years 1990's (Corbetta [4], Bernaud [6], Guo [3]), have allowed now to eliminate satisfactorily this drawback, giving practically the same results as by direct three dimensional approach.

4 Circular symmetric case ($K_0 = 1$)

The question of Figure 2 is equivalent to the following set of equations :

$$\begin{cases} P_i = M(U_i) & \text{ground} \\ P_i = K^*(U_i - U_0) & \text{support} \\ U_0 = ? \end{cases} \qquad (2)$$

Which give support pressure $P_{i.} = P_\infty$ and final convergence $U_{i.} = U_\infty$ of tunnel, provided a supplementary equation for U_0 is given in terms of D_0, and eventually, of $U_{i.}$. Indeed, the problem could be resolved if the relationship $U_{i.}$ vs distance to the face D could be determined (Figure 3).

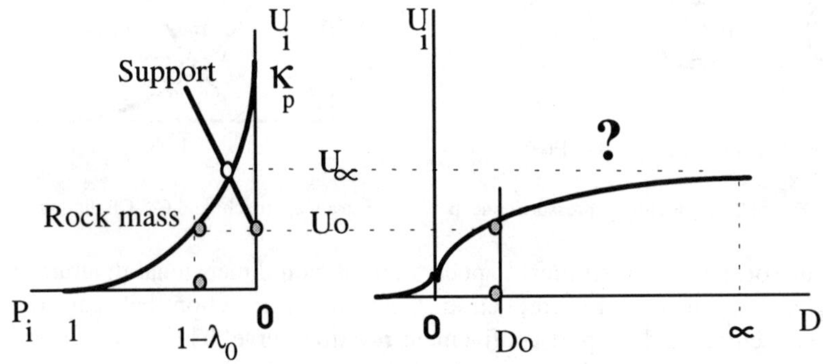

Figure 3: Scheme for determining U_0

4.1 *Self Similarity Principle and approximate solution*

An approximate determination of U_0 refers to the free convergence curve of the tunnel, $U_i^f(D)$, instead of $U_i(D)$ curve so that :

$$U_0 \cong U_i^f(D_0) \tag{3}$$

Historically, this solution was effectively proposed for the elastic case. An analytical expression can be given for the free elastic convergence $U_i^f(D) = F(D)$, for instance (Panet [2]) :

$$F(D) = 1 - 0,71\left(\frac{0,84}{0,84 + D/R}\right)^2 \tag{4}$$

remark that $\lambda_0 = U_0$ only for linear elasticity.

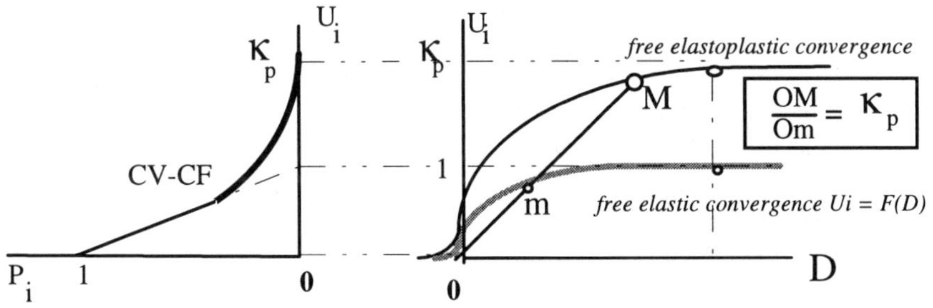

Figure 4: Schematic diagram of self similarity principle

In elastoplasticity, a Principle, called the Self Similarity Principle, based on numerical data evidence for various elastoplatic behaviours (Corbetta [4], Corbetta & Nguyen-Minh [5]), allows us to derive the free elastoplastic convergence curve $U_i^f(D)$ from the normalised elastic one Ui=F(D), by self similarity from the origin (Figure 4). The equation of this curve expresses in terms of elastic function F(D) and the self similarity ratio κ_p obtained from a CV-CF calculation:

$$U_i^f(D) = \kappa_p F(D/\kappa_p) \tag{5}$$

Thus, by a quite procedure to the elastic case, we can easily obtain an approximate value for U_0, much more convenient than tables for λ_0 proposed by AFTES [1].

4.2 Interaction Principle

However, the approximation based on the free convergence curve $U_i^f(D)$ may lead to important errors up to 40% on support pressure (Bernaud [6]). This can be termed as the face support interaction effect, illustrated on Figure 5.

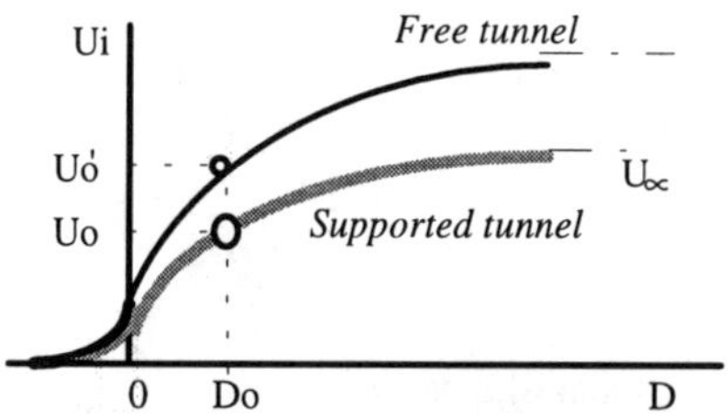

Figure 5: Face support interaction effect

In order to analyse that effect, extensive parametric analyses were carried out on the direct numerical model, including different stability numbers, relative support stiffness K*, and support distances D_0 (Figure 6).

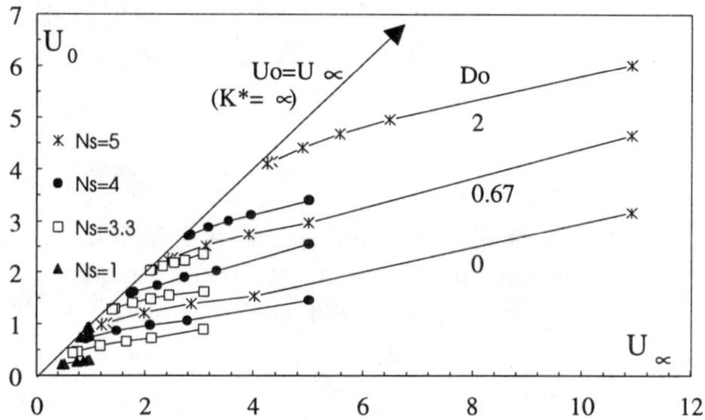

Figure 6: Numerical data for U_0 vs U_∞ (Mises model)

When plotted in an adimensional diagram U_0^* vs U_∞^*, by dividing U_0 and U_∞ by their respective free convergence values :

$$U_0^* = \frac{U_0}{U_0^f} \quad ; \quad U_\infty^* = \frac{U_\infty}{\kappa_p} \tag{6}$$

the results present a definite trend, called the Interaction Principle expressed as follows :

$$\begin{aligned} U_0^* &= \Phi(U_\infty^*) \\ \Phi(x) &= 0.55 + 0.45x - 0.42(1-x)^3 \end{aligned} \tag{7}$$

Where $0{,}23 \leq \Phi(x) \leq 1$ since $0 \leq x \leq 1$.

This curve does not depend on any parameters of the model, e.g. support stiffness K*, distance for setting support D_0, stability number Ns. We shall admit its universality, i.e., its validity for any other rock mass constitutive law. This has been effectively verified by internal and external validations, using other constitutive laws or alternative theories; moreover, interaction principle was proved to work in viscoplasticity (Guo [3]).

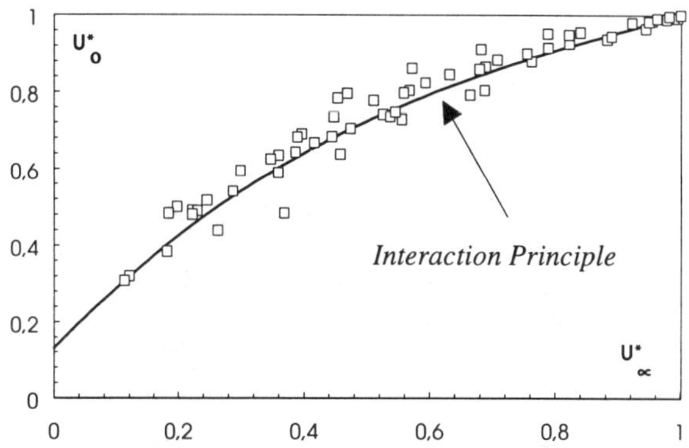

Figure 7: Interaction Principle $U_0^* = \Phi(U_\infty^*)$

4.3 Conclusion

The missing equation for U_0 in system (2) is finally obtained :

- approximately, with $U_0 \cong U_i^f(D_0)$, using Self Similarity Principle (eq (5)),

- with taking account of Interaction Principle (6), combined with the Self Similarity Principle (5), which gives $U_i^f(D_0)$.

5 Examples

5.1 Elastic tunnel

The final convergence of a supported elastic tunnel writes as :

$$U_i = \frac{1+K^* U_0}{1+K^*} \tag{8}$$

Remark $0 \le U_i \le 1$; two different expressions for U_0 can be given, depending on neglecting or not the support interaction effect (eq. (5) , (6) and (7) for U_0). There results a final tunnel closure $U_i = U'_\infty$ or $U_i = U_\infty$, as illustrated graphically, in a (Y, U_i) diagram, by setting Y equal to either member of eq. (8) (Figure 8).

If K* were set to infinity, the second member of eq (8), taking account of interaction, is $Y=F(D_0)\Phi(U_\infty)$, which is function $\Phi(U_\infty)$ reduced by affinity $F(D_0)$.

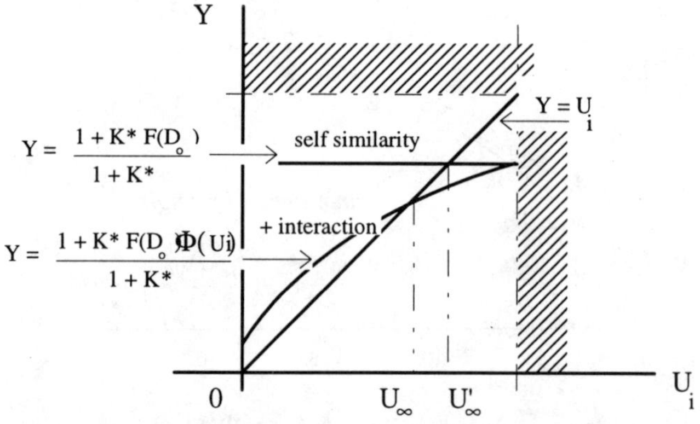

Figure 8: Graphical illustration of the two kinds of solution for elastic tunnel

5.2 *Elastoplastic tunnel (Tresca model, cohesion C)*

The elastoplastic convergence curve writes :

$$P_i = 1 - \lambda = 1 - \frac{1}{Ns}\left(1 + Ln(NsU_i)\right) \tag{9}$$

Elastoplastic behaviour for the supported tunnel implies conditions on U_i and U_0:

$$\frac{1}{Ns} \le U_i \le \kappa_p \qquad ; \qquad \frac{1+K^*U_0}{1+K^*} \ge \frac{1}{Ns} \tag{10}$$

where Ns = P/C is the stability number.
Taking into account support reaction gives :

$$\frac{1+K^*U_0}{1+K^*} = \frac{K^*U_i + \frac{1}{Ns}\left(1+Ln(NsU_i)\right)}{1+K^*} \tag{11}$$

which is the analogous of equation (8). Let us again represent eq (11) in (U_i,Y) diagram, with Y=Y1 (first member) and Y=Y2 (second member) (Figure 9).
Keeping in mind inequality (10) and $F(D_0/\kappa_p)<1$, this diagram clearly shows that a unique solution exists for U'_∞ and U_∞, as well as their difference.

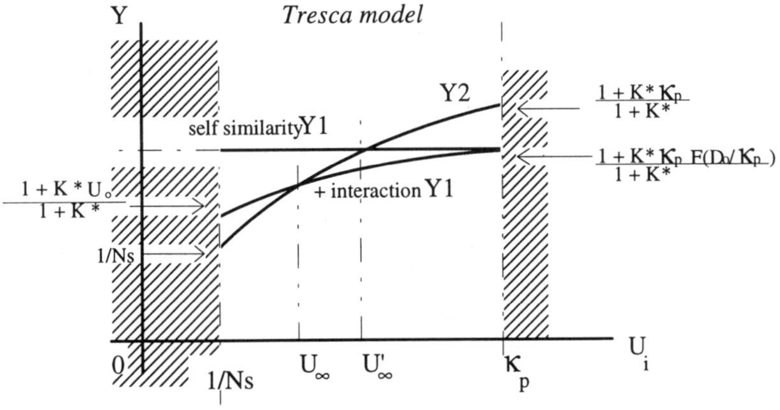

Figure 9: Graphical illustration of the two kinds of solution for elastoplastic tunnel

6 Circular tunnel in anisotropic geostatic stress state ($K_0 \neq 1$)

When rotational symmetry is lost, the fictitious pressure concept can be naturally extended to a fictitious stress, applied at the wall of the gallery. It is currently admitted that this fictitious stress be proportional to the initial stress state, via the deconfining ratio λ :

$$\underline{\underline{\sigma}}^R = (1-\lambda)\,\underline{\underline{\sigma}}_0 \qquad (12)$$

Note that stresses and pressures are now normalised to the mean geostatic pressure P, and convergences and displacements to the elastic convergence of the same tunnel under isotropic geostatic pressure P.

6.1 Validity of fictitious stress concept and deconfining ratio

This concept has however to be verified, and, effectively, we checked that (Guo [3], Nguyen Minh and Guo [10]) :

- For a free tunnel, a $\lambda(D)$ relationship can be exhibited, which relates the deformed section in 2D approach to a tunnel section at a given distance D in the direct 3D calculation. Remind, according to what has been discussed here above, relation $\lambda(D)$ is not sufficient to determine the support activation instant.
- There exists $\lambda=\lambda_0$, for activating support, allowing to find back the final deformed section of the direct 3D approach.

6.2 Mean convergence curve concept

Let us define a mean fictitous pressure $\overline{P}_i$ applied at the wall of tunnel, as the mean flux of the fictitious applied stress, and the mean convergence $\overline{U}_i$ of the tunnel as the mean flux of displacement at the wall (e.g. half surface variation of section).

It is interesting to note, for an elastic tunnel, that the mean convergence curve $\overline{U}_i, \overline{P}_i$:

$$\begin{aligned} \overline{P}_i &= (1-\lambda) \\ \overline{U}_i &= \lambda \quad \text{(in elasticity)} \end{aligned} \qquad (13)$$

is independent on the stress anisotropy ratio K_0, and is identical to that of the tunnel under isotropic geostatic pressure P. This shows that defining mean values according to (13) is adequate to express an intrinsic behaviour of the elastic tunnel under anisotropic stress state. This is no more true in elastoplasticity: the $\overline{U}_i, \overline{P}_i$ curve then characterises the intrinsic influence of the anisotropic stress ratio K_0 (Figure 10).

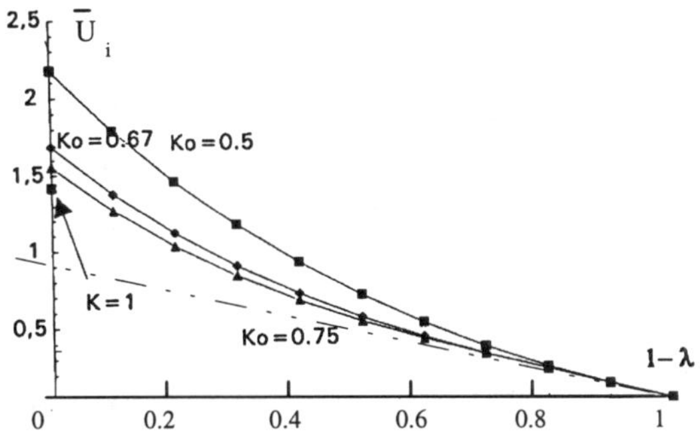

Figure 10: Effect of stress anisotropy on mean convergence curve in elastoplasticity

The "mean convergence curve" concept is all the more interesting, that the bending support stiffness has a negligible influence on the deformation of a tunnel section. This can be understood, since the ratio of normal stiffness to bending stiffness is

$$\frac{K_n}{K_b} = \frac{4}{3}\left(\frac{R}{e}\right)^2 >> 1 \quad \text{(e, support thickness)}.$$

Tunnel closure appears thus to be controlled uniquely by the normal support stiffness.

6.3 Method for détermining λ_0

The interesting properties of the mean convergence curves allow us to expect determining λ_0 by postulating that Interaction Principle still works on it for the anisotropic loading case. However, it must be checked that Self Similarity is still valid, which is actually the case as shown on Figure 11.

Finally, determining λ_0 for the anisotropic stress state is reduced to that of a rotational symmetry problem with a "mean rock mass convergence" curve. Let us remind again, that, except in elasticity, this "mean" curve is not the convergence of the circular tunnel submitted to mean the geostatic stress P, but is sentitive to K_0.

The solution thus obtained in CV-CF method has been proved to agree quite well with the direct tunnel calculation (Guo [3]), which validates at the same time the applicability of Interaction Principle to the non isotropic geostatic stress state.

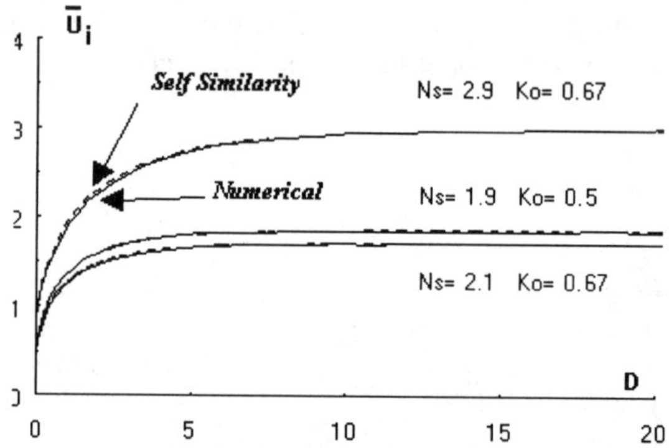

Figure 11: Self Similarity for free tunnel in anisotropic geostatic stresses

7 Conclusion

For plain strain approaches, it has been proved, on a reference three dimensional geometry, that two universal Principles take into account satisfactorily the three dimensional influence of the front face:

- The Self Similarity Principle describes the free convergence behind the face
- The Interaction Principle describes the relationship between convergence of two particular points of the supported tunnel, one on the leading edge of the support, and the other far behind the face.

In the general case, when geostatic stress is anisotropic, these Principles have to be associated with the "mean free convergence" curve to determine accurately the support activation instant.

It is proposed to extend this "mean convergence" concept to analyses of more general tunnel geometries, rock mass structure and rock mass anisotropy, provided the "steady state" conditions prevail for the tunnel under study. This results in a new plane strain approach, which appears as a self-contained CV-CF Method, for treating three dimensional deep tunnel problems, by using practically the same classical software procedures with a minimum added intricacy.

Naturally, if the problem is too far from the reference problem, it may be necessary to treat it by a direct approach, namely, using the steady state method which is well adapted. It may then be expected that the methodology presented herein be extended, provided some modification to the specific cases encountered.

Acknowledgements

The authors are grateful to Dr Habibou Maitournam for his help in steady state calculations

Nomenclature

D, D_0	distance to the face of a unit radius tunnel, and distance to tunnel face for support laying.
P	mean geostatic stress (isotropic geostatic stress)
P_i	fictitious pressure at the tunnel wall
λ	deconfining ratio
U_i	radial displacement at the tunnel wall
U_0, U_0^ℓ	convergence when laying the support and convergence of the free tunnel at distance D_0
U_∞, U_∞^ℓ	stabilised convergence of supported tunnel and stabilised convergence of free tunnel
$\sigma_v, \sigma_h = K_0\sigma_v$	vertical and horizontal principal geostatic stresses (anisotropic geostatic stress state)
K_0	ratio of vertical horizontal principal initial stress
$P = \dfrac{\sigma_v + \sigma_h}{2}$	mean geostatic stress
E, ν	elastic parameters of the rock mass
K_n, K_b	normal and bending support stiffness
$K_n^* = K_n / K_m$	normal relative support stiffness where $K_m = E/(1+\nu)$ is rock mass stiffness

References

1. AFTES . *Tunnels et Ouvrages souterrains* **59** (1983).
2. M. Panet, *Presses de L'ENPC*. Paris (1995).
3. C. Guo, *PhD Thesis, ENPC,* France (1995).
4. F. Corbetta, *PhD Thesis, ENSM* France (1990).
5. F. Corbetta, D. Bernaud, & D. Nguyen-Minh, *Rev. Franç. Géotech.* **54**, 1 (1991).
6. D. Bernaud, *PhD Thesis, ENPC,* France (1991).
7. D. Bernaud & G. Rousset, *Rev. Franç. Géotech.* **60,** 5 (1992)
8. Q.S. Nguyen & M. Rahimian, *J. Mec. Appl.* **5**, no 1 (1981).
9. D. Nguyen-Minh. & C. Guo, *Proc. Eurock'93*, ed.Sousa & Grossman **1**, 171 (1993).
10. D. Nguyen-Minh & C. Guo, *Proc. Eurock'96* ed. G.Barla, **2**, 855 (1996).
11. H. Maitournam, Braham, Dang Van, K. 1995. FAST. LMS, Ecole Polytechnique.
12. H. Maitournam, PhD thesis, ENPC, France, 1989
13. S.Braham, D.Nguyen-Minh, H. Maitournam, C.Guo, Proc. 8th Int. Conf. I.A.C.M.A.C, ed. Siriwardane & Zaman.(1994)

PRACTICE AND THEORY OF THE INTEGRATED TECHNOLOGY OF ROCK REINFORCEMENT

R. SNUPÁREK
Institute of Geonics - Czech Academy of Sciences, Studentská 1768, 70800 Ostrava - Poruba, Czech Republic
E - mail : snuparek@ugn.cas.cz

The paper deals with theory and practice of the two most widespread rock reinforcement technologies - bolting and grouting. The analysis of the deformation and stress - strain distribution in rock mass along fully bonded bolts including practical conclusions is presented. The new integrated system containing both technologies - bolting and grouting is described.

1 Introduction

Employing of rock properties including rock reinforcement for stability control creates the base of recent methods of underground driving and supporting in mining and civil engineering. The most frequent technologies of rock reinforcement are bolting and grouting. Hitherto the two technologies have been executed separately in spite of the fact, that the basic operation for both technologies is drilling of holes with small diameter (mostly up to 40 mm). The practice of separate anchoring and grouting (when both technologies are used) seems to be disadvantageous from both practical and theoretical viewpoint.

2 The combined system of bolting and grouting

Recently the combined reinforcing system has been developed and practically proved. It combines the two basic technologies - bolting and grouting. The system is called Boltex and its scheme is given in Fig.1.

The basic element is hydraulically expansible tube bolt made of the enfolded steel tube of special shape (principle of Swellex). When a liquid medium in high pressure is conveyed into the closed space of the pipe bolt, the bolt expands and presses the wall of the borehole.

The inner end of the bolt is provided with the injection membrane. It is designed in such way so that (after the following increasing of the liquid medium pressure) the membrane cracks and makes it possible to grout rock mass through the bolt. The value of grouting pressure can differ from the pressure of bolt expanding (Fig.2).

The part of the grouting system are packers, preventing the grout from flowing out through the mouth of borehole. There are two types of packers :

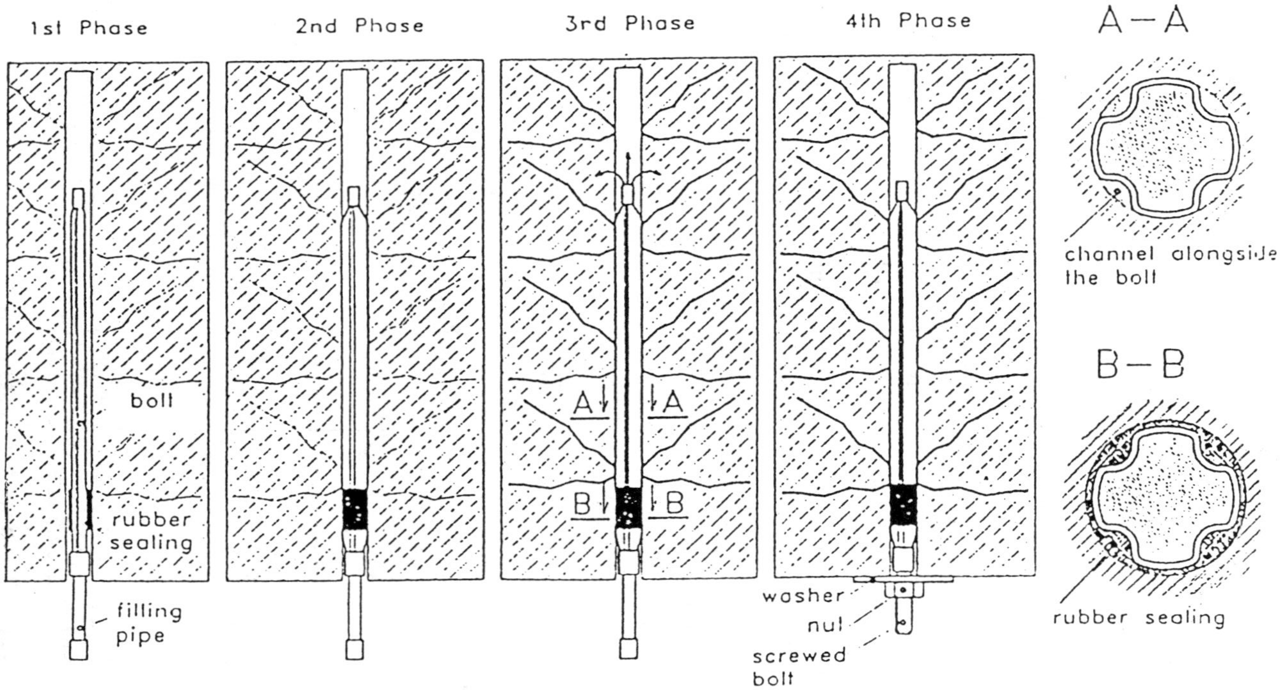

Figure 1: The combined system of bolting and grouting

either rubber sleeve on the pipe bolt, which seals (after expansion of the bolt) the space between the bolt and the borehole wall, or special packers, which are noted with appreciable capacity of expansion and are able to pack the borehole also in damaged parts (caverns, fissures).

The principle of packing is shown on Fig.1. The shape of pipe bolt retains (after expansion) shallow grooves, which serve to transport the grouting material from the inner end of the bolt along the borehole. To prevent grout from flowing out of the hole mouth it is necessary to seal these grooves in proper point. It is provided by raw rubber, which fills the groves and the sleeve from scorched rubber put on the pipe bolt. The length of the packing sleeve is mostly 30 to 50 cm, i.e.much shorter than the length of the bolt. According to the position of the sleeve on the bolt it is possible to change the point of the packing and to influence the active grouting length of borehole. According to the next analysis it is advantageous to locate the packer into the so called neutral point of bolt.

The accessories of bolting and grouting system also contain set of pads, which are situated on the outer part of the bolt. Its construction ensures certain automatic prestression between the pad and the rock.

The bolting and grouting system Boltex enables to use different kinds of grouting materials, e.g. cement or two-part polyurethane resins. Grouting can be realised basically in two ways (see Fig. 2) :

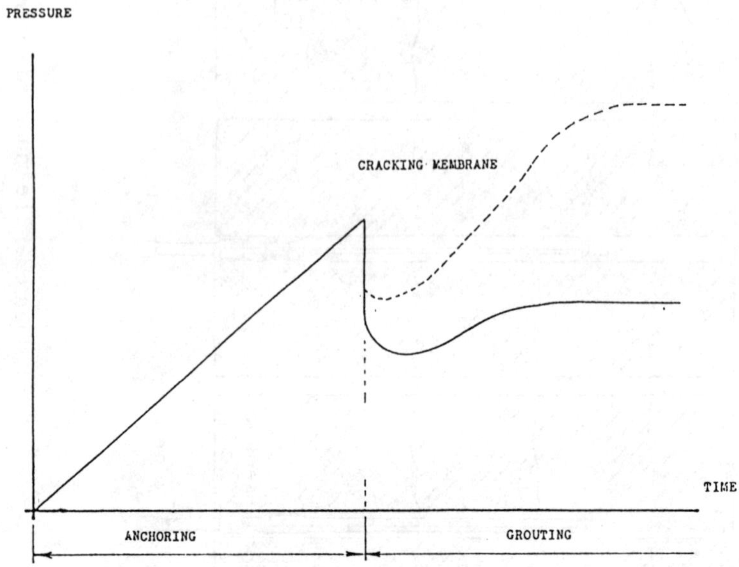

Figure 2: Anchoring and grouting pressure

- as a low pressure injection, when grouting pressure is smaller than pressure necessary for the expansion of bolt. Then the bolt is expanded and the injection membrane is broken by water from bolting pump and in second step proper grouting is realised with low pressure from special injection pump.
- as a high pressure injection, when the expansion of bolt, breaking of the membrane and proper grouting is realised directly with grouting material from only one pump.

It is also possible to detach the procedure of bolting from the process of grouting. In spite of bolting a technology of injection can be realised later, sometimes after weeks or months after excavation according to the course of convergence in the cavern. In some cases the delay can occur because of widening of joints in rock mass.

3 Rock deformation along grouted bolts

The analysis of fully bonded rock bolts is still one of the most up-to date questions in the theory of rock reinforcement. Several results of measuring of bolt stresses were published in the last decades. (T.J.Freeman[1]),(Tao Zhen Yu[3]). Typical results of in situ measurements led to the mechanical model (Fig. 3.) (Sun Xuey[2]).

While working our research programme we tried to compare the mechanical model with the results of mathematical modelling (P.Janas[4]).

During driving the stress - strain pattern in surrounding rock mass is changing due to the distance of the site from driving face. Only in certain distance from the face, when the new stability state occurs, it does not change more. The strain changes correspond to deformations and displacement of rock mass

The FE software package GEM 22 and GEM 32 (R.Blaheta[5]) was used for the analysis of displacement of rock mass surrounding the underground opening.

The mathematic model contains vertical circular underground opening of diameter 10 m in block of rock of size 150x150x110 m. The opening reaches the depth 56 m, i.e.one half the height of the block.Vertical strain of 18 MPa affectes upper bounds of the block. Vertical displacement on the lower bounds and horizontal displacements on lateral bounds were zero.Elastic modulus of rock was 2900 MPa and Poisson´s ratio was 0,25. The value of the primary strain corresponding to the Poisson´s ratio was 6 MPa. The task was solved in elastic regime.

Results of modelling are the values of displacement and strain in particular points of rock mass, corresponding with nodes of modelled mesh. Rock bolts are applicated into rock mass, which is more or less deformed. Rock deformations go on with changing distance from face and bolts are successivly loaded according the way of anchoring, properties of grouting material etc.

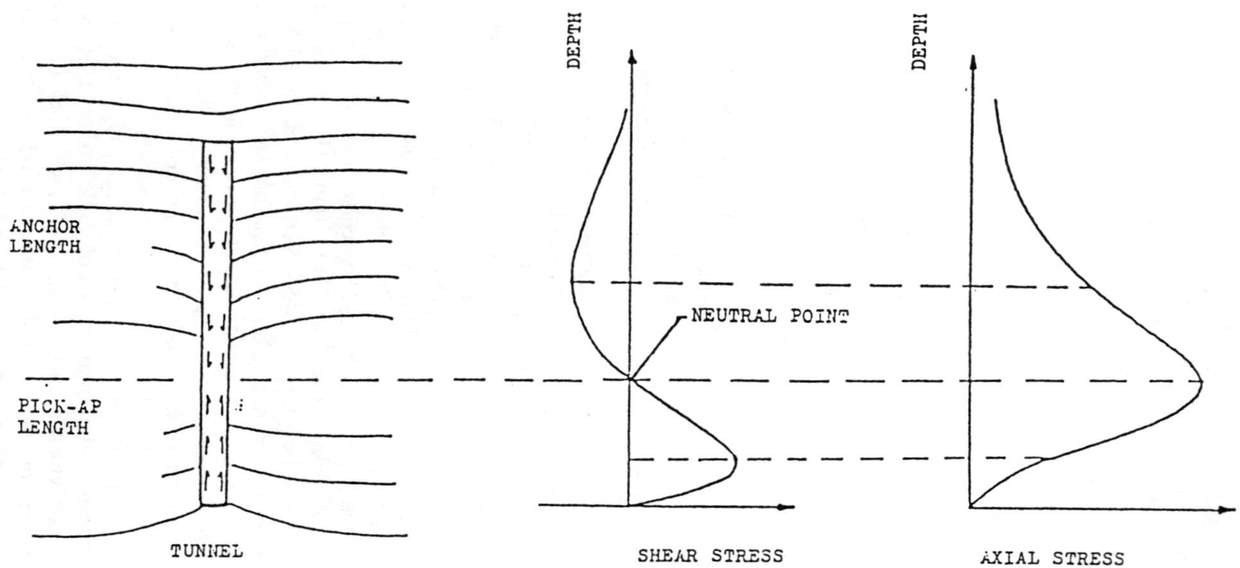

Figure 3 : Stress distribution model for grouted bolts (Sun Xuey 1981)

Analysis of the rock deformation as a base for determination of rock bolt stress results from computed values of the displacement. It was provided as follows :

a) The displacement at the moment of bolt installation was considered as zero.

b) Displacements in particular distances from the driving face were computed by mathematical modelling (GEM), final maximum displacement was calculated by classical Kirsch solution.

c) Changes of the radial displacement from the moment of bolt installation to the final displacement in relation to the distance between considered points of bolt are relative radial deformations.

d) Distances between considered points were voted:

- between nearby nodes, which represented deformation along a fully bonded bolt,

- between the contour of opening and conidered point in rock mass, what represented a point anchored bolt.

e) The values of distance from the face and bolt lengths were expressed as multiples of opening diameter.

The results are presented on Fig. 4. The separate curves express different distances of bolt installation from the face. The sign ".- " means tensile deformation, the sign " + " indicates compression.

The figures show, that the deformation along fully bonded rock bolt depends on the distance of installation site from driving face and it is not influenced by the length of bolts.

While with installation immediately in the face (x = 0 - 0,5 d) there is only tensile deformation along borehole, with distance of installation from the face sleeve on the pipe bolt, which seals (after expansion of the bolt) x = 0,1 - 0,5d , the compressive deformation occurs in the outer part of the borehole and the bolt can be even loaded partly with compression. (In fact , due to prestressing of the outer part of the bolt and lower quality of grouting in the mouth of borehole, it appears mostly as decreasing of tensile stress).

Briefly can state, that 3D FEM model, which was applied, corresponds to the results of in situ measuring of fully bonded bolts and confirms published stress-strain distribution along grouted rock bolts.

The previous results lead to some practical conclusions in the field of bolting technology :

- efficiency of fully bonded bolts decreases with increasing distance of installation site from the face of driving. It is profitable to install rock bolts as soon as possible.

- it is suitable to provide certain prestressing on the outer end even in fully grouted bolts

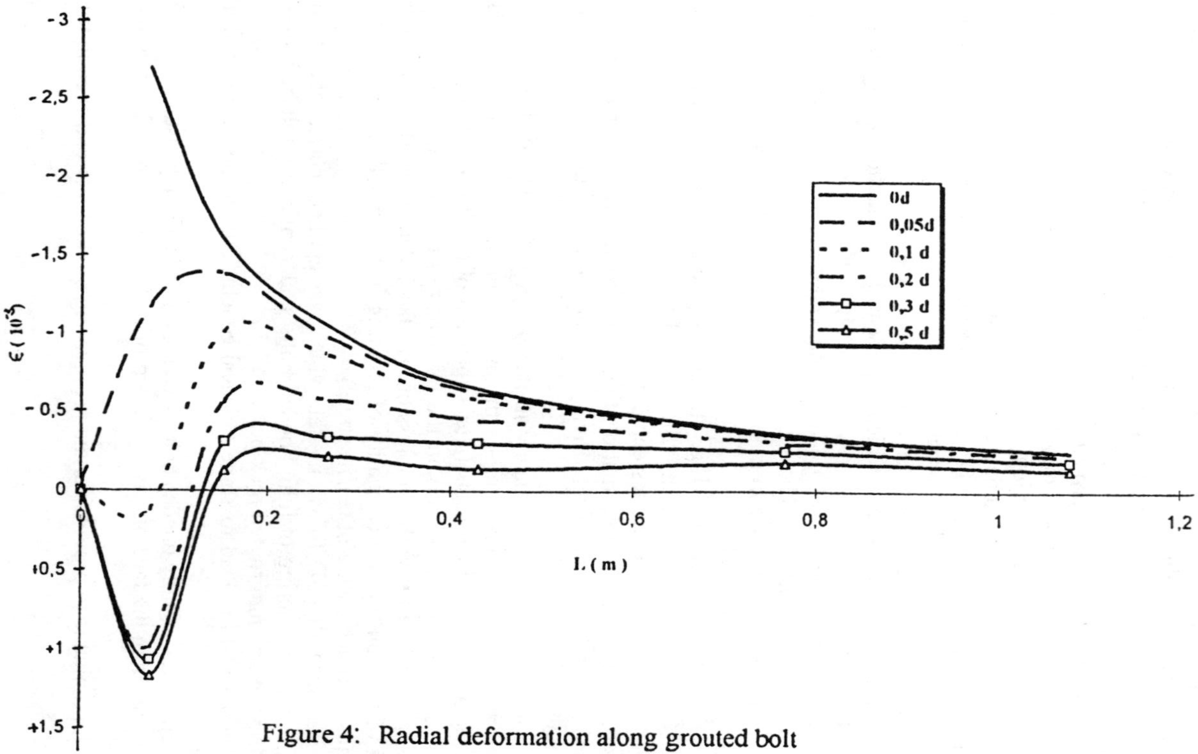

Figure 4: Radial deformation along grouted bolt

4 Stabilising injection of rocks

The reinforcement of rock mass by grouting in tunnelling is used with the goal to ease excavation, to improve the supporting effect of the rock and eventually to reduce the support inside caverns.

A stabilising measure in jointed rocks is to fill joints with grouting material which (after hardening) reinforces the rock structure. Especially wide joints play the main role in the deformation of rock mass. The filling of joints depends on the width of joints in relation to properties (mainly viscosity) of grout.and grouting pressure. The lower viscosity of grout and the higher grouting pressure, the better fulfilling of joints. From the view of grouting pressure, stabilising injection of rocks can be seen as a limiting case between fill grouting and crack grouting. The limit of grouting pressure is influenced by stress state in rock mass and by strength properties of rock in relation to the depth of packers from tunnel walls.

The combined system of reinforcement makes it possible to raise the grouting pressure because the grouting process is performed after strenghtening of rock mass by bolting. The risk of moving blocks of solid rock between joints into a cavern is minimised. The higher grouting pressure improves effectivity of grouting especially in conditions of thin and filled joints.

From the analysis of rock deformation along grouted bolts it is obvious, that the grouting is more effective in the part of surrounding rock mass, where expansion takes part and joints are burst open. Practise proved,that (when using
combined reinforced technology of bolting and grouting) it is profitable to inject rock mass above the depth corresponding to described neutral point The scheme of injection together with bolting is shown in Fig. 5.

5 Experience

The combined reinforcing system has been used first of all in coal mines in conditions of carboniferous rocks like shales and coal.

Hitherto experience of using of the combined bolting and grouting system in coal mines and underground structures are encouraging. It was proved, that bolts reinforce rock mass ring on the perimeter of excavation and prevent from the creation and increasing of fissures, so that the amount of waste grouting material flowing into the excavation decreases. So bolts can partly or fully replace the first (packing) step in two-step injection procedure. The combined technology make it possible to use higher injection pressure. Due to immediately anchored bolts the total amount of grouting material also decreases in 20 - 30 %. This type of bolts

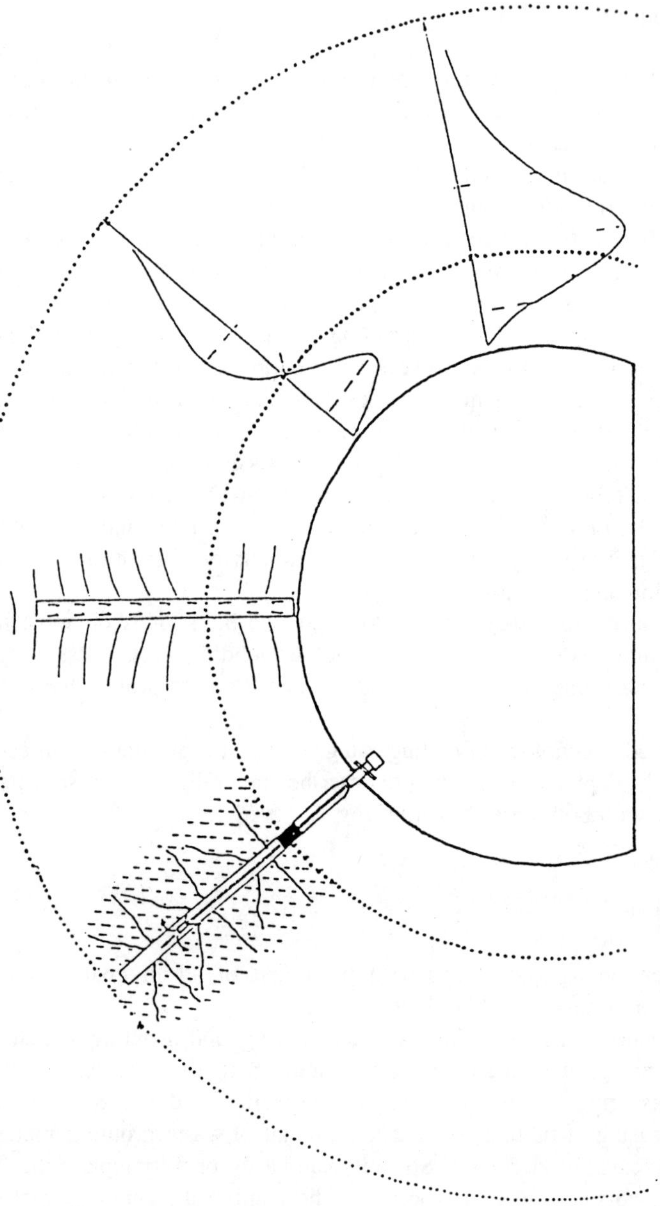

Figure 5: Sheme of combined reinforcing system

can be anchored in weak rocks too.

Practice confirms, that combined bolting and grouting system Boltex brings not only economical effect due to cancellating of certain operations, but it is advantageous from the geomechanical point of view, too.

6 Conclusions

The important advantages of the combined bolting and grouting technologies can be summarized :

- very quick and easy installation
- low requirements on the accuracy of the borehole diameter
- immediate bearing capacity of bolts
- prestressing between pad and rock
- wide range of injection pressure
- unlimited length of bolt column
- arbitrary position of packers along the length of bolt
- brick-box character of the system make it possible to adapt the system to the real conditions of using.

Acknowledgements

The research was provided with the financial aid from the Grant Agency of Czech Republic - project No.105/97/0967 .

References

1. Freeman T.J.: The behaviour of fully bonded rock bolts in the Kieldar experimental tunnel. *Tunnel and Tunnelling* (1978)
2. Sun Xuey . Grouted rock bolt used in underground engineering in soft surrounding rock or in highly stressed regions. In: *Proc. Symposium on Rock Bolting, Abisco,* Balkema, Rotterdam, Netherlands (1981)
3. Tao Zhen Yu, Chen Jie Xian : Behaviour of rock bolting as tunnelling support. In: *Proc. Symposium on Rock Bolting, Abisco*, Balkema, Rotterdam, Netherlands (1981)
4. Janas P. : Mathematic modelling of different bolts in rocks. Report of the Grant Project 103/94/0777, *Report of the Research Mining Institute,* Ostrava Radvanice (1994) (in Czech)
5. Blaheta R., Jakl O., Kohut R., Kolcun A., Slíva J.: GEM 22. Revised User Guide, *Report of the Institute of Geonics,* Ostrava (1993) (in Czech)
6. Snupárek R.,Paloncy L.: Bolting and grouting system Boltex. In: *Proc. International Conf. Underground Structures,* Prague, Czech Rep. (1994)

STABILITY OF ROCK IN TUNNEL AND DESIGN OF SUPPORT

BORIS AMUSIN
E-mail: boris_amusin@hotmail.com

The author has developed a theoretical-experimental method (TEM), which can increase the reliability in designing support parameters, especially for the deep mines. The more precise parameters of stress, strength and creep for the discontinuous rock mass are determined automatically in the computer program "TUNNEL", by inputting the short-term data. This includes simple measurements of the displacements of the actual opening or prototype-opening contour. In the report the correlation between the mechanical properties of the rock mass and the quantity of discontinuities; some important geomechanical peculiarities; the TEM; and the program "TUNNEL" will be discussed.

1. Introduction

Generally poor geological and geomechanical data are obtained from initial investigations carried out before the planning of underground construction. Therefore design parameters of the support of Underground Constructions become unreliable even the design method used were perfect. Traditional design methods based on empirical or theoretical approaches were not reflected real process of interaction support with rock mass.

One of the most widespread defects of empirical methods is its approach for calculating the load on the support. For example, according to the Q-method (Barton [1]), the load on the support is determined depend on Q = rock mass quality and J_r = joint roughness coefficient. If Q and J_r are the same for different depths and ratio of the rock strength to stress state is also constant (SFR = constant) then, the loads on the support will be the same for small depths as well as for large depths. This conclusion however is proved to be false by experimental and theoretical investigations. The Russian empirical U method (Basinski [2]) also concludes the same inaccurate results. This method is based on a statistical generalization of field measurements. It predicts opening stability, depending on the calculation of long-term tunnel contour displacements (U_0) for conditions using a yielding support with minimum capacity (P_{min}). For many of the practical cases using yielding support, the U-method gives us good results. This method however has the same defect as the Q and other empirical methods. It adopts the same assumption regarding to the equality of the loads on concrete support for different depths, if the displacements U_0 or Q-factor for these depths are equal. The assumption is not confirmed by the results of field data and theoretical methods (Basinski [2], Amusin [3], Ardashev [4], Amusin [5]).

Improving of the design methods has become important, especially in recent decades when the productivity, capacity and cost of mines and tunnels have increased, while geomechanical conditions have worsened. Designing mistakes are becoming more and more expensive (Amusin [6]).

Back Analysis Method (BAM) led to improved design of support parameters. For the condition of inelastic rocks however, the number of unknown mechanical parameters are large and this method becomes unreliable.

The present writer is an advocate of the practical use of a somewhat different underground construction design method: the Theoretical-Experimental Method (TEM). The idea of this method is close to BAM, but in this method correlation formulae are used for determining the initial parameters. A system of empirical coefficients is introduced into the original data, which determines the state of stress and the laboratory compressive strength of the rock (Amusin [7]). An axis-symmetrical visco-elasto-plastic solution is used in the TEM as a basis for predicting the rock deformations and failure. It allows the important regularities of deformation and failure of rock mass in time to be considered.

2. Estimation of Mechanical Properties of the Rock Mass

2.1. Strength Characteristics.

For most rocks the Mohr-Coulomb strength condition is preferable. Author introduced into the traditional condition two factors, in order to take into account time dependent (f_t) and post failure deformation (f_p):

$$\sigma_1 - \beta\, \sigma_3 \;\; S_c = S_m f_t \cdot f_p \qquad (1)$$

Where: $\beta=(1+\sin\phi)/(1-\sin\phi)$ (2)

σ_1 and σ_3 are major and minor principal stress; S_m is the strength of rock mass, determined by the empirical formula:

$$S_m = S_c \cdot f_s \cdot f_d \cdot f\alpha \qquad (3)$$

S_c is laboratory compressive strength; factors f_i take into account the scale effect (f_s); the direction of the opening relative to the strike direction (f_d) and dip angle α. Factor f_t is the decrement in strength with time. Determination of the parameters f_i can be found in (Amusin [7]).

An angle of internal friction (ϕ) is another important characteristic of strength of the rock mass. For a small value of ϕ, the size of the inelastic deformation zone can be very large. In this case the influence of the resistance of support will be insignificant on decreasing the displacements of contour. These deformations continue for a long period. For a large value of ϕ, a failure zone will be stabilized in time. The deformations of a contour are determined by the dilatation of the rock. The resistance of support will promote a decrease the displacements. In specialized literature you can find different opinions about the relationship of strength conditions for solid and fractured rocks. Very often the strength conditions for a rock mass are contradictory. It is well known that weak contacts lead to a decreasing internal friction angle. Even when the rock is strong, but the joints between blocks are filled with clay, we must use support with high capacity, because the integrated angle of internal friction of the rock mass will be close to zero. There are no reliable recommendations for determining this parameter for heterogeneous rock masses. Here an indirect method is suggested for determining the effective angle of internal friction for the rock mass.
Using the elasto-plastic solution (Amusin [5]), the author found that when displacements U_r at a distance r (from the center of the opening to the measurement point), are divided by the displacement U_0, the resulting function will be a function from only one parameter. This parameter is the angle of internal friction:

$$\phi = 1/n \, \Sigma \arcsin [(a_i/b_i - 1)/(a_i/b_i + 1)] \qquad (4)$$

Where
$$a_i = \ln (U_i/Uo); \; b_i = \ln (r_0/r_i) \qquad (5)$$

n is number of measurement points; $1 < i < n$; r_0 and U_0 are the radius contour of the opening and the displacement, respectively; r_i and U_i are the distance from the center to the point of measurement and the displacement at this point. Using this idea and data from field measurements, author determined angles of internal friction for heterogeneous rock mass. For most practical situations, values of internal friction angles were much less then in a laboratory (Amusin [7]).

2.2. Elastic Characteristics and Creep Characteristics.

The determination of the elastic characteristics of rock (the modules of elasticity E and Poisson's ratio μ) is now as common as the determination of the compressive strength. These characteristics however, are determined mainly on small specimens. Approach for estimation of elastic and creep parameters of the rock mass are given by author in (Amusin [7]).

3. Theoretical Model of Strata Control in Opening

In the TEM the solution for an axis-symmetrical, visco-elasto-plastic problem (Amusin [5]).is used. It allows one to take into account the effects of post-failure deformations, creep of rocks and their failure in time. In setting up the problem, the following boundary conditions are adopted. An extended circular shaped opening is located at a depth of H. Pressure P_0 corresponding to the support resistance is applied to the internal contour of the opening. The stress field of a virgin rock mass is assumed to be hydrostatic: $Q_h = Q_v = gH$. It is assumed that three zones of non-elastic deformations are formed around the opening. Their size depends on time, stress levels and mechanical properties of rocks mass:
A_1: zone of failed rocks close to the contour. Displacements here take place without dilatation; A_2: zone of intensive fracturing (post-failure) with dilatation; A_3: zone of plastic non-loosening (limit) deformations.

The boundaries of those zones (r_1, r_2, r_3 respectively) are determined by conditions of the continuity of radial stress (σ_3) and tangential strain (ε_θ) from one zone to the neighboring one. The zones A_j correspond with the curves of rock deformations on a rigid press.

The theoretical solution permits one to obtain the deformations for different types of rock. For example, in salts, A_1 and A_2 zones are not formed even for great depths because of considerable limited deformations without failures, which are typical for salt. The openings in this case can be exploited without supports. In brittle rocks (for example, coal) which are not subject to plastic deformations, A_1 and A_2 zones do get developed, and support is necessary for the stability of the opening. The solution permits one to manage without a priory assumptions. For example, according to the theoretical solution, value U_0 does not alone determine the value P_0 as in most of the empirical approaches.

Assuming that for a given rock, any change in depth or stress will change both the strength (S_m) and shear modules (G_m) of the rock mass as follows:

$$Q_v/S_m = \text{constant}; \quad G_m/S_m = \text{constant}; \quad \phi = \text{constant}; \qquad (6)$$

where: $Q_v = gH$, g = the unit weight of rock; ϕ = the angle of internal friction of the rock mass. In this case, the final displacement U_0 remains the same, for the given resistance of support $P_0= 0$ (no support or $P = P_{min}$), independent of the depth. If P_0/Q_v = constant, then the rock mass reaction curve for the different depths also remains constant. For the given support reaction P_0=constant, different curves are obtained for the different depths (H_1 and H_2). The ratio of displacements U_p/U_0 (where U_p is the displacement of the opening contour under the support reaction Po) is lower for the smaller depth, H_1, and the supporting effect is more than for the greater depth, H_2. This peculiarity is very important to the practice of support underground constructions. For example, if for the certain geomechanics conditions we can use concrete support for the small depth, then for the great depth with similar conditions (Uo great depth =Uo small depth) the concrete support will be failured. Similar results were for rock bolt supports in (Lama [8]).

The author (Amusin [7]) found proof of this peculiarity from field measurements of loads on concrete supports. For the same geomechanical condition, the permanent value U_0 was obtained when the support was yielding P_0=P_{min} and the relations H/S_c were constant. Different loads were measured however for the same conditions when supports were made of concrete. Thus, the value U_0 does not determine the value P_0 uniquely and the empirical assumption is not correct.

In the TEM only the initial data: mechanical properties and the Stress State of the rock mass are corrected. For the separate opening in-situ Stress State are assumed: vertical - $Q_v = gH$ and horizontal - $Q_h = \lambda Q_v$. The value of the λ is determined depending on tectonic conditions and the relation H/S_m. In order to take into account 3-dimensional character of problems semi-analytical correction coefficients are introduced into the Stress State. Thus instead of the real depth (H) author used effective depth $\underline{H}$:

$$\underline{H} = HK_i \tag{7}$$

The coefficients K_i allows us to take into account the influence of opening driving rate, types of junctions and the 3-dimensional character of the problem for chambers (Amusin [9]). For example, the influence of the opening driving rate (V, m/day) and distance from the face (L, m) are taken into account by the coefficient K_v:

$$K_v = 1 - \exp(-2.6L/B) \tag{8}$$

where $L = V \cdot t$; B = width opening; t= elapsed time of support installation. This allowance for the time (t) makes it possible to take into account, simultaneously, the distance from the face and the changes of rock deformations in time. While calculating the strata control in chambers, coefficient K_{ch} is introduced into Eq. (7) as follows:

$$K_{ch} = [1 - \exp(-1.3L/B)]^2 \tag{9}$$

where L and B are the length and width of the chamber, respectively.
For a junction of two openings or for parallel openings author introduced in Eq. (7) a coefficient K_{junct}:

$$K_{junct} = 1 + 1 / [\, a + r_b / r_o]^2 \tag{10}$$

where $r_b = L_p \, tg\,(\zeta / 2)$; L_p = the distance from the center of the junction to the calculated cross-section; r_o = half of the opening width, close to the calculated one; a = the empirical coefficient (for the cross openings: $a = 1.15\,(1 - \zeta / 100)$); ζ = the angle between two openings. For the parallel openings: a = 1 and $\zeta = 90^o$, Lp = a pillar between two openings.

The loads on the support in the TEM are determined by taking into account: the time of the support installation; creep of concrete; sizes and thickness of support; thickness and nonlinear properties of backfill. In order to take into account the heterogeneous structure of the rock mass, different stress states and properties of support, separate calculations are carried out for the roof, floor and sides (Amusin [9]).

4. The Automated Design System "TUNNEL".

The benefit of using a computer is clearly evident when an opening is located in unstable and very unstable rocks. In these cases, the reliable support parameters can only be determined by using a computer program for the structural and strength calculations. The use of the "TUNNEL" program will greatly increase the accuracy of the solution, as it enables the quick calculation of various possible cases of the support construction and the use of optimization criteria (Amusin [6], Lama [8], Amusin [9]). The input data in the system are the following: the thickness and the type of rocks in the core, their uniaxial compressive strength Sc, azimuth of strike direction, rock dip angle, the opening depth (H), type of opening (extended separate and parallel openings, shaft, junction of openings, chambers and two openings in different planes) and the sizes of cross-section. The support is modeled in the calculated scheme by the polygonal beam system with hinges.

The peculiarity of the calculation technique is in accounting for the cohesion and the tangent interactions of the concrete support with the rock mass and in estimating stability for the block support. The limit loads on the latter are determined by means of successive calculations while taking into account geometric non-linearity. Resulting from these calculations, the forces and the bending moments are determined for the given conditions and support rigidity. The ratio of bearing capacity to normal force in the weak support section determines the safety factor.

The "TUNNEL" system provides the solution of support optimization and the opening construction technology in a new way. The well-known criterion for optimization support consists of creation of support installation technology that results in minimum loads on the support with permissible contour displacements. The criterion appears to be erroneous rather often since the supporting costs are determined by the required minimum bearing capacity of the support while the latter not only on the maximum normal load but also on its distribution and the contact conditions (Amusin [9]). When the criterion is applied, the possible methods of reducing the loads on the support (by means of reducing the support stiffness and increasing the thickness and the yielding of the backfill, and increasing the time lagging in the course of erection of the rigid support, etc.) appear to be equal. However, if the support is designed based on the static and the strength methods, the effect of the above factors on bearing capacity may differ considerably. Therefore, for optimization of the technology of the opening and of the support parameters, it is proposed to use the criterion of maximum ratio of the support bearing capacity to the normal force in the weak elements of construction. The different kinds of the support with varying installation construction technology, which are rational for the preset conditions, are proposed to be analyzed on the basis of this criterion (Amusin [9]). It appeared, for example, that under some conditions (double track opening, width = 6m), at the full contact of the support with the rock mass, the support capacity can be twice as much as the same support installed with the backfill up to 0.5m in the roof. In other cases (single-track opening, width =3 m) the use of a yield backfill or a certain delay of the support installation from the face prove more efficient than the installation of the support with full contact. In that case loosening of the contact does not result in a decrease of the bearing capacity of the support.

The using of the correlation formulae and the system of the empirical coefficients allows for an opportunity to correct the discrepancy between the assumed and actual schemes of calculation. As a result the TEM can be used for a wide range of geomechanical conditions and can bring the calculated parameters to the real value. It is also very important to use the "TUNNEL" system to predict the initial Stress State and mechanical properties of the discontinuities rock mass. These parameters are calculated in the program from back analysis of short-term measurements of the contour displacements within time span of one month. For the design of large openings and junction measurements of displacements can be carried out in small prototype openings. Some examples of case histories are given in (Amusin [10]).

References

1. Barton N., Lien R., Lunde J. (1974). "Engineering Classification of Rock mass for the Design of Tunnel Support," *Rock Mech.*, **V 6**, No 4, 189-236.
2. Basinski Y.M., Ivanov E.A. (1972). "Dependence of Rock Pressure Phenomena in Permanent Workings Deep Shaft of the Central Donbass from Main Geological and Mining Parameters", *Tr. VNIMI*, Leningrad, No 85, 79-84 (in Russian).
3. Amusin B.Z., Ardashev K.A., Koshelev V.F., Linkov A.M. (1979). "Post limiting Rock Deformation in Support Design Calculations" *Soviet Mining Science.***V 15**, No 5, 529-532.
4. Ardashev K.A., Amusin B.Z., Koshelev V. F. (1985). "Numerical Methods of Analysis of Underground Construction," *5th Int. Conference on Numerical Methods in Geomechanics* /Nagoya/ 1-5 Apr., 1077-1084.
5. Amusin B.Z.(1977). "Prediction of the Stability of Main Workings with Allowance for Gradual Fracture of the Rock in the Zone of Inelastic Deformation." *Soviet Mining Science.***V 13**, No 5, 460-466.
6. Amusin B.Z., Ardashev K.A., Basinski Y.M. (1984)."Geomechanics Principles of Planning Mine Workings." *ISRM Symposium on Design and Performance of Underground Excavations*, Cambridge, 3-6 Sept. 1984, Publ. London British Geomech. Society, 81-87.
7. Amusin B. Z. (1994). "A Method for Design Support of Underground Construction", *Proc. Symposium "Underground Openings for Public Use", Gjovik, Norway* 14-17 June, 65-77.
8. Lama R.D., Amusin B.Z. (1992). "Rigid or Yielding Roof Bolts: At the Face or away from the Face". *X1 Intern. Conf. on Ground Control in Mining*, Wollongong, NSW, July 1992.
9. Amusin B.Z., Bulichev I.N., Koshelev V.F. (1989). "Influence of Rubble Stowage Chamber and Contact Conditions on the Carrying Capacity of a Support System." *Soviet Mining Science.* **V 25**, No 3, 243-248.
10. Amusin B.Z., Shick V.M., Lama R.D. (1992). "Using Back Analysis to Estimate Geomechanical Field Parameters for the Design of Support Systems for Tunnels"."*Tunneling and Underground Space Technology*", V7, No 3, pp.281-284.

Drilling Model

EVALUATION OF DRILLING PARAMETERS OF A PDC BIT

H. GEOFFROY [a] , D. NGUYEN MINH, H. MAITOURNAM, J. BERGUES
LMS-Ecole Polytechnique,
91128 Palaiseau, France
E-mail: duc@athena.polytechnique.fr

C. PUTOT
IFP, 3-4 Avenue Bois-Préau,
92506 Rueil-Malmaison, France
E-mail: claude.putot@ifp.fr

A drilling model for PDC bit is presented herein within steady bit advance hypothesis, with taking into account influence of wear and its evolution. First, friction mechanism on cutter wear flats, is studied by experimental and numerical approach. It is shown to correspond to a relationship between the normal applied pressure to the cutter wear flat, and its subsequent dip angle α into the rock surface. It is shown that frictional wear is directly influenced by bit profile. Typical results are exhibited for elliptic profiles and compared with literature.

1 Introduction

Polycrystalline Diamond Compact (PDC) bits were introduced in petroleum industry in the early 1970's. A PDC bit is composed of a multiplicity of individual cutters mounted on the surface of the bit body (Figure 1). The high rate of penetration (ROP) of these bits is due to shearing action of cutters instead of crushing, producing faster penetration rates with less weight on bit (WOB) than roller cone bits. A sharp PDC bit will typically drill two to three times faster than the best tricone in soft formations, but such performance will gradually decrease with increasing wear.

In normal drilling conditions, e.g. under moderate temperature condition not exceeding $350°C$, abrasion is the main cause of wear, resulting in growth of a wear flat on the cutter base (Glowka[2]). Friction on wear flat can dissipate up to 50 % of the total energy for drilling. For this reason, a better understanding of bit performance has to consider the worn tool. Although many studies have dealt with this subject in literature, the frictional mechanism generated by wear on cutters has still been overlooked.

This study is devoted to model the frictional mechanism on a single worn cutter in relation with rheology of the rock, in order to integrate it in a global drilling model. Illustrative examples are given for some bit geometries.

---

[a]Now in LGM-ENTPE, 69518 Vaulx-en- Velin, France

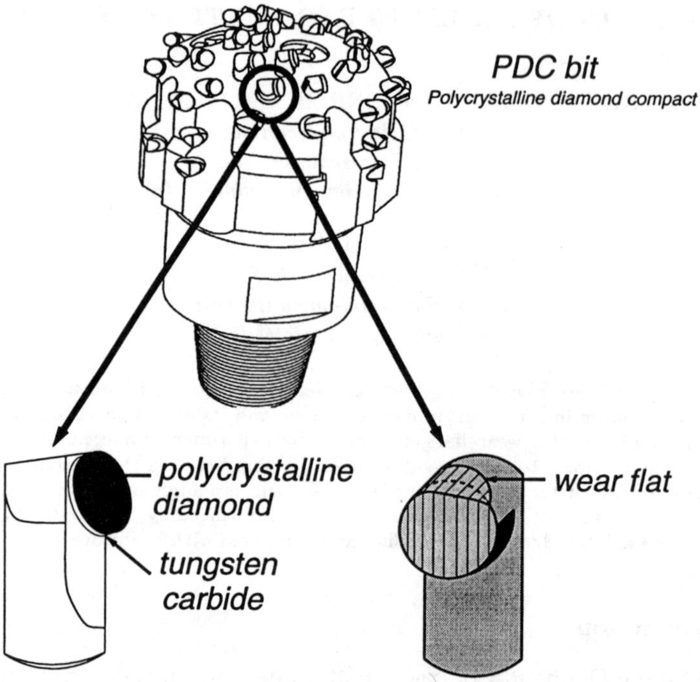

Figure 1: A PDC bit

2 Mechanisms involved by a worn PDC cutter

A cutter model with wear flat is shown on Figure 2. Let n and t be the local cutter axes in a plane containing the displacement vector of the cutter, and normal to the wear flat. The forces applied to the cutter are usually considered to share into two forces related to independent mechanisms: cutting force on the cutting face (superscript c), and a frictional one on the wear flat (superscript f) (Detournay [3], Sellami [4], Kuru[5], Nguyen Minh [6]). Their coordinates in (n, t) plane (subscript n, t) are :

$$F_t = F_t^c + F_t^f \text{ and } F_n = F_n^c + F_n^f \tag{1}$$

Nguyen Minh[6], Sellami[4] considered cutting as a yield load mechanism. Indeed, this concept is somehow equivalent to specific energy one ϵ_0 (Detournay [3], Kuru[5]), defined as the amount of energy to remove a unit volume of rock, with

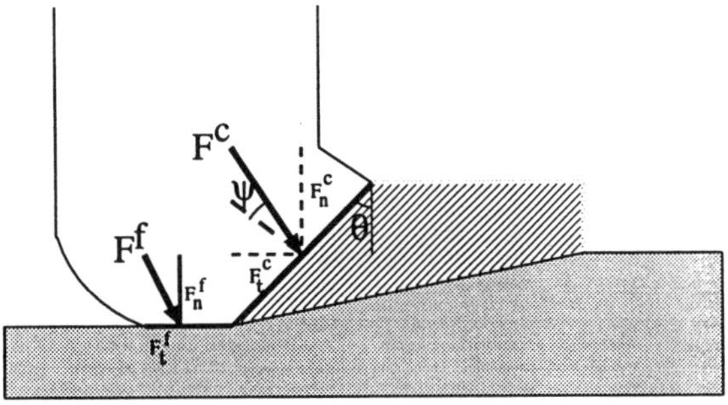

Figure 2: Forces acting on the worn cutter.

ϵ_0 of the order of the rock compressive strength in atmospheric conditions. We have :

$$F_t^c = \epsilon_0 A_c \text{ and } F_n^c = \zeta F_t^c \tag{2}$$

A_c is the cross-sectional area swept the cutter in its movement. ζ is given by $\zeta = \tan(\theta + \psi)$. θ is the back rake angle and ψ the interface friction angle.

Things are not so clear for the frictional mechanism. Kuru[5], sets the mean normal pressure on wear flat to point load strength, Sellami[4] to the normal cutting stresses, and Detournay to the uniaxial compressive strength.

Alternatively, on the basis of experimental observations, Nguyen Minh [6], proposes a mechanism related to elastoplastic deformations. The author performed drag tests on a flat based cutter on a model material standing for a plastic rock (sand mixed with paraffin). The cutter penetrates into the horizontal rock surface with a dip angle α, increasing with the normal applied load unless the surface bearing capacity of the sample is exceeded. This behaviour would fully represent the friction mechanism on the wear flat, but needs to be verified, namely by experimental confirmation on actual rocks and by numerical validation.

This is the first objective this study, which aims at finding out, for a given elastoplastic rock behaviour, a relationship between α and the applied forces on the wear flat. For a flat nose bit, with cutter wear flats parallel to bottom face, angle α represents the dip angle of the helical trajectory of each cutter (Figure 3).

With a typical penetration rate ranging from 15 m/h to 60 m/h, and average 120 rev/mn rotary speed, penetration per revolution δ range from 2 to 8 mm. So, for a 216 mm diameter bit, we have typically $3.10^{-3} < \tan\alpha < 6.10^{-2}$

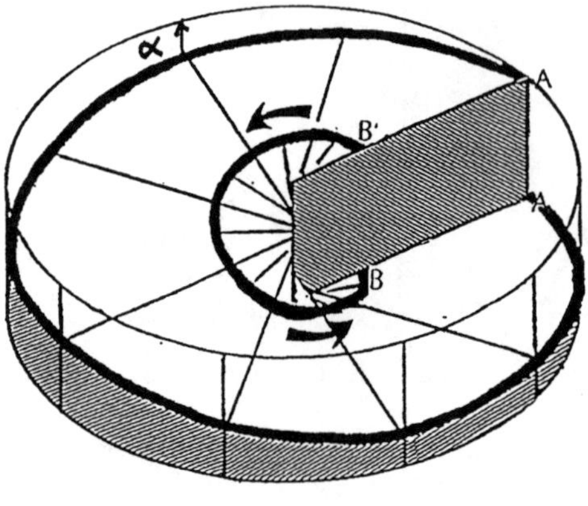

Figure 3: helical trajectory of a cutter on a flat nose bit, and α angle

3 Study of the frictional mechanism

3.1 Experimental analysis

In order focus on parameter α, tests were performed with constant normal applied load (Figure 4). With regard to the small α angle involved, it would not be possible to run classical tests with constant depth of cut. A circular sample is rotated by an electric motor, allowing measurement of an average penetration slope of the cutter on a great distance, e.g. on several rounds. The cutter is mounted on a ram fixed on a rigid frame. The ram can be placed at different radii in order to simulate different cutter positions on the bit. Forces applied on the cutter are recorded by a three dimensional piezoelectric force gage, and depth of penetration by LVDT. The rock tested is a chalk from Liege (Belgium). Its complex behaviour exhibits important plastic deformations due to the gradual breakage of the porous structure, both under hydrostatic and deviatoric loadings. This rock was selected because of its good homogeneity and small grain size, and its low strength.

Two different tests were performed with iron cutter models, one with a sharp circular cylinder with back rake angle 20° and 13 mm diameter, to identify cutting mechanism, the other with a flat based parallelepiped 10 mm long, 3 mm large and 6 mm thick with 0° rake angle θ to enhance friction mechanism. (Geoffroy [7]). Depth of cut (d) versus displacement of the cutter

Figure 4: Testing apparatus

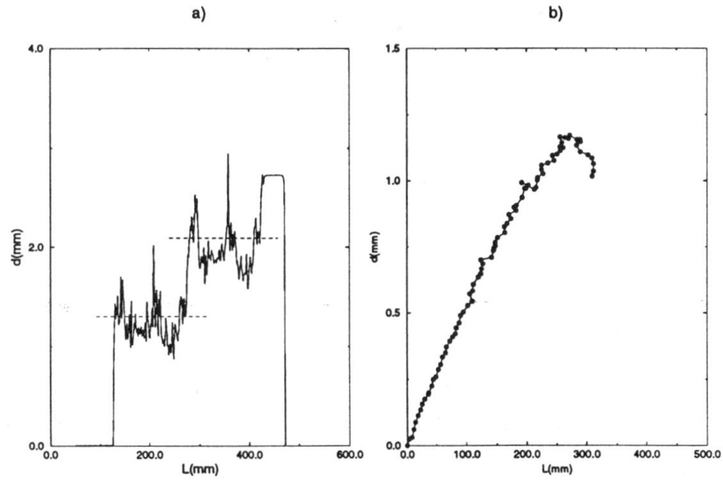

Figure 5: Depth of penetration d versus distance L. a) Cutting mechanism, b) Frictional mechanism

(L) curves show the difference between the two mechanisms (Figure 5). The sharp cutter stabilizes rapidly to a definite depth, whereas the flat based parallelepiped penetrates gradually into the rock with a mean dip angle α, which has to be related to the mean applied normal stress.

3.2 Numerical analysis of the frictional mechanism

Formulation

Within the hypothesis of two independent mechanisms of the worn tool - cutting and friction - the frictional contact between the cutter wear flat and the rock can be idealised by a plane strain problem, with a constant velocity moving rigid slider inclined by angle α on the surface of a half plane (Figure 6).

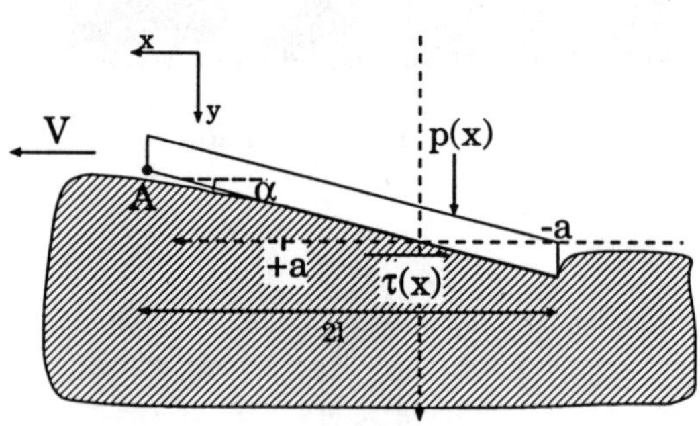

Figure 6: Wear flat model

This is a frictional unilateral contact problem, with unknown contact length $2a$, normal pressure condition $p(x) \geq 0$ and shear stress related to normal pressure by a Coulomb's law $\tau(x) = \mu p(x)$ on contact area $[-a, a]$. This problem is resolved by means of an algorithm of contact described in more details in (Geoffroy [7]). It has been validated, using an analytical elastic solution of this problem (Adachi [8]).

The steady motion of the slider is simulated, using a specific steady algorithm in plane strain finite element code. This algorithm allows, in a reference moving with the cutter, to treat the stationary problem on a fixed geometry (Maitournam [9]).

Two models were used to describe elastoplastic rock mass behaviour.

A Von Mises law, as a perfect elastoplastic model, is intended to confirm the α vs F_n^f relationship revealed by tests, and to verify whether the "frictional mechanism" may be related to a deformation of elastoplastic type, and to show how transition to the limit point load occurs.

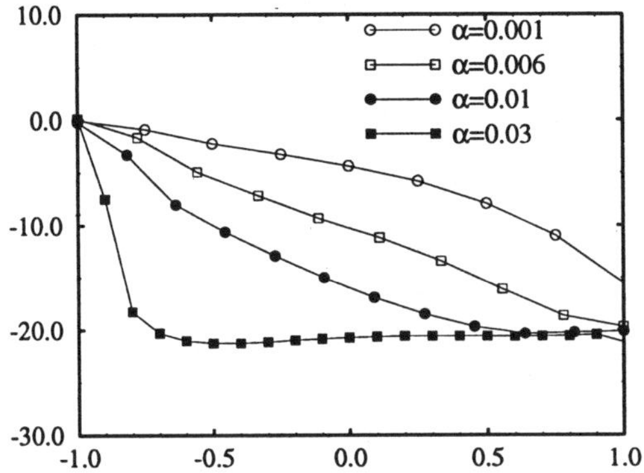

Figure 7: Evolution of the distribution of normal pressure beneath the punch for different values of penetration angle α.

A Cam-Clay model precises the plastic volumetric strains inside the rock medium, taking into account strain hardening and softening, pressure dependency, dilatancy and pore collapse of chalk.

Indeed, an approximate solution of the frictional problem is obtained in two steps. First, the smooth slider is considered, $2a$ and $p(x)$ are calculated for a prescribed α angle ($10^{-3} \leq \alpha \leq 5.10^{-2}$ rd). Then, for the frictional case ($\mu = 0.5$), pressure distribution $p(x)$ already obtained in the former calculation and a shear stress distribution $\tau(x) = \mu p(x)$ are prescribed on the contact length $2a$, so that a mean α angle can be evaluated.

Numerical simulation data

Let P be the mean normal pressure on the slider contact area.

With Von Mises model, P initially increases with α until 2.10^{-2} rad, then reaches a stationary value, which appears to be close to the theoretical bearing capacity of a flat punch (Figure 8). In the same way, as α increases, the distribution of normal pressure $p(x)$ under the punch tends towards a uniform distribution when limit point load is approached (Figure 7).

But, with Cam-Clay model, P increases continuously, due to strain hardening, and an ultimate bearing capacity does not appear in this case. For a

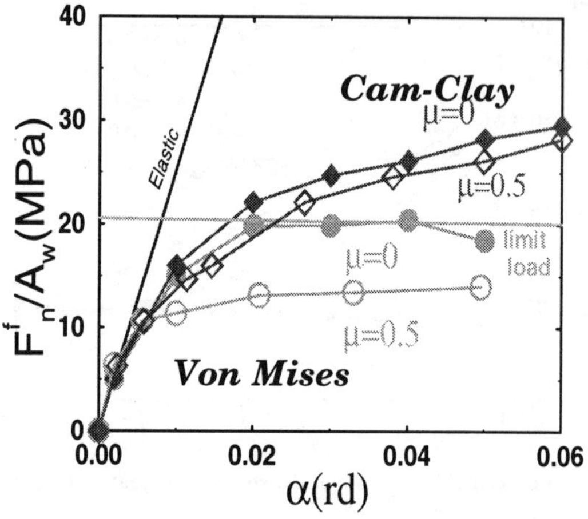

Figure 8: $P_0(MPa)$ versus α : Von Mises and Cam-Clay laws

given mean applied normal stress P, α is larger if friction increases.
Plastic volumetric strains under the slider have been represented for Cam-Clay law in the smooth slider case (Figure 9). Contractant zones appear beneath the punch (pale zone, water drop shape). Behind the punch, near the surface, a thin contractant zone appears too, overlying a more important dilatant region in the rockmass (dark zone) . However, with frictional contact, all the domain turns to be contractant.

3.3 Conclusion

We have shown that friction mechanism on cutter wear flat can be understood as an elastoplastic driven mechanism which expresses as a relationship $P = f(\alpha)$ between the mean applied normal pressure P on the wear flat surface, and its subsequent dip angle α. This angle will be locally determined by the movement of the wear flat relative to rock surface. To sum up, simplified forms can be proposed, which take into account an eventual existence of a limit load σ_0:

$$f(\alpha) = \sigma_0(1 - \exp(-\alpha/\alpha_0)) \text{ if a limit point load exists} \tag{3}$$

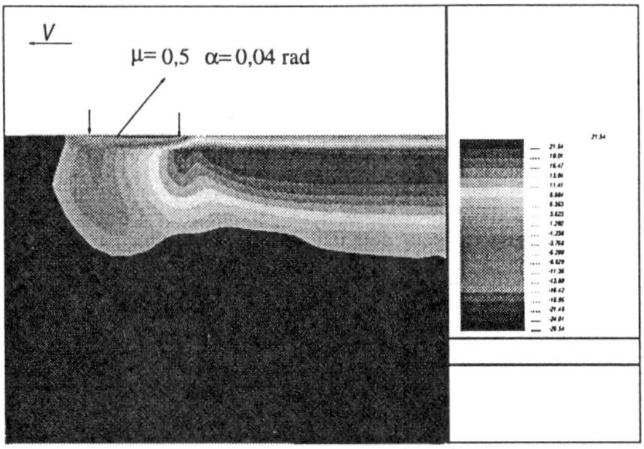

Figure 9: Cam-Clay law ; plastic volumetric zones ; $\mu = 0.5$; $\alpha = 0.05$ rd

$$f(\alpha) = \sigma_0(1 + \beta_0\alpha - \exp(-\alpha/\alpha_0)) \text{ otherwise} \tag{4}$$

The parameters are function of the rheology of the rock and could be obtained numerically or experimentally. Remark that the slope at the origin is related to elasticity and should be proportional to the elastic modulous (Adachi[8]), which should imply a relationship between σ_0, α_0, β_0. Indeed such a relationship has to be ignored with such a simple formulation.

4 A drill bit model

It is now possible to propose a drill bit model with taking into account wear effect and wear evolution, provided a relevant wear law be choosen. The geometry of the bit has first to be precised.

4.1 geometry of the bit

Let us define the bit profile (curve $z(y)$), as the locus of all the PDC centres when projected on a same fixed meridian plane (y, z), moving along with the bit on direction z (Figure 10). This profile is supposed to be a smooth curve $z(y)$. A cutter on the bit is then defined by a point on the bit profile and its angular position θ. The bit is supposed to proceed steadily, with constant rotation and advance rate.

As no viscous effects or cutting evacuation problem is considered herein, the movement of the bit can be simply described by its depth of penetration δ per revolution in the bit axis direction ($\delta = \dfrac{ROP}{N}$).

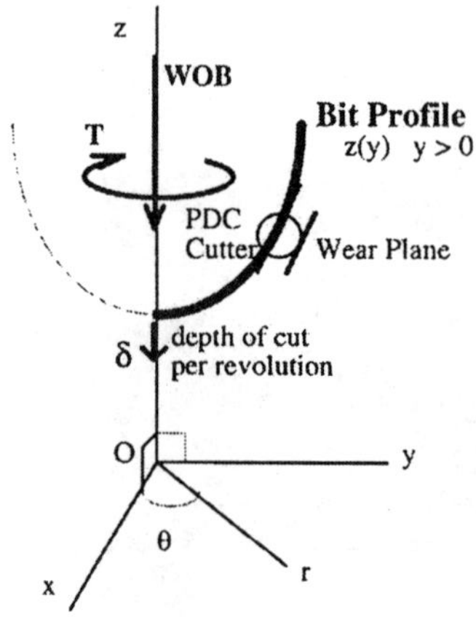

Figure 10: Schematic diagram of bit profile in meridian plane

4.2 Statement of a wear law

Describing wear evolution needs defining orientation of the wear flat, and the volumetric wear rate.

It is usually admitted that wear flat develops parallel to the wall of the rock e.g. practically tangent to the axisymmetric volume defined by the bit profile, which is supported by experimental evidence (Warren, [11]).

Volumetric wear law is usually proposed in the general form (Glowka, [10]):

$$\frac{dV_w}{dl} = C^* \frac{F_n}{A_w^m} \quad \text{with } 0 \leq m \leq 1 \tag{5}$$

Where V_w is the volume of PDC cutter removed by wear, l the linear displacement of the cutter on the rock surface, F_n the total normal force applied to the cutter, A_w the wear flat area, and C^* a wear coefficient.

Instead of this global law, since it is possible to separate the forces on the cutter, it seems reasonable to postulate a more rational law in terms of thickness ξ lost by wear on the cutter wear flat and shear stress F_t^f/A_w applied on it, as follows:

$$\frac{d\xi}{dl} = \frac{1}{C}\frac{F_t^f}{A_w} = \frac{\mu}{C}\frac{F_n^f}{A_w} \tag{6}$$

This law is equivalent to (5) when setting $m = 0$, and noticing that $dV_w = A_w d\xi$

4.3 Dip angle of cutter wear flats

It now possible to precise angle α as a function of position of the cutter on bit profile and depth of penetration δ, according to the hypothesis of wear flat parallel to bit profile (Figure 10):

$$\sin\alpha = \vec{n}.\vec{\tau} = \frac{\delta}{(4\pi^2 y^2 + \delta^2)(1 + z'^2)^{1/2}} \tag{7}$$

$\vec{n}$ is the unit external normal to bit profile and $\vec{\tau}$ is the unit tangent to the helicodal trajectory of a cutter passing through the bit profile.

5 Bit wear evolution

Using expression (4), rate of wear expresses as:

$$\frac{d\xi}{dl} = \frac{\mu}{C}\frac{F_n^f}{A_w} = \frac{\mu\sigma_0}{C}(1 - \exp(-\alpha(\delta, y)/\alpha_0)) \tag{8}$$

expressing distance increment dl in terms of the rotational frequency N gives finally :

$$\frac{d\xi}{(2\pi\mu N\sigma_0 dt)} = y(1 - \exp(-\alpha(\delta, y)/\alpha_0)) \tag{9}$$

This equation appears as an intrinsic cutter wear rate, which is function of cutter position on bit profile. For a given rock, this quantity depends on the bit profile and on the bit penetration per revolution δ, but not on the location and the number of cutters.

6 Global forces on the bit

Global forces result from local integration of local forces on cutters. With similar notations as for the single cutter, Torque and Weight On Bit write :

$$(T) = (T)^c + (T)^f \text{ and } (W) = (W)^c + (W)^f \tag{10}$$

It is interesting to note with Detournay[3], that the global cutting forces applied to the bit do not depend on bit geometry, provided the specific energy be supposed constant for all cutters :

$$(T)^c = \frac{1}{2}\epsilon_0 \delta R^{*2} \text{ and } (WOB)^c = \zeta\epsilon_0\delta R^* \tag{11}$$

where (δR^*) is the rock surface swept by the bit in one revolution in a meridian plane and R^* radius of the "bit cutter line profile", defined by Detournay[3]. Indeed, radius R^*, which includes existence of external cutters can be replaced by the profile radius R_b with the maximal relative error of $\frac{d_c}{2R_b}$.

$(W)^f$ and $(T)^f$ result from summing up friction forces on cutters wear flats :

$$(T)^f = \Sigma(yF_\tau^f) \text{ and } (W)^f = \Sigma F_n^f \tag{12}$$

The cutter friction forces can be expressed as :

$$F_n^f = A_w f(\alpha) \text{ and } F_t^f = \mu F_n^f \tag{13}$$

where A_w is the wear flat area, which shape is a truncated ellipse, given the cylindrical circular shape of the PDC cutter. This surface can be expressed in terms of the wear depth ξ in the normal direction to the surface of the rock wall ([5]):

$$A_w = \frac{d_c^2}{2\sin\theta}\left(\frac{1}{2}\arccos\sqrt{(1-4(\omega-\omega^2))} - \sqrt{(\omega - 9\omega^2 + 8\omega^3 - 4\omega^4})\right) \tag{14}$$

$$\omega = \frac{\xi}{d_c\cos\theta} \text{ with } \theta, \text{ rake angle, } d_c, \text{ cutter diameter}$$

ξ is given from integration of (9)

7 Illustrative examples

For illustration, some simple bit geometries have been considered. Intrinsic wear distribution on the bit profile, according to equation (9), is given in Figure 11 for an elliptic bit profile $y^2/R^2 + z^2/S^2 = 1$, with ellipticity $R/S = 0.57; 1; 1.74$ corresponding to elongated elliptic, circular, oblate elliptic profile and for two different bit penetretion δ. Notice that for higher δ, a higher wear rate occurs for external cutters, which is currently predicted by other models in literature (Glowka[10]). But for lower δ, a reverse tendency can be noticed for the sharp elliptical profile.

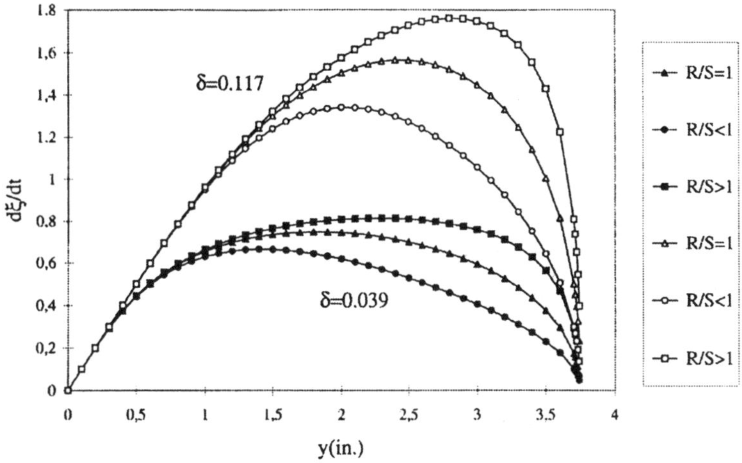

Figure 11: Intrinsic wear distribution on the bit profile

The model allows to find again some results observed on laboratory tests (Warren [11], Glowka [10]). Wear flat effect results in the upward concave curve T vs. WOB (Figure 12), illustrating the better efficiency of the bit as WOB increases, i.e., as depth of cut increases.

8 Conclusion

It can be considered that a reasonable understanding of mechanisms on the cutting edge and wear flat of a PDC cutter has been now achieved, so that a better modelling of PDC bit performance and better insight in wear laws can be expected. In this study, the elastoplastic nature of the frictional mechanism has been demonstrated, which expresses as a law relating the frictional stresses on wear flat contact to the movement of the bit, relatively to the rock wall.

A simplified law for this mechanism has been proposed, together with a reasonable wear law for the cutter wear flat. This has allowed to propose a PDC drill bit model which wear evolution is governed by the frictional mechanism which can be called abrasion. This model shows that wear evolution depends on the profile and on the depth of penetration per revolution. In this case, the cutter distribution on the bit has no influence with respect with the frictional wear, unless another wear law be envisaged with respect to the cutting effects.

The next step will be the validation and experimental identification of the

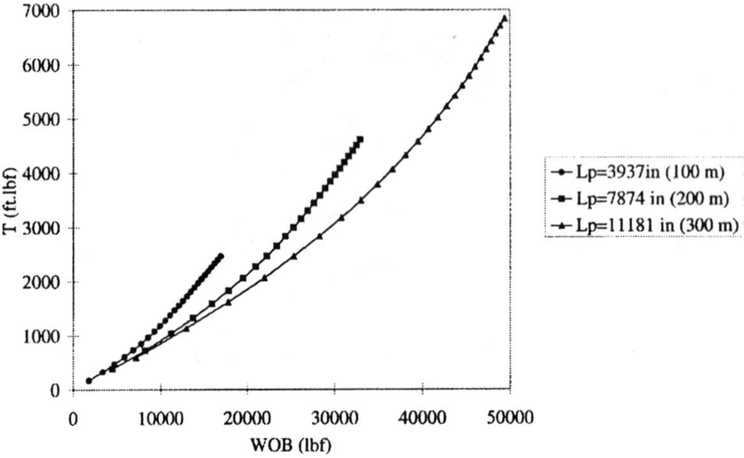

Figure 12: T versus WOB

parameters of the model, which will finally allow for an optimisation of the geometry of the bit.

References

1. R. Feenstra, *SPE*,1988,675-684.
2. D.A. Glowka, *SPE 14222*, 1985.
3. E. Detournay and P.Defourny, *Int. J. Rock Mech. Min. Sci. and Geomech. Abstr.*, **29**(1),13- 23,1992.
4. H. Sellami, PhD thesis, ENSMP, France, 1987.
5. E. Kuru, PhD thesis, Louisiana State Univ. Baton Rouge, La, 1990.
6. D. Nguyen, Minh, PhD thesis, Université de Paris VI, France, 1974.
7. H. Geoffroy, PhD thesis, Ecole Polytechnique, France, 1996.
8. J.I. Adachi, Master's thesis, University of Minnesota, USA, 1996.
9. M.H. Maitournam, PhD thesis, ENPC, France, 1989.
10. D.A. Glowka, Sandia Report 86-1475, Sandia national laboratories, 1987.
11. A. Sinor and T.M. Warren, *SPE drilling Engineering* 128-136, 1989.

Rock Classification

AN INTRODUCTION OF THE CHINESE STANDARD FOR ENGINEERING CLASSIFICATION OF ROCK MASSES (GB50218 - 94)

YUNMEI LIN
Department of Cicil Engineering, Northeastern University,
Shenyang 110006, PR China
E-mail YMLin@ramm.neu.edu.cn

From 1995-07-01 in China executed a forced standard for engineering classification of rock masses(GB 50218 - 94). 10 rock mechanics experts have worked on the forming of this standard about 8 years. This paper deals with some results concerning Chinese Standard[1] and new development of rock classification in China

1 Introduction

A large number of rock classification systems have been put forward in the world (Coates,D.F.[2]). However, in most of them, the choice of criterions and class number is made subjectively. Considering the efficacy of a system will be remarkably influenced by the selected parameters, a series of key questions must be answered before the building of a standard. For example:

1. Which is the optimal set of classification criterions and how could be obtained this optimal set?
2. Which is the appropriate number of classes?
3. How can be worked out a standard fulfilling high percentage of hits and accepted by audience in Chinese rock mechanics and engineering field?

After analyzing all kind of rock classification methods, which could be found in the world, finally it was decided to build a system on the basis of statistical data with modern mathematical approaches, such like the hierarchical cluster analysis, dynamic cluster analysis and fuzzy cluster analysis etc. Therefore the most significant advantage of this standard, which has over the traditional classification systems is its objectivity.

2 The procedure for building of Chinese Standard

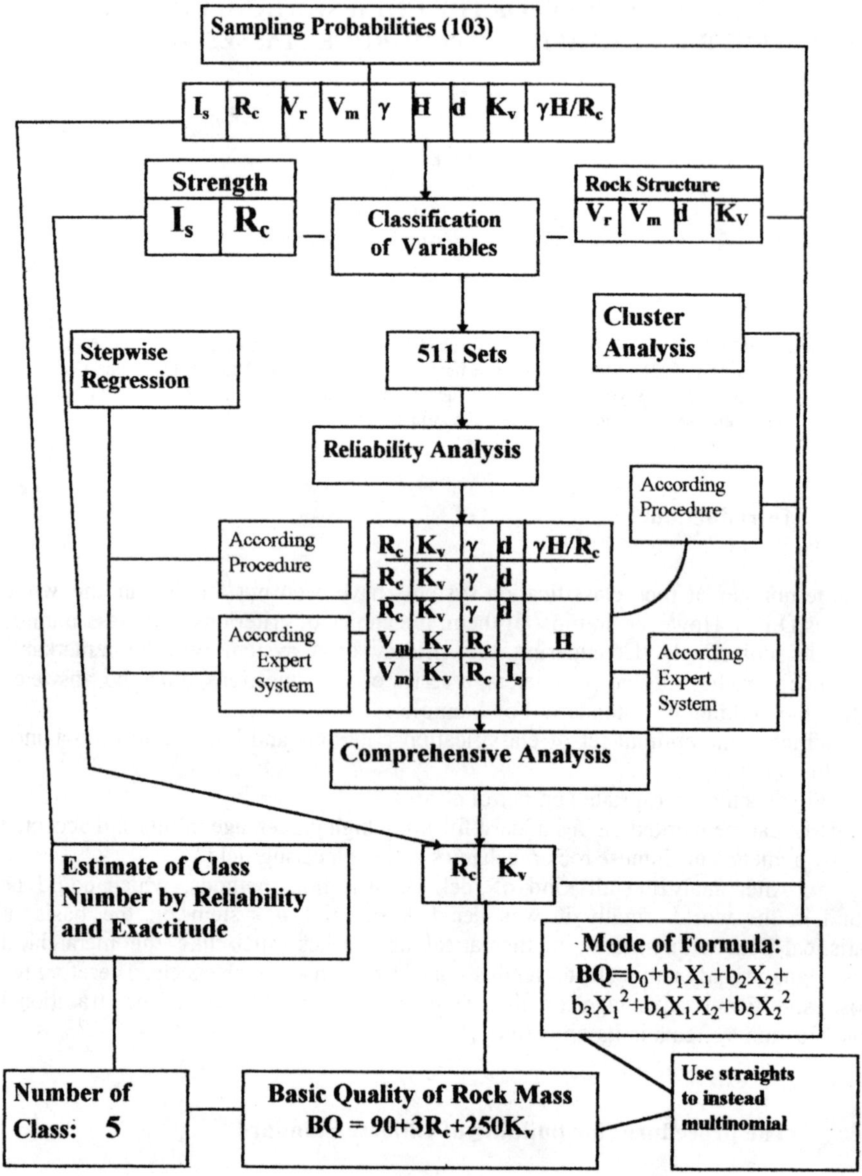

figure 1: Procedure for building of Chinese Standard

In order to clearly express the process of building of Chinese Standard, a diagram shown in Fig.1 was used to illustrate each step of the procedure, which can be described as follows:

2.1 Building of database

Collected all the data in the whole country and built a database of 460 rock engineering with comprehensive information in different aspects.

2.2 Selection of total sampling probabilities

The total sampling probabilities included 103 rock engineering, which were selected from the database according the requirement to the laboratory and *in-situ* measurement results. That is: every engineering must simultaneously have complete measurement of following parameters:

1. The point load strength index $\mathbf{I_s}$
2. The uni-axial compressive strength $\mathbf{R_c}$
3. Sonic velocity of rock $\mathbf{V_r}$
4. Sonic velocity of rock masses $\mathbf{V_m}$
5. unit weight of rock γ
6. Depth of rock engineering $\mathbf{H}$
7. Average spacing of various joint sets $\mathbf{d}$
8. Completeness coefficient $\mathbf{K_v}$
9. Stress ratio $\boldsymbol{\gamma}\mathbf{H/R_c}$

2.3.Classification of variables

By using 11 kind of mathematical methods (Yunmei Lin etc[3]), containing correlative analysis, 8 hierarchical cluster analysis, dynamic cluster analysis and fuzzy dynamic cluster analysis, the 9 variables shown above are divided into three families as follows:

1.Strength family containing two variables: point load strength index and the uni-axial compressive strength;

2.Rock structure family containing 4 variables: sonic velocity of rock and rock mass, completeness coefficient, as well as the average spacing for various joint sets;

3. Stress family containing 3 variables: unit weight of rock, depth of rock engineering and stress ratio.

2.4 The optimization of three elements of classification

Normally, a table should be proposed for any classification system, which basically constitutes by three major elements as follows:

1. Criterion for classification;
2. Number of classes;
3. Standard of classification or some estimated formulae for calculation.

The way of optimization of three elements are described in the following paragraphs.

3 Selection of optimal classification criterion

Among three elements, the criterion for classification must be studied at first. In this regard, the contention of academic point of view is very lively, and most of authors insist to classify rock with their favorable criterion without sufficient mathematical analysis. In order to avoid above pointed defect, the Chinese Standard does not previously select single variable (for example: $\mathbf{R_c}$ - the compressive strength or **RQD**) or any combination of variables as the criterion of system, but only studies the total sampling probabilities to look for the optimal criterion matching with the statistical rules.

There are all together 511 sets of parameters which may be combined from the 9 variables. In order to know which is the optimal combination of parameters three methods, namely the reliability analysis, stepwise regressive analysis and stepwise discriminative analysis were used to choose the optimal sets, meanwhile, let each analysis gave the sequence of variables in the selected set, so that people can understand which parameter has the first important meaning and which is the second etc.

In the reliability analysis (Yunmei Lin[4]) suppose each rock engineering as the object for judgement and the every set of parameters as an independent judge. No matter every judge has his own standard for judgement, but under the judgement of numerous experts, the final result definitely can reflect an appropriate suitable identification for the judged object. And the qualified level of each judge (here the level might be understood as the reliability of each set of parameters) could be estimated by the average probability obtained in the process of judgement. This average probability is defined as reliability function. The process of reliability analysis is to calculate and compare the magnitude of reliability function of each set of variables.

During the application of different criterion, the ith rock engineering might be identified as different but adjacent class. Therefore the probability of ith rock engineering to be divided into k^{th} class can be expressed by following equation:

$$P_k(i) = \sum_{md=1}^{ms} (\delta_{k,md})/ms \tag{1}$$

Where : $\delta_{k,md} = 1$,while i^{th} is judged as the kth class;
$\delta_{k,md} = 0$,while i^{th} is not judged as the kth class;
ms - the total number of sets;
md - the serial number of sets, md =1,2,...,ms;
i - i^{th} rock engineering;
k - k^{th} class

Then the average probability or the reliability function RF can be obtained as following formula:

$$(RF)_{md} = 1/n \sum P_k(i) \tag{2}$$

Where:

n - number of rock engineering judged in the total sampling probabilities.

The result of reliability analysis for all sets of parameters shows that the best set of parameters is the combination of parameters: $\mathbf{R_c}$, $\mathbf{K_v}$, $\boldsymbol{\gamma}$, **d** , $\boldsymbol{\gamma}\mathbf{H/R_c}$ (Fig.1).

The result of stepwise regressive analysis based on the procedure of dynamic cluster analysis is the combination of parameters: $\mathbf{R_c}$, $\mathbf{K_v}$, $\boldsymbol{\gamma}$, **d** (Fig.1).

The result of stepwise regressive analysis according expert system is the combination of parameters: $\mathbf{V_m}$, $\mathbf{K_v}$, $\mathbf{R_c}$, **H** (Fig.1).

The result of stepwise discriminative analysis on the basis of procedure of dynamic cluster analysis is the combination of parameters: $\mathbf{R_c}$,$\mathbf{K_v}$, $\boldsymbol{\gamma}$, **d** (Fig.1).

The result of stepwise discriminative analysis according expert system is the combination of parameters: $\mathbf{V_m}$,$\mathbf{K_v}$,$\mathbf{R_c}$, $\mathbf{I_s}$ (Fig.1).

From above selected 5 sets of parameters and the position of each parameter appeared in the ranges, it could be clearly understood, that the most important parameters are the uni-axial compressive strength $\mathbf{R_c}$ and the completeness coefficient $\mathbf{K_v}$. Both they also satisfy following important requirements:

1. same with the opinions of experts according the physical consideration;
2. represent two main families: strength and rock structure;
3. have available substitutive parameters $\mathbf{I_s}$ and $\mathbf{d}$, which could be used to instead the main parameters in the condition of lack of $\mathbf{R_c}$ and $\mathbf{K_v}$. For the convenience of use by engineers in rock project, the standard has offered the corresponding substitutive equation between $\mathbf{I_s}$ and $\mathbf{R_c}$,as well as between $\mathbf{d}$ and $\mathbf{K_v}$.

based on above comprehensive considerations these two parameters have been selected finally as the basic quality parameters of rock masses and estimated as the optimal criterions in the Chinese National Standard.

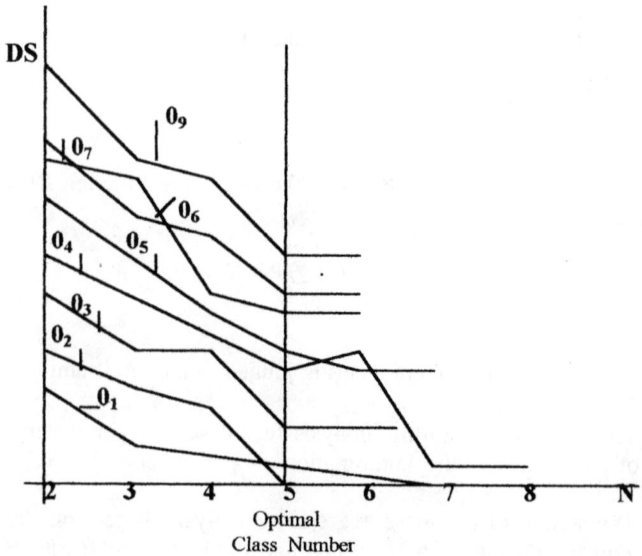

Figure2: curves of **DS** versus class number **N** under the condition of various combination of parameters

0_9 - 9 parameters 0_7 - 7 parameters 0_6 - 6 parameters
0_5 - 5 parameters 0_4 - 4 parameters 0_3 - 3 parameters
0_2 - 2 parameters 0_1 - single parameter

4 Selection of the number of classes

After the determination of classification criterions, the next step is to find the optimal number of classes. This problem could be solved also by the dynamic cluster analysis . In this method, there is a function named **DS** which could be calculated during amending the classes of sampling probabilities. Because function **DS** is closely related with the class number, so it can be used as an indicator for the selection of reasonable number of classes. Fig.2 shows the relationship between **DS** and different number of classes. An optimal number is determined by finding the point on the **DS** versus **N** plot where the gradient significantly reduces. Anyway, the reliability reduced due to the increase of class number, however the exactitude should be increased while the number of classes is increased. Finally 5 classes have been chosen as an appropriate number of classes on the basis of both reliability and the exactitude.

5 Mathematical mode for classification

The last subject is to find out a mathematical mode for the selected two parameters combined as an index **BQ**, which can be easily used and remembered for estimating the basic quality of rock mass.

A lot of mathematical modes has been applied in the process of research. Finally, the multinomial is selected from 5 modes to build the **BQ** formula, which can be expressed as follows;

$$Y = b_0 + b_1X_1 + b_2X_2 + b_3X_1^2 + b_4X_1X_2 + b_5X_2^2 \tag{3}$$

The regressive formula has very complex mode as follows:

$$\mathbf{BQ} = -41.0451 + 1.9212\,R_c + 546.1130\,K_v - 0.0064\,R_c^2 + 1.0492\,R_cK_v - 276.9786\,K_v^2 \tag{4}$$

No matter this precise formula give satisfactory results for classification of rock mass, it can't be used in the standard due to its complexity and un-popularization. Therefore three straight lines have been used to instead this multinomial curve(Fig. 3). The basic quality index of rock masses should be calculated with the quantitative index of classification criterion $\mathbf{R_c}$ in MPa value and $\mathbf{K_v}$ by using the formula of the main straight line(**A,B,C,D** lines in Fig.3),that is:

$$\mathbf{BQ} = 90 + 3\mathbf{R_c} + 250\mathbf{K_v} \tag{5}$$

Before using **BQ** formula, the parameters $\mathbf{R_c}$ and $\mathbf{K_v}$ must be checked by following two restrictive formulae:

1. If $\mathbf{R_c} > 90\,\mathbf{K_v} + 30$; then let:
   $\mathbf{R_c} = 90\,\mathbf{K_v} + 30$; (1,2,3,4 lines in Fig.3)
   This value of $\mathbf{R_c}$ and $\mathbf{K_v}$ should be substituted into formula (5) for calculation of the value of **BQ**

2. If $\mathbf{K_v} > 0.04\,\mathbf{R_c} + 0.4$; then let:
   $\mathbf{K_v} = 0.04\,\mathbf{R_c} + 0.4$; (5,6 lines in Fig.3)

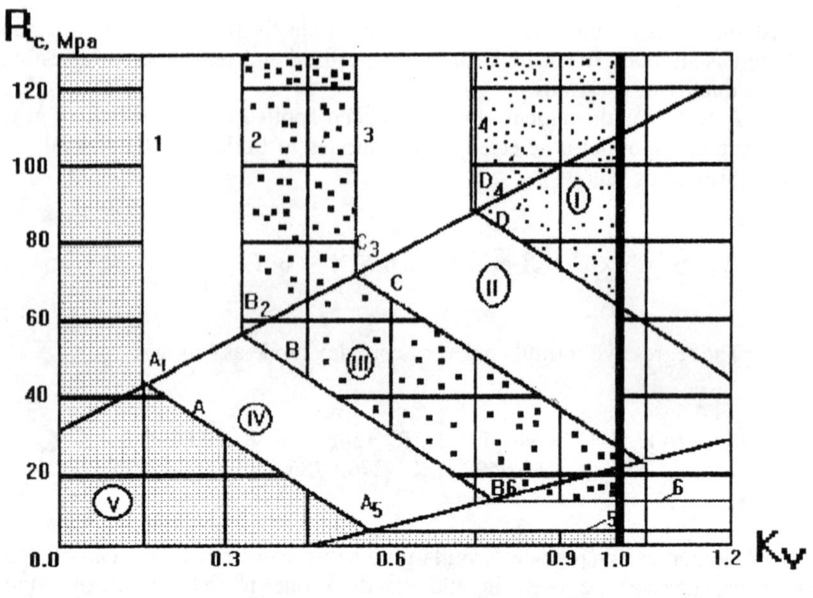

Figure3: Approximation of three straight lines to the multinomial curve

Table1: The Basic Quality Table For Estimation Of Rock Mass

Class of Basic Quality	The Qualitative Characteristic of Basic Quality of Rock Masses	The Index of Rock Mass Basic Quality, **BQ**
I	Strong Rock; Intact Rock Mass	550
II	Strong Rock; Less intact rock Mass. Less Strong Rock; Intact Rock Mass.	550 - 451
III	Strong Rock; Less Fractured Rock Mass. Less Strong Rock or Interlayer of Soft and Strong Rock; Less Intact Rock Mass. Less Soft Rock; Intact Rock Mass.	450 - 351
IV	Strong Rock; Fractured Rock Mass. Less Strong Rock; Less Fractured to Fractured Rock Mass. Less Soft Rock or Interlayer of Soft and Strong Rocks, with Predominant Soft Rocks; Less Intact to Less Fractured Rock Mass. Soft Rocks; Intact to Less Intact Rock Mass.	350 - 251
V	Less Soft Rock; Fractured Rock Mass. Soft Rock; Less Fractured to Fractured Rock Mass. All Kinds of Very Soft Rock ; All kinds of Very Fractured Rock mass	250

This value of $\mathbf{K_v}$ and $\mathbf{R_c}$ should be substituted into formula (5) for calculation of the value of **BQ.**

With the calculated value of **BQ** the class of basic quality of rock masses can be easily estimated according the Table 1.

The class of rock mass basic quality, determined in accordance with Table 1 may be used as the class of rock mass in preliminary class determination of engineering rock mass.

6 Class determination of engineering rock mass

For detailed class determination of engineering rock mass necessary corrective parameters should be taken into consideration on the basis of classification of rock mass basic quality. These are, depending on the features of different engineering types, the state of underground water, the initial rock stress state, the combination of relative orientation of the structure axis or strike with the occurrences of the main weak joint sets.

The influence of surface water should be taken into consideration also for rock mass of slopes.

The influence of the following factors should be considered during class determination of engineering rock mass, if they became the main factors, effecting rock mass stability. These are expandability and solubility of rock mass, the existence of rock structural planes, which are long in relation to the area of rock engineering and with high degree of continuity.

There are a series of appendices included in the Chinese Standard, which can be used to correct the index of rock mass basic quality **BQ** for various factors in the determination of engineering rock mass.

7 Conclusions

1. The Chinese Standard for engineering classification of rock mass is a forced standard and the first standard of rock classification in the world;
2. This classification system was built on the base of statistical data of 460 rock engineering by using 11 kind of advanced mathematical analysis, containing: correlative analysis, 8 hierarchical cluster analysis, dynamic cluster analysis and fuzzy analysis;

3. The most significant advantage of this standard, which has over the traditional classification systems is its objectivity in the determination of three elements of classification, the index of rock mass basic quality and the influence of corrective parameters for engineering rock mass;
4. The contributions and significance of this standard are unifying the method of rock classification in whole China and forming a firm foundation for investment and management of enterprises, because standard make every Chinese rock engineering comparable.

References

1. Standard for engineering classification of rock mass, the national standard of People's Republic of China(**GB 50218 - 94**), issued by THE NATIONAL DEPARTMENT OF TECHNICAL MONITORING AFFAIRS AND THE MINISTRY OF CONSTRUCTION, PRC,(1994)
2. Coates, D.F., Handbook on Mechanical Properties of Rocks, Publication Trans. Tech.,(1978)
3. Yunmei Lin etc, Numerical analysis methods and its application in rock mechanics, Northeastern University Press, Shenyang,(1989)
4. Yunmei Lin, Theory and practice of rock classification, Metallurgical industrial press, Beijing,(1994)